… UNIVERSITY OF NORTH TEXAS

Date Due

A charge is made for overdue books

DEC
JAN 2011 C R
      2010 C R
JUN
    JUL 05 2011 C R

(Form 14-8/97)
PS56483-11/05

The Expression of Negation

The Expression of Cognitive Categories

ECC 4

Editors
Wolfgang Klein
Stephen Levinson

De Gruyter Mouton

The Expression of Negation

edited by
Laurence R. Horn

De Gruyter Mouton

ISBN 978-3-11-021929-6
e-ISBN 978-3-11-021930-2

Library of Congress Cataloging-in-Publication Data

> The expression of negation / edited by Laurence R. Horn.
> p. cm. − (The expression of cognitive categories ; 4)
> Includes bibliographical references and index.
> ISBN 978-3-11-021929-6 (hardcover : alk. paper)
> 1. Grammar, Comparative and general − Negatives. 2. Negation (Logic)   I. Horn, Laurence R.
> P299.N4E97 2010
> 415−dc22
>
>   2010013733

Bibliographic information published by the Deutsche Nationalbibliothek

The Deutsche Nationalbibliothek lists this publication in the Deutsche Nationalbibliografie; detailed bibliographic data are available in the Internet at http://dnb.d-nb.de.

© 2010 Walter de Gruyter GmbH & Co. KG, Berlin/New York

Cover design: Frank Benno Junghanns, Berlin
Printing: Hubert & Co. GmbH & Co. KG, Göttingen
∞ Printed on acid-free paper

Printed in Germany

www.degruyter.com

Contents

Introduction 1
Laurence R. Horn

Typology of negation 9
Östen Dahl

The Acquisition of negation 39
Christine Dimroth

On the diachrony of negation 73
Johan van der Auwera

Multiple negation in English and other languages 111
Laurence R. Horn

Quantifier-negation interaction in English:
A corpus linguistic study of *all...not* constructions 149
Gunnel Tottie and Anja Neukom-Hermann

Negative and positive polarity items: An investigation of the
interplay of lexical meaning and global conditions on expression 187
Jack Hoeksema

Negation as a metaphor-inducing operator 225
Rachel Giora, Ofer Fein, Nili Metuki, and Pnina Stern

Negation in Classical Japanese 257
Yasuhiko Kato

Negation in the new millennium: A bibliography 287
Laurence R. Horn

Contributors 331
Index of subjects 333
Index of languages 335
Index of persons 337

Introduction

Laurence R. Horn

Negation is a sine qua non of every human language but is absent from otherwise complex systems of animal communication. While animal "languages" are essentially analog systems, it is the digital nature of natural language negation, toggling between 1 and 0 (or T and F) and applying recursively to its own output, that allows for the essential properties of our own linguistic systems. In many ways, negation is what makes us human, imbuing us with the capacity to deny, to contradict, to misrepresent, to lie, and to convey irony.

The apparently simplex nature of logical negation as a one-place, two-valued operator that reverses truth and falsity belies the profoundly complex and subtle expression of negation in natural language. Not only do we find a plethora of negative adverbs, verbs, copulas, quantifiers, and affixes, but the interaction of negation with other operators (including multiple iterations of negation itself) can be exceedingly problematic, extending (as first explored in detail by Otto Jespersen) to negative concord, negative incorporation, and the widespread occurrence of negative polarity items whose distribution is subject to principles of syntax, semantics, and pragmatics. Situated at the core of the mental faculty of language, negation interacts in significant ways with principles of morphology, syntax, and logical form, as well as with processes of language acquisition and sentence processing, whence the prominent role played by work on negation in the recent development of grammatical and semantic theory. The semantics of negation has been under close investigation since Parmenides, Plato, and Aristotle (cf. Horn 2001a). Much of this work, as reflected by key passages in Aquinas, Kant, Bergson, and Russell, has focused on the asymmetry between negative and affirmative sentences, often resulting in attempts to define negation out of existence whether through subsumption under falsity, incompatibility, positive difference, dissimilarity, or true disbelief; negation, however, survives these attempts at elimination, as befits its status as the Rasputin of the propositional calculus (Horn 2001a: 59).

The key modern landmark in the study of the meaning and expression of negation is Jespersen's monograph, "Negation in English and other languages" (1917). This magisterial, though flawed, work ranges from morphology to logic to what would now be elucidated through the application of (neo-)Gricean pragmatics, but is celebrated in particular for its exposition of the cyclical process of successive weakening, strengthening, and reanalysis that has been known since Dahl 1979 as JESPERSEN'S CYCLE:

> The history of negative expressions in various languages makes us witness the following curious fluctuation: the original negative adverb is first weakened, then found insufficient and therefore strengthened, generally through some additional word, and this in its turn may be felt as the negative proper and may then in course of time be subject to the same development as the original word.
>
> ...The negative notion, which is logically very important, is ... made to be accentually subordinate to some other notion; and as this happens constantly, the negative gradually becomes a mere proclitic syllable (or even less than a syllable) prefixed to some other word. The incongruity between the notional importance and the formal insignificance of the negative may then cause the speaker to add something to make the sense perfectly clear to the hearer. (Jespersen 1917: 4–5)

This process, re-examined in van der Auwera's chapter (and in his companion article, van der Auwera 2009), has been widely attested in a variety of languages, especially (but not exclusively) within the Romance, Germanic, Greek, and Celtic families of Indo-European. The reinforcers that fill Jespersen's 'add[ed] something' role fall into two general classes, one involving indefinites of either positive or negative morphological character within the scope of negation, as with Latin *non* < *ne-oenum* 'not one' or Eng. *not* < *ne-a-wiht* [lit. 'not ever a creature'] and the other exemplified by minimizers denoting small entities or negligible quantities from various domains *(a crumb, a hair, a red cent, a shred, an iota)*. These postverbal indefinites and minimizers may gradually oust the original prosodically weakened proclitic negative as essentially occurred earlier with *not* and is occurring currently with Fr. *pas* ('a step'). Jespersen's cycle plays a key role in the development of negative polarity and negative concord, two linked phenomena that lie at the heart of contemporary work on the syntax and semantics of negation, as explored in the papers of this volume. (Valuable overviews of variation and change in the expression of negation in English form the heart of two recent books, Anderwald 2002 and Mazzon 2004.)

The chapters in the present book examine the patterning of negative utterances in natural languages across time and space, spanning such foundational issues as how negative sentences are realized cross-linguistically, how negation is acquired by children, how it is processed by adults, and how its expression changes over time. Other chapters offer focused empirical studies of negative polarity items, pleonastic negation, the scopal interaction of negatives with quantification, and detailed examinations of the form and function of negation and negative polarity in specific languages.

Östen Dahl's chapter recapitulates work on the typology of sentence negation dating back to his own trailblazing article (1979) from three decades ago and including the more recent extensive surveys of Dryer (2005) and especially Miestamo (2005a, b, c; 2006; 2007). A starting place for much of this work is to define a notion of "standard negation", excluding lexical negatives as in English *un*-adjectives (but extending to negative affixes that do express canonical clausal negation as in Turkish or Japanese). Dahl sorts out both definitional and substantive issues in the forms expressing standard and non-standard negation – negative particles, negative verbs, and varieties of affixal negation – and their interaction with word order, finiteness, and other aspects of syntax and morphology.

Dahl's study of the typology of negation touches on the role of diachrony and grammaticalization encapsulated in the developments of Jespersen's cycle. This is the topic of Johan van der Auwera's chapter. Like Dahl, van der Auwera focuses on Miestamo-style standard negation, realized by an operator taking sentential scope, typically in the form of a verbal predicate in a declarative clause exploiting the general strategy made available within a given language, although the development of non-standard negation (and of prohibitives in particular) is also considered. In tracking the Jespersen cycle – or what may be more fully designated as the Bréal-Gardiner-Meillet-Jespersen cycle – van der Auwera marshals extensive cross-linguistic data to determine the plausibility of different possible analyses of the motivation for and the details of the reanalysis involved in the relevant shifts. Other features of this study include the interaction of negation with verbal aspect, subordination, and finiteness, the genesis of negation in existential and non-verbal clauses, and the derivation of prohibitives, in which negation is incorporated within the scope of directive illocutionary force.

Constraints on the lexical incorporation of negation in logical operators are touched on this chapter and in other recent work by van der Auwera (2001), Jaspers (2005), Seuren (2006), and Horn (2001a, to appear). The primary riddle is to predict the asymmetry of values that can be mapped on the Square of Opposition, e.g. the nonexistence of a lexicalized *nall* (= 'not all') operator alongside *all, some, no(ne)* or the nonexistence of any connective *nand* alongside *and, or, nor*. This asymmetry is motivated along Gricean lines in Horn (to appear) and work cited therein, but there are complexities in the data that support alternative explanatory options, as van der Auwera observes; one problem is the discrepancy between the total non-occurrence of *nall, *nand, or *nalways (= 'not always') alongside the limited but clearly possible instantiation of the corresponding O-vertex modals, such as English *needn't*.

The expression of negation can also be mapped through its ontogeny. Christine Dimroth's contribution points out the centrality of negation within the study of both first and second language acquisition. Since the inception of work on L1 acquisition, much attention has been devoted to mapping the appearance and frequency of the various categories and subcategories of negation in child language – denial, nonexistence, disappearance, unfulfilled expectation, metalinguistic objection, and (last and in developmental terms possibly least) falsity; as Dimroth notes, the ordering differs for the early (one-word) and the later (clausal) stages, but the general tendency is a shift from negation as a sign of rejection to negation as a truth-functional operator. Dimroth relates the acquisition of negation to the development of other aspects of grammar, and several of the same topics come up with ontogeny as with phylogeny: the development of clausal vs. constituent negation, negative concord and polarity, and the interaction of negation with word order and finiteness. Unlike children acquiring their first language, L2 learners can use their head start to bypass the problem of mastering the functions of negation, but problems arise from the variation among language types in the formal expression of the relevant categories.

The chapter by Rachel Giora and her colleagues focuses on the processing of negative utterances, a topic dealt with in much earlier empirical work (see Horn 2001a: Chapter 3 for a summary). Giora et al. summarize earlier research in their chapter, much of it by their group, demonstrating that negated propositions and concepts are in general mentally retained rather than simply discarded from the discourse model. Suppression varies with the discourse goals and the real and assumed intentions of the speaker and hearer(s). The more relevant the information in the scope of negation is taken to be, the more likely it is to persist, as measured by psycholinguistic studies reported on here. As a result, the processing of negation and affirmation is often less asymmetrical within an actual discourse context than it is in isolation (and in artificial experimental paradigms). But a robust distinction emerging from the work by Giora et al. is that negative utterances tend to induce a figurative or metaphorical interpretation largely absent from the processing of the corresponding affirmative. Negation thus functions cross-linguistically as a metaphor-inducing operator, a property derived from its suitability across diverse languages to serve as a marker of rejection.

Following the first four contributions devoted to general properties of negation – its expression across languages, its development across time, its acquisition by first and second language learners, its processing by adults – the next four chapters examine more specific aspects of the expression of negation.

Gunnel Tottie and Anke Neukom provide a closely observed corpus-based study of the interaction of negation with universal quantification. The *All that glitters* phenomenon (cf. Horn 2001a: §4. 3, §7. 3) has offered a target to irate prescriptivists and a challenge to logicians and descriptive linguists since Tobler (1882) and Jespersen (1917): why do speakers in English, French, German, and other languages express negated universals ('not all') by universalizing a negative *(All...not)*? And when is this reading more or less likely to be associated with that expression? Drawing extensively on the substantial extant computerized historical and synchronic corpora, Tottie and Neukom point out that in addition to the two well-known readings of *all...not* sequences, the NEG-Q or 'not all' reading and the more "logical" but statistically less frequent NEG-V or 'all not' reading, a third, collective reading for the quantifier must be admitted, illustrated by Shakespeare's "All the perfumes of Araby will not sweeten this little hand" or Carroll's "All the king's horses and all the king's men could not put Humpty Dumpty together again". Data from the 100 million word British National Corpus of written and transcribed English shows that NEG-Q readings predominate more in the written sample and, along with the collective reading, in contrastive and formulaic utterances ("All is not lost/well"). Other factors favoring one reading or another include the role of context – syntactic, collocational, and extralinguistic. In addition to the question of why a NEG-Q or a NEG-V reading prevails in a given context, Tottie and Neukom acknowledge a second salient question: why does a speaker use the *all...not* construction to express the former (as opposed to *not all...*) or the latter (as opposed to *no/none*), given the possibility of misinterpretation? While *all... not* may win out over *not all* for reasons of syntactic markedness (cf. Horn 2001a: 488 ff.), the authors point out that it may be chosen over the negative quantifier because of the role of presupposition: an *all (not)* universal is more likely to be understood with existential import and/or to evoke old, thematic information than is the corresponding proposition with *no* or *none*.

Jack Hoeksema's chapter seeks to characterize the constraints on the inventory and distribution of negative (and to a lesser degree, positive) polarity items. NPIs are items that occur only in the scope of expressions that have the semantic value (but not always the formal character) of overtly negative elements. Hoeksema begins with what Ladusaw (1996), in his overview of the semantics of negation and polarity, calls the LICENSOR QUESTION for NPIs. Following Michael Israel's lead, Hoeksema explores the role played by the lexical semantics of the members of classes of NPIs and the collocational restrictions affecting their grammatical and felicitous occurrence, emphasizing the nature of the mapping between the intended meaning of the (NPI-enriched) utterance and the form of that utterance. His microanalysis of particular NPIs in Dutch, English, and German across a variety of syntactic categories and of the conditions on their occurrence is informed by both extensive corpus work and subtle intuitive judgments. (One limitation to corpus-based methodology here and elsewhere in research on negation is the relative paucity of exemplification of many of the crucial constructions, as Hoeksema notes.) While drawing on approaches to polarity licensing based on downward entailment (Ladusaw), implicature (Linebarger), and non-veridicality (Giannakidou), as well as the roles played by morphological blocking, semantic bleaching, and focus, Hoeksema argues that any explanatory account must extend beyond local conditions satisfied by a given NPI to consider global conditions on utterance meaning, however that meaning is conveyed.

Just as Tottie and Neukom's chapter addresses the motivation for a speaker's choice of an unlikely vehicle *(all...not)* to express a given meaning ('not all') and Hoeksema's chapter addresses the motivation for a speaker's choice of complicating her utterance with the addition of excrescent polarity-restricted elements whose distribution is limited and whose contributions to meaning are elusive if not obscure, so too my chapter on multiple negation asks why a speaker would go out of her way either (i) to express a positive assertion through two mutually destructive negations or (ii) to garnish an implicitly negative predication by adding a pleonastic or expletive negation in an embedded clause within the scope of an exclamative, a comparative or a verb expressing fear, denial, doubt, or prohibition. (Not coincidentally, these and other contexts licensing pleonastic negation, e.g. adverbs like *à moins que* 'unless', *avant que* 'before', and *depuis* 'since' in French, share the downward entailing semantics of standard NPI triggers.) The motivation for "hypernegations" of either type (i) or type (ii) must be sought outside the domain of truth-conditional semantics – in rhetoric or in conventional implicature. The chapter extends the purview of these "illogical" negatives to a variety of constructions in non-standard or colloquial English as well as to the lexical domain.

Several of the above chapters focus on the development of and variation in the expression of negation in Indo-European languages, particularly those in the Germanic and Romance families, from a largely typological and descriptive perspective. To fill out the story, Yasuhiko Kato travels back a millennium to the Heian period of *The Tale of Genji* and contemporary texts to examine the form and function of negation in Classical Japanese, looking back to the Old Japanese of previous centuries and forward to the modern language. While revisiting the topics familiar from several other chapters in the volume – the formal properties of sentence negation, double negation, negative imperatives, metalinguistic negation, negative polarity items, and the interaction of negation with focus and irrealis modality – Kato's study of of a language temporally distant and genetically unrelated to the others under investigation in the volume offers a useful perspective. His contribution also focuses on the implications of the grammar of negation in Classical Japanese for current theories of generative syntax.

The volume concludes with an extensive bibliography covering work on negation and polarity in the 21st century, encompassing publications appearing in the first decade of this century. Most of the articles and books included should be relatively accessible; others can in general be downloaded from the authors' or publishers' web sites. Without committing myself to either side in the controversy over when the millennium and century technically began, I arbitrarily chose January 1, 2000 rather than January 1, 2001 as the starting date for the bibliography. (An extensive, if somewhat less comprehensive, listing of work on negation and polarity appearing during the immediately preceding decade can found in the reissue edition of my *Natural History of Negation,* Horn 2001a: xxxix–xlvii.)

References

(Note: other work cited in the preface is listed in the general volume bibliography.)

Dahl, Östen
    1979   Typology of sentence negation. *Linguistics* 17: 79–106.
Jespersen, Otto
    1917   *Negation in English and Other Languages*. Copenhagen: A. F. Høst.
Ladusaw, William
    1996   Negation and polarity items. In *Handbook of Contemporary Semantic Theory*, S. Lappin (ed.), 321–342. Oxford: Blackwell.
Tobler, Adolf
    1882   Il ne faut pas que tu meures "du darfst nicht sterben." *Vermischte Beiträge zur französischen Grammatik 1,* 3rd ed., 201–205. Leipzig: S. Hirzel, 1921.

Typology of negation

Östen Dahl

Introduction

Modern language typology goes beyond mere classification or taxonomy. It can be defined as the systematic study of cross-linguistic patterns and cross-linguistic variation, in other words, of similarities and differences among languages. It is not in any way opposed to linguistic theory; on the contrary, typologists share with other linguists the ultimate aim to understand human language as a general phenomenon, but emphasize the necessity of seeking a secure empirical basis for generalizations in cross-linguistic data, and see cross-linguistic patterns as an important key to theoretical understanding.

Negation has in a way been a "low-hanging fruit" for typologists, since few grammatical descriptions fail to provide at least some basic information about negation in the language under study. In addition, negation has some features that makes it relatively unique among linguistic items, whether lexical or grammatical: it has a comparatively straightforward basic meaning which varies little among languages at the same time as it tends to have grammatical properties that set it off from other items in the language. The easy availability of basic information about negation is somewhat deceptive, however, and may have had the adverse effect of restricting the view to the most salient phenomena; accordingly, much of the cross-linguistic variation remains to be mapped, and many relevant questions have not been answered or even asked.

An important and much-cited forerunner to modern works on the typology of negation is Jespersen (1917). In the 1960's and 1970's, the period of early generative grammar and post-Greenbergian typology, negation figured prominently in many works (the most famous being Klima 1964), but there were no general typological surveys of negation before Dahl (1979) and Payne (1985)[1]. Of these, only Dahl's paper is based on an explicit sample, comprising 240 languages, although it is rather a "convenience

---

[1] The six-year gap between the publication years is a bit misleading; pre-publication or working paper versions of the papers existed around 1978 and they may seen as having arisen independently and roughly at the same time.

sample" with significant areal and genetic bias. There are also differences in focus, in that Payne is relatively brief on issues of word order, which are treated in some detail by Dahl, while Payne on the other hand goes more deeply into other topics, e.g. the relation of negation and quantifiers. In spite of this, the two treatments have much in common; the typological classifications of negation differ mainly in details, and have also stood the test of time in the sense that the classification presented a quarter of a century later in Dryer (2005a) can be seen as a synthesis of them. Other notable works are Dryer (1988), which focuses on word order issues based on a balanced sample; Croft (1991), which discusses the evolution of negation; Bernini & Ramat (1996), with a stress on European languages, and Kahrel (1996), largely devoted to negation and quantification (under the rubric "term negation"). Kahrel & van den Berg (1994) contains chapters on negation constructions in 16 languages from many different regions and language families. A recent important work is Miestamo (2005c), who presents a new proposal for the typological classification of clausal negation based on a stratified sample of 297 languages. Miestamo also contributed two of the 142 maps in the recent typological atlas, Haspelmath et al. (Miestamo 2005a,b). Several other maps in the same atlas also treat negation or phenomena related to it: Dryer (2005a) on negative morphemes, Haspelmath (2005b) on negative indefinite pronouns and predicate negation, van der Auwera et al. (2005) on prohibitives, and Zeshan (2005) on irregular negatives in sign languages.

The notion of "standard negation"

One complicating factor in the study of negation is that it is often not expressed in a homogeneous fashion across clause-types. In accordance with what was said in the Introduction, there has been a strong tendency for typological studies to concentrate on what has been seen as the basic negation constructions in languages. Thus, Dahl (1979) says that his study concentrates on negation "in simple indicative sentences with a verbal predicate". Payne (1985) introduces the notion of "standard negation", which he defines as "that type of negation which has as one function the negation of the most minimal and basic sentences". In English, Payne says, such sentences are "those involving weather predicates of zero valency, but requiring the dummy syntactic *it*", e.g. *It is raining*. Without using the term "standard negation", Dryer (1988: 2005a) specifies the domain as "simple clausal negation" and "the expression of negation in declarative sentences". Miestamo (2005a, b, c) follows Payne in using the term "standard negation" but defines it in a more extensional way as "the basic way (or ways) a language has for negating declarative verbal main clauses". The scholars mentioned thus seem to have at least partly independently arrived at more or less the same delimitation of the domain of study. The only difference lies in the restriction to verbal predicates, which is mentioned explicitly by Dahl and Miestamo only; on the other hand, both Payne and Dryer also tend to focus on negation of verbal predicates in their discussion. Curiously, however, none of the authors discusses at any length why, for instance, the declarative verbal sentence *It is raining* should be considered more basic than the imperative *Come!* or than the copular sentence *He is here* and its verbless counterparts in other languages. What is striking is that the clause-type that is argued to be basic in the definition of "standard negation" is identical to the type of predication argued to be prototypically associated by finiteness (Anderson (2007)), and that deviations from that prototype are often connected with a change in choice of negation construction. This is not the only way in which negation is linked up with finiteness, as we shall see below.

The choice of the term "standard negation" is perhaps not wholly fortunate, since it implies that anything that is not used in simple indicative sentences is "non-standard", but it is hard to come up with something very much better, so I will use it in the following. I will also use the term "negator" as a convenient way of the referring to words and morphemes that express negation, and accordingly, "standard negator" is what expresses standard negation.

Since "standard negation" is sentential or clausal, it follows that it does not include e.g. English prefixes such as *un-*, *in-*, and *dis-*, which belong to word formation rather than syntax. Although the term "affixal negation" is sometimes used for the latter (Zimmer 1964), it should be noted that the criterion is not whether the negator is a word or an affix; standard negation is expressed affixally in many languages, as we shall see below. For this reason, I shall use the term "lexical negation" instead.

Classifications

To bring some order in the apparent chaos with regard to the ways negation is expressed in human languages, we need some way of classifying negation constructions. Issues of classification are indeed prominent in the typological literature on negation. There is an obvious danger for these issues to detract attention from other, more directly empirical questions. I want to point to one general problem with typological classifications that is relevant here: the dependence of classifications on how expressions are analyzed grammatically, and on how various theoretical concepts are defined. For instance, if we cannot tell what is the subject and what is the object of transitive sentences in a language, we can neither assign the language to a Greenbergian basic word order type nor determine whether its case system is ergative or accusative. Analogous examples from the typology of negation will be discussed below.

As noted above, Dahl (1979), Payne (1985) and Dryer (2005a) all present classifications of negation constructions which largely coincide, sharing a focus on the status of negative markers. Thus, three major types of negation are identified by them all, although the terminology varies to some extent: (i) morphological or affixal negation; (ii) negative particles; (iii) negative verbs. Double negative particles are treated as a type of their own in Dryer (2005a) and in Dahl's main text; Payne sees them as a variation of type (ii) and the same policy is followed in Dahl's Appendix A. What Payne calls "secondary modifications" are not used in his classification and are not mentioned at all by Dryer (2005a) but are used to cross-classify syntactic types in Dahl's Appendix A. Payne adds a further type, negative nouns, not mentioned by the other authors.

A different kind of classification is proposed by Miestamo (2005a, b, c), where the key distinction is that between "symmetric" and "asymmetric" negation, based on whether there are structural differences ("asymmetries") between affirmative and negative sentences that go beyond the addition of one or more negative marker(s). Asymmetric negation is divided into three subtypes, "A/Fin", involving asymmetries in "the finiteness of verbal elements", "A/NonReal", involving marking of negative clauses as nonrealized ("irrealis"), and "A/Cat", involving changes in grammatical categories such as tense/aspect, mood, and person. The types are not mutually exclusive at the construction level, since one and the same construction may exhibit asymmetries of more than one type, and one could thus say that Miestamo classifies asymmetries rather than negation constructions. Miestamo (2006) discusses the relationship between his classification and the notion of complexity, where complexity is understood in the information-theoretic sense as depending on the length of the length of the description a phenomenon requires. He notes that asymmetric negation is "generally more complex than symmetric negation" (2006: 312). This statement could in a way be turned on its head. The logician Haskell Curry (1961) made a distinction between two levels of grammar, which he called "tectogrammatics" and "phenogrammatics", where the first concerns "the study of grammatical structure in itself" and the latter – how grammatical structure is represented in terms of expressions. Since Curry was working within the framework of categorial grammar, tectogrammatics was for him basically equivalent to a categorial grammar representation. At this level, standard negation is arguably universally always "S/S" or "VP/VP", i.e. an operation that derives a new sentence/verb phrase out of an old one. Phenogrammatically, however, negation varies widely. In Dahl (2004), where the same notion of complexity is employed as in Miestamo's work, Curry's distinction is taken as a point of departure for the discussion of "phenogrammatical complexity". Minimal phenogrammatical complexity is there said to equal "unrestricted concatenation", i.e. realization according to the rule "Concatenate the input expressions in any order" (p. 52).

This sounds fairly similar to Miestamo's characterization of symmetric negation when he says that what it does is to "simply add a negative marker to the corresponding affirmative" (e.g. 2005c: 351). However, since negative markers usually have a fixed position in the sentence, symmetric negation has to be compatible with restrictions on the concatenation operation. Moreover, in many cases the negative marker is not concatenated with the expression it operates upon but is rather spliced in at a specific position in the middle of it, as in the following example from Paez (Colombia; Paezan) where the negative morpheme -*me:*- occurs after the aspectual suffixes but before the subject marker of the verb (Miestamo 2005c: 11, quoting Jung 1989: 102–104):

(1)   a.   *u'x-we-ts-thu*
           go-IMPF-PROG-DECL.1SG
           'I'm going'

      b.   *u'x-we-ts-me: -th*
           go-IMPF-PROG-NEG-DECL.1SG
           'I don't go/I'm not going'

Also, the marker may be added as a clitic or an affix rather than as a separate word, which may influence prosody and word order. Thus, in various languages (e.g. in the Slavic and Iranian branches of Indo-European), negative prefixes form a prosodic unit with the following verb and receive word stress. In English, we can see that when a verb is moved to the front of the sentence, the suffix -*n't* but not the free morpheme *not* has to move with it, yielding *Isn't he here?* vs. *Is he not here?* Miestamo also subsumes discontinuous or double negation such as French *ne...pas* in symmetric negation, although this deviates from the characterization above at least with regard to the number of negative markers. All this means that symmetric negation can also have a significant degree of phenogrammatic complexity, and it may not be totally obvious what is to count as the "structural change" that is necessary to qualify as asymmetric negation. The symmetric/asymmetric distinction is also analysis-dependent in the sense described above. Most pertinently, it presupposes that we can determine which morphemes are negative markers in a construction. This is not so difficult as long as there is only one of them, but if there are two or more, there is often a choice between treating them both as negative markers, which in Miestamo's system means that we are dealing with symmetric negation, or treating one of them as a negative marker and the other as something else, i.e. the result of some structural change that motivates calling the construction asymmetric. Thus, Miestamo (2005c) regards both *ne* and *pas* in the French *ne...pas* construction as negative markers, but as he himself notes (2005c: 415), Kahrel (1996) argues for an analysis where *ne* is treated as a marker of non-reality, which should yield a classification as "A/Non-Real" in Miestamo's system.

Morphological (affixal) negation

In morphological negation, negation is expressed morphologically, most often as an affix, normally on a verb or an auxiliary. Turkish is a stock example, where the standard negator is a suffix *-mV-* (the vowel varies due to vowel harmony):

(2) a. *Oku-yor-um*
    read-PROG-1SG
    'I am reading'

 b. *Oku-mu-yor-um*
    read-NEG-PROG-1SG
    'I am not reading'

It may be noted that the function of the negator affix in (2b) is that of sentential negation rather than of what has been called "affixal negation" in English (prefixes such as *un-*), here called "lexical negation".

While Payne (1985) and Dryer (2005a) regard morphological/affixal negation as a type on a par with e.g. negative particles, Dahl opposes morphological negation to all other types, lumped together under the heading "syntactic negation". In Dryer (1988) and Miestamo (2005c), neither of which makes use of the distinction between bound and free marking in their classification, this is criticized. Dryer's critique seems to be restricted to Dahl's reluctance to extend his word order classification to bound morphemes, but Miestamo (2005c) claims that Dahl's classification is "not ideal for bringing out the essence of the cross-linguistic variation in the expression of" standard negation. Partly, this seems to rest on a misunderstanding of the intended criteria for morphological negation[2]. Another problem brought up by Miestamo is in fact one that is not exclusive to the morphological/syntactic distinction but will tend to appear with any attempt to reduce what is essentially a continuum to a set of discrete classes, namely that cases that end up on different sides of a borderline are treated as totally distinct even if they are in fact quite similar to each other. Miestamo's example (actually discussed already in Dahl 1979) is Polish vs. Czech: the orthographic criterion applied in Dahl (1979) makes Polish *nie* syntactic but the very similar-looking Czech *ne-* morphological[3]. However, as I shall discuss in more detail below, quite analogous problems show up in Miestamo's own classification. Dahl (1979) found that suffixal negation was more common than prefixal – the proportions were about 1.75: 1 in his sample, although he admitted the possibility of the sample being biased. Bybee (1985: 177) thinks that the latter was in fact the case, since in her balanced sample of 50 languages, there was in fact a 'slight preference for prefixal negation', with seven clearly prefixing and five clearly suffixing out of 15 languages with morphological negation in a sample of 50 languages. However, the data in Miestamo (2005c), taken from a balanced sample much larger than Bybee's, seem rather to confirm Dahl's claim, actually with as much as three times as many suffixes as prefixes among the clear cases. This would of course

---

[2] Miestamo argues that Dahl's treatment obscures the similarity between the negative constructions in Suena (Papua New Guinea, Trans-New Guinea) and Apalaí (Brazil, Carib) (which both involve the addition of an extra auxiliary), assuming that the latter would be seen as morphological. This assumption is based on an analogy from Dahl's treatment of "a similar construction in Chukchi". However, Dahl lists Chukchi (Russia; Chukotko-Kamchatkan) as having both morphological and syntactic negation; after checking the sources it now seems to me that the latter was wrongly classified and should have been type S22 (auxiliary+modification of finite verb) rather than S11 (negative particle). The definition of morphological negation was intended to exclude any multi-word construction; the classification of Chukchi negation as morphological was thus intended to apply only to the cases where there was no auxiliary (these cases do not seem to have a counterpart in Apalaí).

[3] It may be noted that differences in orthographic practice between Polish and Czech *ne-* do have some foundation in the spoken language: in Czech (a language with initial stress), *ne-* takes over the word stress from the following verb, which means that the negator and the verb form a clearer prosodic unit than in Polish (a language with penultimate stress), where the word stress is not moved.

be in accordance with the general preference for suffixing at least in inflectional morphology, but it should be noted that the tendency is still a bit weaker for negation than it is for some other categories, such as tense and aspect – thus, in the sample of Dryer (2005b), there were about four times as many languages with suffixes than prefixes marking those categories.[4]

Payne (1985: 226) declares (without motivation) that in morphological negation, "the negative morpheme must be considered to form part of the derivational morphology of the verb", Dahl (1979: 81) (also without further argumentation) says that morphological negation "is an inflectional category of the verb". Frequently, morphological negation interacts rather intimately with tense-aspect, mood and person/number; this can be seen as an argument for seeing it as inflectional rather than derivational, at least in the languages where this is the case. However, although Dahl does discuss at length the borderline cases between morphological and syntactic negation, he does not raise the question whether it is reasonable to see these as inflectional affixes, rather than as results of cliticization of free markers (Dryer 1988: 116). It may be noted here that the position of negative markers relative to other inflectional morphemes varies quite extensively between languages. Going through the languages classified as having bound negative markers in Miestamo (2005c), I found that in about almost half of those, the negation marker was the outermost morpheme in the word (judging from the examples given). Such negation constructions would be candidates for an analysis in terms of clitics. They are, however, much more frequent in prefixal negation; both negative prefixes and negative suffixes tend to precede other inflectional markers, meaning that prefixal negation is mainly word-initial, whereas negative suffixes either directly follow the verb stem or show up in the middle of a sequence of inflectional morphemes (cf. (1) above). In addition, fusion of negative affixes with other markers seems to happen only to the right of the verb stem. In other words, not only are bound negative markers more often suffixal than prefixal, but suffixal negation is also much more integrated into verbal morphology than prefixal negation. It should also be pointed out that the distinction between standard negation and lexical negation gets a bit blurred in some languages where

---

[4] Since the world's languages seen as a population may be biased in one way or other due to historical factors, numbers of actual languages having a certain property may be misleading. What we would like to know is how often the property shows up in an ideal sample of human languages. Although such a sample is an obvious fiction, the ultimate aim must be to assess the strength of cross-linguistic tendencies.

the negator is not integrated into the inflections of the verb. Thus, in some Slavic languages, the same prefix *ne-* is used for both, which means that the difference may be wholly neutralized or depend only on the order of elements, as in Czech *nebyl zdravý* 'was not healthy' vs. *byl nezdravý* 'was unhealthy'.

Morphological negation tends to be expressed affixally, rather than by other means. Not too infrequently though, tone changes also enter into the picture in various languages, in most cases probably in connection with affixation. A particularly complex example of this is found in Igbo (Miestamo 2005c: 275, quoting Green & Igwe 1963), where negation is expressed by a combination of tone and a "flip-flop" vowel prefix, present in imperfective negated and perfective affirmative sentences. However, at least in some cases, affirmative and negative verb forms may be marked by tone alone. Dahl (1979), quoting Becker-Donner (1965), mentions examples from Mano (Liberia, Niger-Congo) and Bond (2006), quoting (Barnwell 1969: 63, 80), has examples from Mbembe (Nigeria, Niger-Congo) such as

(3) a. *mɔ́-tá*
       3.FUT-go
       'He will go.'

     b. *mɔ̀-tá*
       3.NEG-go
       'He won't go.'

Another morphological process to be mentioned is reduplication, which was mentioned in Dahl (1979) as marginally appearing in Tabasaran (Dagestan, North-East Caucasian: Khanmagomedov 1967). Bond 2006 mentions two African languages – Eleme (Nigeria, Niger-Congo) and Banda-Linda (Central-African Republic, Niger-Congo: Cloarec-Heiss 1986), but in neither reduplication seems to be the predominant way of expressing negation.

The phenomenon of "paradigmatic asymmetry" (Miestamo 2005c: 52), i.e. the lack of a one-to-one correspondence between affirmative and negative paradigms, is common. It may take different forms: one is neutralization of other inflectional categories in negative paradigms. For instance, in Tamil, one single negative verb form corresponds to affirmative past, present, and (optionally) future verb forms (Schiffman 1999):

(4) naan poo-r-een 'I go' naan pooha-lle 'I don't go'
naan poo-n-een 'I went' naan pooha-lle 'I didn't go'
naan poo-v-een 'I will go' naan pooha-lle (or) naan pooha-maaTTeen
'I won't go'

Some Dravidian language have similar systems although there is no overt negative morpheme, that is, negation is expressed by dropping the tense marker. as in the following example from Old Kanarese (Master 1946: 142):

(5) kēḷ-v-en 'I hear' kēḷ--en 'I do not hear'
kēḷ-gu-m 'I will hear' kēḷ--en 'I will not hear'
kēḷ-d-en 'I hear, I shall hear' kēḷ-en 'I do not hear'

This, then, is a rather glaring counterexample to the generalization that morphological negation is always affixal.

Paradigmatic asymmetry can also involve a misfit between tense-aspect categories in the affirmative and negative paradigms without there being a straightforward neutralization. A much cited example is Swahili, where the choice between the different negative tenses depends on factors apparently not relevant in the affirmative, such as whether an event is expected or not (Contini-Morava 1989). A similar phenomenon is found in some Northern Swedish vernaculars, where the perfect can be negated in two ways, the usual syntactic one as in (6a) and with the prefix *o-* 'un-' prefixed to the main verb, as in (6b), with the latter carrying the additional meaning of expectedness ('not yet'). The examples are from Northern Westrobothnian (Marklund 1976):

(6) a. *i hæ eint skrive breve*
I have not written letter
'I haven't written the letter'

b. *i hæ oskrive breve*
I have un-written letter
'I haven't written the letter (yet)'

The identification of morphological and affixal negation is thus a slight over-simplification, although it may be true that there is no language in which negation is consistently marked by non-segmental means. As noted by Horn (1989: 472–473), the marking of negation behaves rather differently than that of polar questions, which is often expressed exclusively by intonation or word order.

How frequent is morphological negation? Estimates vary between 30% in Bybee (1985) and 45% in Dahl (1979). Dryer (2005a) gives 33% for negative affixes and Miestamo (2005c) has 40% for bound negative morphemes – the latter figure also includes a number of constructions in which the negative morpheme is bound to the verb but there is also some kind of added auxiliary.

This means that the proportion of one-word negation constructions in an ideal sample would probably closer to the figures given by Dryer and Bybee, Dahl's sample being rather unrepresentative in this regard. In any case, if as much as a third of all languages have morphological negation, that is still a notable fact, in view of the quite limited number of things that can be grammaticalized in verb morphology, in particular in the form of inflections.

Negative particles

'Negative particles' are negators that are characterized by two features: (i) they are independent words rather than affixes – as we have seen, a somewhat fuzzy condition; (ii) they are not inflected. This is arguably the most common type of standard negation. In the sample in Dryer (2005a), negative particles are found in about half of the languages – an exact figure does not make sense since there is both variation and many unclear cases. In Miestamo's classification, most negative particle constructions fall under symmetric negation.

A straightforward example of a negative particle would be Indonesian *tidak* as in

(7)   *Saya tidak tidur*
      I   not   asleep
      'I am not asleep'

From the syntactic point of view, the most interesting general property of negative particles is their placement in the sentence, a problem which will be discussed in detail below. There are, however, a couple of variants of the negative particle construction that demand special treatment. The first is the double particle construction, well-known from (written) French, where the negated counterpart of *Jean chante* 'Jean is singing' is *Jean ne chante pas*. This construction is found also in a number of other Romance and Germanic varieties, further in Celtic, Mayan and West African languages of different families. (See the section "Classifications" for an account of earlier treatments.)

Historically, such constructions in attested cases arise from the addition of a particle whose original function was to reinforce the negation in the French case, *pas* comes from a noun meaning 'step'. later development may, as in spoken French and some earlier stages of the Germanic languages, lead to the disappearance of the original particle (Jespersen 1917), and thus a return to the original simple particle construction. This kind of process, referred to in Dahl (1979) as 'Jespersen's cycle', might be seen as a result of a conflict between a tendency to grammaticalize negation, leading to, among other things, the loss of prosodic autonomy and independent stressability, and the pragmatic need of giving emphasis to the negated character of the sentence (Horn 1989: §7.1).

Both Dahl (1979) and Payne (1985) talk of double negative particles as being in general positioned on each side of the verb. There is at least one counterexample to this generalization, viz. Afrikaans, whose double particle construction is remarkable in two respects: (i) the particles both follow the verb and (ii) they are identical: *hy skryf nie 'n brief nie* 'he is not writing a letter'.

The other thing that can happen is that the verb in the negated sentence takes another form than in the corresponding affirmative sentence, most often one which is also used in various non-asserted clause types. For instance, in Mawng/Maung (Australia, non-Pama-Nyungan) where the particle *marig* is combined with one of two "irrealis suffixes", which are also used e.g. in contexts labelled "potential" and "hypothetical" (*ŋiudba* 'I put': *marig ŋiudbaŋi* 'I did not put'). For Miestamo (2005c), who seems to be alone among the other analysts to take these cases seriously, such cases belong to the subtype "A/NonReal", where there is a marking "that denotes non-realized states of affairs". Miestamo finds this type in about an eighth of all languages in his sample.

Negative verbs

There are two varieties of this type of construction: higher negative verbs and auxiliary negative verbs. The first type, in which negation is expressed by a verb with a sentential complement, is relatively uncommon but is attested in Malayo-Polynesian languages and at least one North American language, Squamish (Canada, Salishan; Kuipers 1967). An illustrative example from Payne (1985): in Tongan (Tonga, Malayo-Polynesian), the negative counterpart of *Na'e 'alu 'a Siale* 'Charlie went' is

(8) Na'e   'ikai [$_s$ke   'alu 'a   Siale $_s$]
     ASPECT NEG   ASPECT go   CASE Charlie
     'Charlie did not go'

where *ke* is an aspect marker which shows up in subordinate clauses only. Payne also provides other arguments for the claim that there is a clause boundary in (8). Higher negative verbs constitute the clearest counterexamples to the generalization made in Dahl (1979) that standard negation does not create syntactically complex sentences.

Auxiliary negative verbs are a more common type, but considerably less frequent than negative particles. In this type, the negative element takes over all or some of the inflectional categories characterizing finite verbs. The standard example is Finnish (Fenno-Ugric):

(9) a. *Pekka lukee*
       P.   read.PRS.3SG
       'Pekka is reading'

    b. *Pekka ei   lue*
       P.   NEG.3SG read
       'Pekka is not reading'

where *ei* is a negative auxiliary which agrees with the subject but does not have more than one tense and *lue* is the stem form of the verb. This illustrates the tendency for negative verb paradigms to be more or less defective – there are, though, examples of full sets of forms, as in Evenki (Russia, Tungusic; Nedjalkov 1994). Categories that are lacking in the negative auxiliary may instead be marked on the main verb, as tense in Finnish (the past of (9b) is *Pekka ei lukenut* 'Pekka was not reading'). Estonian (Fenno-Ugric), with its uninflected negative 'auxiliary' combined with various non-finite forms of the main verb, is an example of a degenerate auxiliary construction which comes close to a negative particle construction, possibly representing a general tendency for negative verbs to fossilize.

In Dryer (2005a), slightly less than 5% of the languages in the sample were labeled as clear cases of negative auxiliaries; interestingly, however, as many as 6.5% were classified as "negative word, unclear if verb or particle". This group includes languages which lack verbal inflectional morphology but also ones that Payne treats as higher negative verbs. Negative auxiliaries show clear areal patterns, being quite frequent in Northern Eurasia (which caused them to be overrepresented in the sample of Dahl 1979).

In Miestamo's system, negative verbs generally fall under the type A/Fin/NegVerb.

Non-negative auxiliaries in negation constructions

In some languages where negation is marked by an affix, this has the effect of making the verb non-finite, and a (non-negative) auxiliary has to be added. This is identified as a separate type only in the work of Miestamo (or rather as a sub-type, labeled "A/Fin/Neg-LV"). Hixkaryana (Brazil, Carib; Derbyshire 1979: 48) would be an example.

(10) a. *ki-amryeki-no*
1SUBJ-hunt-IMMPST
'I went hunting.'

b. *amryeki-hira w-ah-ko*
hunt-NEG   1SUBJ-be-IMMPST
'I did not go hunting.'

However, what is finite and non-finite is often a tricky question. Consider Japanese, discussed in Dahl (1979). The negated formal past verb form *kaimasen desita* 'did not buy' looks like the Hixkaryana example, in that negation is expressed by the suffix *-en* on the main verb (following the formality marker *-mas-*), and past tense is marked on the following auxiliary. In the present tense, on the other hand, *kaimasen* appears on its own, and would thus appear to be finite. This situation, which is found in various languages, can be compared to the use of copulas in marked tenses/persons in languages which do not employ them otherwise. Miestamo (2005c) classifies as much as 11 per cent of all languages as being of this type, but it should then be noted that the construction is often identifiable only in restricted cases.

A second type of negation construction where an auxiliary plays a role without carrying negative meaning by itself is also only classified separately by Miestamo (2005c) (labeled "A/Fin/Neg-FE"). In this type, a negative affix is attached to an auxiliary not present in the affirmative. This is less common; Miestamo finds it only in five languages (2 per cent of the sample). One of these languages, Korean, was in Dahl (1979) lumped together with English and a couple of other languages as having a "dummy auxiliary construction". Although Korean and English both employ auxiliaries with the original meaning 'do', what seems special to English is the possibility for the negator to be a free morpheme rather than an affix. It is actually hard to find any close parallels to the English situation, which is perhaps somewhat ironic in view of the central role dummy *do* constructions played in the early development of generative grammar. Miestamo treats English as belonging to the type A/Emph, that is, negative constructions that involve marking "that expresses emphasis in non-negatives". (He does not mention the use of *do* in questions, where it does not seem to express emphasis.) This is also an uncommon type and the other examples provided are not very clear.

Standard negation and word order

The position of many linguistic elements is largely predictable from the basic word order patterns of the language in question, as was shown by Greenberg (1963). Some linguists have tried to formulate Greenbergian principles also for the placement of negative morphemes, although in radically different ways. Thus, Lehmann (1974) claimed that negation would be preverbal in VO languages and postverbal in OV languages while for instance Bartsch & Vennemann (1972) thought that negation would behave like other adverbial modifiers, resulting in an order opposite to what Lehmann proposed. Another approach was taken by Jespersen (1917: 5), who claimed that there is a tendency to place negators "first, or at any rate as soon as possible, very often immediately before the particular word to be negatived (generally the verb)". The logical structure of this claim is actually a bit complicated – it can be seen as the disjunction of three different statements: (i) negators are placed initially; (ii) negators are placed "as soon as possible"; (iii) negators are placed immediately before the negated word, generally the verb. Here, it turns out that (i) does not receive strong support: sentence-initial placement of negators in non-verb-initial languages is not very common (it is even hard to find clear cases, Kiowa (USA, Kiowa Tanoan; Watkins & McKenzie 1984: 214) is perhaps the best example). What "as soon as possible" is supposed to mean is not quite obvious, but (iii) indeed seems to be empirically supported. It is also complex, however, and can be seen as the logical conjunction of the following three distinct claims: (a) the placement of the negator is generally defined relative to the verb; (b) the negator is in direct contact with the verb; (c) the negator tends to precede the verb. In fact, all of these receive fairly unanimous support from typological surveys of negation constructions. Thus, judging from the figures in Dryer (1988), negators are placed either directly before or directly after the verb in 80–90 per cent of all cases, and in both VO and OV languages, syntactic negators overwhelmingly precede verbs, the ratio between preverbal and postverbal placement being something like 3:1 in a hypothetical ideal sample. In other words, there is a preverbal tendency which is fairly independent of the order between object and verb – although it appears to be strongest for verb-initial languages in which there are only a handful of examples of post-verbal placement of negation. In other words, there is a "canonical" position for syntactic negators immediately before the verb which is relatively independent of Greenbergian basic word order. A few questions remain here, however:

1) What about morphological negation? We saw above that suffixal negation appears to be more common than prefixal negation. Might it be that there are more pre-verbal particles because the post-verbal ones have become suffixes (Dryer 1988: 114)? Moreover, it turns out that there is a positive correlation between verb-final word order and morphological negation: in OV languages, there are slightly more bound than free negators, whereas in VO languages, only about a fourth have morphological negation. Thus, it might even be the case that there is a Greenbergian correlation after all – negation tends to be postverbal in OV languages but this tendency is hidden by the fact that many postverbal negators attach to the verb as suffixes. Indeed, if we lump together particles and affixes, as is done e.g. by Dryer (1988), there is a correlation between the position of negators and the position of the object relative to the verb, in the way Lehmann suggested. But this correlation is far from perfect: in fact, Dryer still finds a slight preponderance for preverbal negation even in SOV languages. Also, we may remember in this connection that morphological negation is more often prefixal than comparable inflectional categories. This speaks in favour of an independent tendency for preverbal placement of negators.

2) What counts as a verb? More specifically, what happens if the sentence contains both a main (lexical) verb and an auxiliary? This question has been somewhat neglected in the literature. Thus, in his discussion of universals of negative position, Dryer (1988) does not mention auxiliaries at all except when saying that even if "English requires an auxiliary in negative sentences", this is "a relatively idiosyncratic quirk" which can be ignored. In Dahl (1979), on the other hand, it was proposed that the position of negators was typically defined relative to the "finite element" of the sentence – that is essentially an auxiliary whenever present or else the finite verb, and that uninflected negators tended to be placed before the finite element and as close as possible to it. In the case of verb-non-final languages, auxiliaries usually precede main verbs, and here the attested orders are Neg Aux Verb and Aux Neg Verb, or in the case of double negation, Neg Aux Neg Verb. When a negation is placed after the auxiliary, it tends to attach to the left, i.e. to the auxiliary, rather than to the main verb. The order Aux Verb Neg

does not seem to occur. These facts speak in favour (although perhaps not too strongly) of the hypothesis that it is the finite element that determines negation placement, . As for verb-final languages, Dahl's proposal suggests that in a language where the auxiliary follows the main verb, the preferred order would be Verb Neg Aux. Choosing the verb rather than the auxiliary as pivot, which Dryer appears to do, predicts the order Neg Verb Aux. In fact, both orders occur, sometimes in one and the same language. Thus, in Hindi the following sentences are both grammatical (Vasishth 1999)

(11) a. *raam roṭii nahĩĩ khaataa   thaa*
        Ram bread NEG   eat-IMP-PART-MASC be-PAST-MASC
    b. *raam roṭii khaataa   nahĩĩ thaa*
        Ram bread eat-IMP-PART-MASC NEG   be-PAST-MASC
    'Ram did not (use to) eat bread.'

If auxiliaries historically derive from main verbs, the order Verb Neg Aux would be expected, so it is possible that the Neg Verb Aux order is an innovation in the languages where it occurs. In that case, it may be that in combinations of verbs and highly grammaticalized auxiliaries, there is a tendency to let the position of the negation be determined by the whole "finite cluster" rather than by the auxiliary alone.

3) Not all kinds of syntactic negation necessarily obey the same word order principles. In fact, it is not immediately clear how to apply a principle relating negators to a verb or an auxiliary if the negator itself is an auxiliary and thus presumably the finite element in the sentence. Accordingly, unlike Dryer (1988), who treated negative auxiliaries in the same way as negative particles, Dahl (1979) claimed that negative auxiliaries are not subject to the preverbal tendency but rather follow Greenbergian word order principles, meaning that they would follow the lexical verb in verb-final languages, like auxiliaries in general. The problem in evaluating this claim is that negative verbs are concentrated to certain areas and families, and may therefore not be numerous enough in balanced samples to show significant tendencies. Thus, in Miestamo's sample, there are 12 languages classified as "A/Fin/ NegVerb" with OV order. Of these, the negative auxiliary precedes the main verb in 5 languages and follows it in 7. This is a higher proportion than for negative particles in OV languages, where the ratio between pre-verbal and post-verbal placement is 2.67: 1 in the same sample, but the difference is not statistically significant. It may also be noted that if the position of negative verbs were totally determined by Greenbergian principles, we would

expect all such verbs to be sentence-final in verb-final languages, which is rather far from being the case.

In a number of languages of different word order types (particularly common in West and Central Africa), inflectional categories usually connected with finiteness such as tense and subject markers show up on an auxiliary-like element which may be non-contiguous with the verb. Not infrequently, negation is also marked on this element. Superficially, then, such languages look as if they contradicted the thesis that negation is in direct contact with the verb. Thus, Dryer (1988: 123) lists six languages with the order "SNegOV": Yaqui (Mexico, Aztec-Tanoan), Bambara (Mali, Niger-Congo), Mandinka (Senegal, Niger-Congo), Vai (Liberia, Niger-Congo), Berta (Ethiopia, Nilo-Saharan) and Songhai (Mali, Nilo-Saharan). If we disregard Berta (which appears to be a mistake[5]), four out of the five remaining ones – the three Niger-Congo languages and Songhai – are SOV languages of the type just described. Likewise, Dryer (2009) notes a number of languages in Central Africa which have the order SVONeg, but which also exhibit the order SVOAux. All these would be counterexamples to the contact thesis but only if we assume that it applies to the relation between negators and main verbs.

On the other hand, as was noted in Dahl (1979) and demonstrated on a larger sample in Dryer (2009), sentence-final placement of negators is common in all the three major language families (Afro-Asiatic, Niger-Congo, Nilo-Saharan) in Central Africa, also in languages which do not have auxiliaries in final position. In the same area, double negative particles are also common, often with the second negator in sentence-final position. Dahl (1979) suggested a connection between these tendencies, in that they would both be explained as results of Jespersen's Cycle, if it is assumed that sentence-final negators arise from adverbial elements used to reinforce negation. This hypothesis seems yet to await confirmation.

"Non-standard negation"

As was noted above, most typological work on negation has focused on standard negation, i.e. the negation constructions used in main verbal declarative clauses – the structures prototypically associated with finiteness. It is quite common – in the case of imperatives one should perhaps even say "normal" – for negation in other constructions to deviate more or less

---

[5] Dryer (2005a) lists Berta as having a negative affix, and this is also in accordance with the information in Cerulli (1947).

completely from standard negation. However, the focus on standard negation means that there are in several cases no systematic typological surveys to rely on, therefore I have to largely restrict myself to illustrative examples here. (See also Horn 1989: 447–462, for a general discussion of differentiated expression of negation.)

Negative imperatives. The least complex way of negating an imperative should be one where the same marker as is used in declaratives is added to a positive imperative. This kind of construction is indeed found in some languages, and is dominant in Europe, but it constitutes a minority – in fact only 23 per cent of the sample in van der Auwera et al. (2005), where two kinds of "asymmetries" are distinguished: (i) differences in negation strategy between (indicative) declaratives and imperatives; (ii) differences in the verbal construction used in positive and negative imperatives, by them labeled "prohibitives"[6]. Almost exactly two thirds of the languages in van der Auwera et al. (2005) use different negation strategies for declaratives and imperatives. For instance, Classical Greek used *ou* in declaratives and *mē* in imperatives. Two fifths of the languages display differences in the construction used in positive and negative imperatives, and almost thirty per cent show both kinds of asymmetries.

Negation in sentences with non-verbal predicates. Sentences where the predicate is not a lexical verb but e.g. a noun, an adjective or a locative phrase often exhibit special ways of expressing negation, even if the claim put forward in Eriksen (2005) that "nominal predicates may never be directly negated" appears a bit too strong. Languages differ as to whether they use copulas with non-verbal predicates and as to what constructions demand copulas. In copula-less constructions, special negators are often used, as in Indonesian (Malayo-Polynesian), where *bukan* replaces the standard negator *tidak* with nominal predicates, e.g.

(12) *Itu bukan jeruk.*
 this NEG orange
 'This is not an orange'

---

[6] In view of the pervasive differences between positive and negative imperatives, it may be a good idea to use a special term such as "prohibitive" for the latter. It should be noted, however, that negative imperatives can express other speech acts than prohibitions, notably warnings – these are sometimes formally differentiated, as in Russian, where negative imperatives take the perfective aspect only if they express a warning rather than a prohibition.

In copular constructions, the ordinary copula is sometimes replaced by a special negative one, as in Czech:

(13) a. *Jan je   doma*
        Jan COP.3SG at_home
        'Jan is at home'
    b. *Jan není   doma*
        Jan NEG.COP.3SG at_home

In addition, optional copulas may show up more frequently in negated sentences, although unequivocal examples of this are hard to come by.

Negation in existential sentences. Existential constructions show similarities to non-verbal predication – often the existential verb is identical to the copula. There are parallels also in negative constructions. Thus, suppletive existential negative verbs are common, e.g. Turkish *var* 'exist' vs. *yok* 'not exist' or Russian *est'* 'exist' vs. *net* 'not exist' (only used in the present, for the other tenses the standard negator *ne* is combined with forms of *byt'* 'to be'). Another possibility is represented by Polish, where the existential *jest*, identical to the copula, is replaced by *ma* 'has' in negated sentences, although the negator is the standard one (*nie*).

Sometimes, the standard negator and the negative existential are identical (Croft 1991: 11). In Sirionó (Bolivia, Tupían), a negated existential sentence can be constructed simply by suffixing the the standard negator *-ä* to a noun, as in

(14) *tikise-ä   tuchi*
     machete-NEG INTENSIFIER
     'we have no machetes at all'   (Priest & Priest (1980: 96))

Existential sentences normally involve quantifying over an indefinite set of entities, which means that they tend to display phenomena such as polarity-sensitivity (see next section).

One reason for negative existentials being lexicalized is their high frequency in spoken language. In certain spoken Russian genres, *net* 'not exist' occurs twice as often as the affirmative existential *est'* (Ljuba Veselinova, pers. comm.). Obviously, the lack of this or that is a favourite topic in most human societies, but one should make the reservation that there may be alternative ways of expressing the existence or presence of something in affirmative sentences that influence such figures.

Negation in embedded clausal structures. Embedded clauses, whether finite or non-finite, is another type of context in which negation behaves differently from standard negation in many languages. Examples discussed in Payne (1985) include Yoruba and Welsh, in which subordinate clauses employ different negators than main clauses. Classical Greek would be another example, where many finite as well as non-finite embedded constructions use the negator *mē* also found in negative imperatives (see above) rather than the standard negator *ou*.

Payne also notes that non-finite negation in English is preverbal rather than postverbal – analogously, the two parts of the French *ne...pas* construction show up before infinitives (*ne pas chanter*) rather than wrapped around them, as with finite verb forms. In many Germanic languages there are differences in the placement of negation in main and subordinate clauses, although most syntactic theories would ascribe those to the effect of finite-verb movement in main clauses.

Quantification, polarity-sensitivity and focus

Negation and quantification. Negation often shows up in combination with quantification, as in *No man is an island*. Languages vary in the ways such combinations are realized, and even within one language more than one construction may be possible, as when English *nobody* varies with *not anybody*. Surveys of the typological variation are given in Kahrel (1996) and Haspelmath (2005). The main criterion for Haspelmath's classification is whether negative indefinite pronouns such as 'nobody' and 'nothing' cooccur with the marker of standard negation. Consider e.g. the following examples from Russian and German:

(15) a. N*ikto   ne   prišel*
       nobody NEG come.PST.M.SG
       'Nobody came'
    b. *Niemand kam*
       nobody   come.PST.3.SG
       'Nobody came'

In Russian, the negator *ne* is obligatory in (15), whereas the addition of the standard negator *nicht* to the German sentence would render it ungrammatical. An overwhelming majority (170 languages) of the 206 languages in Haspelmath's sample go with Russian here, and only about five per cent (11 languages) consistently behave like German. About the same number

(13 languages) show "mixed behaviour": this includes languages such as Spanish and Italian, which are like German when the negated pronoun is in subject position and like Russian in other cases, and English, where there is often a choice e.g. between *nobody* and *not anybody*.

Kahrel (1996) gives a more fine-grained classification in which indefinite pronouns that cooccur with standard negators are classified into three types: "NEG plus indefinite", "NEG plus special indefinite" and "NEG plus zero quantification". In the first type, the same indefinite pronouns are used under negation as in affirmative sentences, in the second, polarity-sensitive items such as English *anybody* are used under negation (see further below). The third type is the one exemplified by Russian, where a standard negator is combined with "inherently negative quantifiers". In his 40-language sample, Kahrel found that the type "neg plus indefinite" was represented in about two thirds of the languages, the other two types being much less frequent. Haspelmath rejects the notion of "inherently negative" indefinite pronouns, which means that the following two sentences from English and French both represent the "Russian" option, in spite of the fact that *rien* would usually be thought of as corresponding to English *nothing* rather *anything*:

(16) a. *I have not said anything*

　　b. *Je n'　　ai　　　　　　rien　dit*
　　　I　NEG　have.PRS.3SG　nothing　say.PP

In addition to the types discussed so far, both Haspelmath and Kahrel define a separate "negative existential construction" type, in which the translation of *Nobody came* would be something like "there isn't a person who came". This type is found in 6 per cent of the languages in Haspelmath's sample, particularly often in Oceanic languages. Counting also languages in which this is not the primary strategy, Kahrel found it in 7 languages of 40 (18 per cent).

Polarity-sensitivity. Polarity-sensitivity, which has been mentioned a couple of times above, is the phenomenon by which certain expressions or forms are restricted to occur either inside or outside the scope of negation. (See Hoeksema's paper in this volume.) In addition to the pair *some* and *any*, polarity-sensitive items in English include idioms like *would rather* (affirmative polarity) and *lift a finger* (negative polarity), and sentential operators such as *also/either*. From other languages one may mention examples of strengthening adverbs such as Swedish *mycket* 'very' which cannot be used in negative contexts but must be replaced with *särskilt* 'particularly' or some synonym. Other noticeable examples are found in the case systems of e.g. the Slavic and Fenno-Ugric languages where the genitive or the partitive is often used in negative sentences instead of the accusative or (less often) the nominative, or in mood systems, e.g. that of French, where the complements of negated propositional attitude verbs tend to be in the subjunctive rather than the indicative.

Even if polarity-sensitivity has been discussed from a cross-linguistic point of view in works such as Wouden (1997), no systematic typological survey of polarity-sensitivity has been undertaken, to my knowledge, and most studies have been done on European languages (including the papers in Hoeksema et al. 2001). In general, polarity-sensitivity seems to be linked to quantifying contexts and scalarity phenomena. Thus, negative polarity items are often what Horn (1989: 400), following Bolinger (1972), calls "minimizers", i.e. expressions like *the slightest chance* which denote a minimal quantity of something, which are often used to add emphasis to negative statements (and are consequently often involved in Jespersen's Cycle). Polarity-sensitivity is thus clearly a phenomenon with a semantic or pragmatic basis but can sometimes be grammaticalized (as in the case of case in Slavic and Fenno-Ugric).

Negation and focus. This is a topic that has to be treated more briefly here than it deserves. It is obvious that negation often behaves in special ways in focus constructions. Thus, in many languages, a negator can be moved from its standard position to the focused constituent, as in the following Russian examples:

(17) a. *V sledujuščij raz ja ne pridu.*
      in next   time I NEG come.NON-PST.3SG
      'Next time I won't come' (unmarked focus)

  b. *V sledujuščij raz pridu ne ja.*
      in next   time come.NON-PST.3SG NEG I
      'Next time it won't be me who comes' (focus on subject)

In the absence of a systematic survey, it is hard to tell how widespread this phenomenon is. In grammars, it is often seen as a case of constituent negation, but this obscures the relationship to focus; thus (13)(b) is the denial of the affirmative focused sentence (18):

(18) *V sledujuščij raz pridu ja.*
      in next   time come.NON-PST.3SG I
      'Next time it will be me who comes' (focus on subject)

Diachrony and grammaticalization

In language typology, diachronic aspects have grown in importance as the field has developed; it has become increasingly clear that the processes which linguistic patterns arise and spread in languages are important for the understanding of why languages look the way they do. Particularly important are processes of grammaticalization, by which lexical items are recruited and become integrated into grammatical constructions as markers.

In the case of negators and negative constructions, a relatively small number of diachronic sources have been proposed. The most celebrated one involves "Jespersen's Cycle", which was discussed above in the section on negative particles. (See van der Auwera's chapter in this volume for elaborations.)

Another source for standard negators is negated existential verbs (Croft 1991). Croft in fact proposes another diachronic cycle, by which languages go through the following stages: (i) the standard negator is used also in existential sentences; (ii) a special negative existential arises by fusion of the standard negator and the existential verb or by some other process; (iii) the negative existential starts being used as a standard negator; (iv) negated existential sentences employ a combination of the previous negative existential and the non-negated existential verb – at which point the original stage is reached. Negative existentials are plausible sources for negators in many Indo-Aryan languages (cf. the Hindi example (11)).

A third source, apparently not widely attested, would be verbs such as 'lack', 'leave', and 'refuse' (examples mentioned in the literature are mainly from African languages (Givón 1973; Marchese 1986).

As an object of grammaticalization, negation displays some not quite usual features. Thus, although Jespersen's Cycle is often adduced as a paradigm case of grammaticalization, what happens here is actually somewhat different from many other cases of grammaticalization, where a previously optional element becomes obligatory and backgrounded. Grammaticalized elements, such as tense markers, tend to express redundant information that is either presupposed or "incidental" to the basic message. This is hardly possible with negation, since the very point of a negated sentence is typically the fact that it is negated – that is, the denial of some proposition. Whereas languages may well lack inflectional categories such as tense, aspect, case and definiteness, it is hard to imagine a language in which negation has zero expression, i.e. where negative sentence are identical to their affirmative counterparts. In spite of this, what we observe in language after language is that negation morphemes tend to be unstressed, phonetically reduced and eventually fuse with the finite verb or auxiliary of the sentence, seemingly without any change in the semantics or pragmatics of the negation morpheme. This process of reduction and tightening cannot be accounted for by a change in the informational or rhetorical value of the negation morpheme, but must be explained in some other way. We may note that when we emphatically assert or deny some proposition, this is done in many languages by assigning extra stress to the finite element of the sentence. In English, this will often be the dummy auxiliary *do*. In the Slavic languages, negation morphemes (*ne, nie* etc.) usually form a prosodic word unit with the verb, although this is not always reflected in the written language. It is thus normal that it is this unit as a whole, rather than the negative morpheme as such, that receives extra emphatic stress. This process, by which prosodic prominence is shifted from the negation morpheme to the finite element, is probably what is behind the reduction of the former and its integration into the latter. In Jespersen's Cycle, something extra happens, viz. some strengthening morpheme (such as *pas* 'step' in French) is bleached and finally reinterpreted as a standard negation morpheme. Here, the original negation morpheme does become redundant and may even disappear totally. In an alternative but somewhat similar development, a negator may be replaced by negative indefinite pronoun such as 'nothing' or adverb such as 'never'. Thus, Latin *non* derives from *ne oenum* 'not a thing', and words deriving from *never* are used as negations in some English-based creoles.

It is also relevant to point out here that negation is not particularly unstable in comparison to many other grammatical items. It is somewhat striking that whereas at least some historical sources of negators have been identified, negators themselves more seldom appears as the source in grammaticalization processes. Heine & Kuteva (2002: 216) list only one type of process that starts with negation and ends with something else, viz. the genesis of question markers from negators. Thus, probably for reasons similar to those mentioned in the previous paragraph, a negator does not normally undergo "semantic bleaching" (unless it is combined with some other element in Jespersen's Cycle). Furthermore, it should be emphasized that Jespersen's Cycle and other similar processes are not determined by fate; the fact that the Proto-Indo-European preverbal negator *ne* survives after 6,000 years or so in more or less unmodified form in most Slavic and Iranian and some Indo-Aryan languages shows that negators are not doomed to undergo change or be replaced.

Concluding remarks

I said in the beginning that typologists who have studied negation have had a tendency to focus on the most salient, or supposedly most basic, phenomena, and at several points in the chapter I have noted the absence of systematic typological surveys. In particular, there has so far been relatively little attention paid to the principles by which languages with more than one negative construction choose between them.

In fact, there is a wealth of potential topics for research papers or even doctoral theses relating to the typology of negation, some of which I haven't even had a chance to mention above, such as lexical negation, negated questions, the interpretation of scope ambiguities, the interaction with modality, etc.

References

Anderson, John M.
    2007   Finiteness, mood, and morphosyntax. *Journal of Linguistics* 43: 1–32.

Barnwell, Katherine
    1969   A grammatical description of Mbembe (Adun dialect), a Cross River language. University College London.

Bartsch, Renate and Theo Vennemann
    1972   *Semantic Structures*. Frankfurt am Main.: Athenäum.

Becker-Donner, Etta
    1965   *Die Sprache der Mano*. Wien: H. Böhlaus Nachf.

Bernini, Giuliano and Paolo Ramat
    1996   *Negative sentences in the languages of Europe: a typological approach*. Empirical approaches to language typology 16. Berlin/New York: Mouton de Gruyter.

Bolinger, Dwight L.
    1972   *Degree words*. Janua linguarum. Series maior 53. The Hague: Mouton.

Bond, Oliver
    2006   A questionnaire for negation: capturing variation in African languages. http://www.soas.ac.uk/linguistics/research/research-projects/negtyp/39436.pdf.

Bybee, Joan L.
    1985   *Morphology: a study of the relation between meaning and form*. Typological studies in language (TSL) 9. Amsterdam/Philadelphia: John Benjamins.

Cerulli, Enrico
    1947   Three Berta Dialects in Western Ethiopia. *Africa* XVII: 156–169.

Cloarec-Heiss, France
    1986   *Le Banda-Linda de Centrafrique*. Cambridge: Cambridge University Press and Éditions de la Maison des Sciences de l'Homme pour la Société d'Études Linguistique et Anthropologique de France.

Contini-Morava, Ellen
    1989   *Discourse pragmatics and semantic categorization: the case of negation and tense-aspect with special reference to Swahili*. Discourse perspectives on grammar 1. Berlin/New York: Mouton de Gruyter.

Croft, William
    1991   The Evolution of Negation. *Journal of Linguistics* 27: 1–27.

Curry, Haskell B.
    1961   Some logical aspects of grammatical structure. Structure of Language and Its Mathematical Aspects: Proceedings.

Dahl, Östen
    1979   Typology of sentence negation. *Linguistics* 17: 79–106.

Dahl, Östen
    2004   *The growth and maintenance of linguistic complexity*. Studies in language companion series, 71. Amsterdam: John Benjamins.

Derbyshire, Desmond C.
    1979   *Hixkaryana*. Lingua descriptive studies (LDS), 1. Amsterdam: North-Holland.

Dryer, Matthew S.
    1988   Universals of negative position. In *Studies in Syntactic Typology*, Michael Hammond, Edith A. Moravcsik and Jessica R. Wirth (eds.), 93–124. Amsterdam: John Benjamins.

Dryer, Matthew S.
    2005a   Negative Morphemes. In *The World Atlas of Language Structures*, Martin Haspelmath, Matthew S. Dryer, David Gil and Bernard Comrie (eds.), 454–457. Oxford: Oxford University Press.

Dryer, Matthew S.
    2005b   Position of Tense-Aspect Affixes. In *The World Atlas of Language Structures*, Martin Haspelmath, Matthew S. Dryer, David Gil and Bernard Comrie (eds.), 282–285. Oxford: Oxford University Press.

Dryer, Matthew S.
    2009   Verb-object-negative order in central Africa. In *Negation Patterns in West African Languages and Beyond*, Typological Studies in Language 87, Norbert Cyffer, Erwin Ebermann, and Georg Ziegelmeyer (eds.), 307–362. Amsterdam: John Benjamins.

Eriksen, Pål Kristian
    2005   On the Typology and the Semantics of Non-Verbal Predication. Department of Linguistics and Scandinavian Studies, University of Oslo.

Givón, Talmy
    1973   The time-axis phenomenon. *Language* 49: 890–925.

Green, Margaret M. and Georgewill E. Igwe
    1963   *A Descriptive Grammar of Igbo*. London: Oxford University Press.
Greenberg, Joseph H.
    1963   Some universals of grammar with particular reference to the order of meaningful elements. In *Universals of Language*, Joseph H. Greenberg (ed.), 73–113. Cambridge, MA: MIT Press.
Haspelmath, Martin
    2005   Negative Indefinite Pronouns and Predicate Negation. In Haspelmath et al. (eds.), 466–469.
Haspelmath, Martin, Matthew S. Dryer, David Gil and Bernard Comrie (eds.)
    The World Atlas of Language Structures. Oxford: Oxford University Press. (Online version 2008.)
Heine, Bernd and Kuteva, Tania
    2002   World lexicon of grammaticalization. Cambridge: Cambridge University Press.
Hoeksema, Jack, Hotze Rullmann, Victor Sanchez-Valencia and Ton van der Wouden (eds.)
    2001   *Perspectives on Negation and Polarity Items*. Amsterdam: John Benjamins.
Horn, Laurence R.
    1989   A natural history of negation. Chicago: University of Chicago Press.
Jespersen, Otto
    1917   Negation in English and Other Languages. Reprinted in *Selected Writings of Otto Jespersen* (1962): George Allen and Unwin, London.
Jung, Ingrid
    1989   *Grammatik des Paez, ein Abriss*. University of Osnabrück.
Kahrel, Peter
    1996   *Aspects of negation*. University of Amsterdam.
Kahrel, Peter and van den Berg, René
    1994   *Typological studies in negation*. Amsterdam: John Benjamins.
Khanmagomedov, B. G.-K.
    1967   Tabasaranskij jazyk. In *Jazyki narodov SSSR – Iberijsko-kavkazskie jazyki*, E. A. Bokarev and K.V. Lomtatidze (eds.), 545–561. Moskva: Nauka.
Klima, Edward S.
    1964   Negation in English. In *The structure of language: readings in the philosophy of language*, Jerry A. Fodor and Jerrold J. Katz (eds.), 246–323. Englewood Cliffs, NJ: Prentice-Hall.
Kuipers, Aert H.
    1967   *The Squamish language: grammar, texts, dictionary*. Janua linguarum. Series practica 73. The Hague: Mouton.
Lehmann, Winfred P.
    1974   *Proto-Indo-European syntax*. Austin: University of Texas Press.

Marchese, Lynell
1986   Tense/aspect and the development of auxiliaries in Kru languages. Summer Institute of Linguistics publications in linguistics, publication no. 78. Dallas, TX; [Arlington]: Summer Institute of Linguistics, University of Texas at Arlington.

Marklund, Thorsten
1976   *Skelleftemålet: grammatik och ordlista: för lekmän – av lekman*. Boliden: self-published.

Master, Alfred
1946   The zero negative in Dravidian. *Transactions of the Philological Society* 1946: 136–155.

Miestamo, Matti
2005a   Subtypes of Asymmetric Standard Negation. In *The World Atlas of Language Structures*, Martin Haspelmath, Matthew S. Dryer, David Gil and Bernard Comrie (eds.), 462–465. Oxford: Oxford University Press.

Miestamo, Matti
2005b   Symmetric and Asymmetric Standard Negation. In *The World Atlas of Language Structures*, Martin Haspelmath, Matthew S. Dryer, David Gil and Bernard Comrie (eds.), 458–461. Oxford: Oxford University Press.

Miestamo, Matti
2005c   *Standard Negation: The Negation of Declarative Verbal Main Clauses in a Typological Perspective*. Berlin/New York: Mouton de Gruyter.

Miestamo, Matti
2006   On the Complexity of Standard Negation. In *A Man of Measure: Festschrift in Honour of Fred Karlsson on His 60th Birthday* [Special Supplement to SKY Journal of Linguistics 19], M. Suominen, A. Arppe, A. Airola, O. Heinämäki, M. Miestamo, U. Määttä, J. Niemi, KK Pitkänen, and K. Sinnemäki (eds), 345–356.

Nedjalkov, Igor
1994   Evenki. In Typological studies in negation, Peter Kahrel and René van den Berg (eds.), 1–34. Amsterdam: John Benjamins.

Payne, John R.
1985   Negation. *Language Typology and Syntactic Description* 1: 197–242.

Priest, Perry and Anne Priest
1980   *Textos sirionó*. Riberalta, Bolivia: Instituto Lingüístico de Verano.

Schiffman, Harold F.
1999   *A reference grammar of spoken Tamil*. Cambridge: Cambridge University Press.

van der Auwera, Johan and Ludo Lejeune
2005   The Prohibitive. In *The World Atlas of Language Structures*, Martin Haspelmath, Matthew S. Dryer, David Gil and Bernard Comrie (eds.), 290–293. Oxford: Oxford University Press.

Vasishth, Shravan
 1999 Word order, negation, and negative polarity in Hindi. In *OSU Working Papers in Linguistics*, Vol. 53, Robert Levine, Amanda Miller-Okhuizen and Antony Gonzalves (eds.), 108–131. Columbus: The Ohio State University, Department of Linguistics.

Watkins, Laurel J. and Parker McKenzie
 1984 *A grammar of Kiowa*. Studies in the anthropology of North American Indians. Lincoln: University of Nebraska Press.

Wouden, Ton van der
 1997 *Negative contexts: collocation, polarity and multiple negation*. Routledge studies in Germanic linguistics 1. London: Routledge.

Zeshan, Ulrike
 2005 Irregular Negatives in Sign Languages. In *The World Atlas of Language Structures*, Martin Haspelmath, Matthew S. Dryer, David Gil and Bernard Comrie (eds.), 560–563. Oxford: Oxford University Press.

Zimmer, Karl E.
 1964 *Affixal negation in English and other languages: an investigation of restricted productivity*. Word Monograph 5. New York: Linguistic Circle of New York.

The Acquisition of Negation

Christine Dimroth

1. Introduction

The acquisition of negation has fascinated researchers since the early child language studies in the beginning of the last century and has been a fruitful topic of inquiry ever since. This might have to do with the clash between the perplexing complexity of negation in natural languages on the one hand and the observation that children start to use negation early and apparently without major problems on the other. A word for negation is typically one of the first words children learn, and, often to the regret of their care-givers, it is also "one of the most consistently used words throughout the single-word utterance period" (Pea 1980: 178). These early negation words differ from most other early words in that they are inherently relational in nature. As early as 1928, Stern and Stern already noticed the head start of *no* and equivalents, which typically also precede their affirmative counterpart (*yes*) in child language. Once children leave the one-word stage and start to master more complex syntactic constructions, negation plays a leading role again. It is one of the first clausal operators that children productively use.[1]

The first language (L1) acquisition of negation is not an immediate development, however. Children's early negation words do not cover the whole array of negative meanings found in adult language. The language specific lexical differentiation of negation words and their integration into fuller utterances takes time.

Questions regarding the acquisition of negation as one of the first clausal operators are particularly interesting since we are dealing with developing systems. It is not the case that children just have to learn how to add negation to otherwise complete sentences. They have to learn how to express negation given the current state of their learner language.

This is equally true for second language (L2) acquisition later in life. The main difference between the two types of acquisition lies in the avail-

---

[1] In many languages, additive particles like *also* are attested equally early (Dimroth 2009; Penner et al. 2000).

ability of knowledge about the function of negation. Adult L2 learners are experienced language users who "know" about the different functions of negation from their L1, but are often struggling with their formal expression in the target language.

Children do not only have to understand how to negate sentences, they must also learn that negation can affect (or have scope over) different parts of sentences and leave others unaffected. Adult L2 learners, on the other hand, might again "know" these principles from their L1. What both groups have to learn is what the meaning is of different negation words, which ones are used in given syntactic contexts and how the scope of negation is signaled in the target language. L1 and L2 learners thus have different starting points, and the developmental patterns observed vary as a function of this knowledge, and also as a function of the specifics of the (source and) target language.

Whereas the literature on the role of negation in L1 acquisition has covered the entire developmental process, studies on the L2 acquisition of negation are often more selective. In particular, studies have focused on the interaction between negation and other properties of sentence grammar that seem to pose related problems to the learner. For example, it has been shown that the acquisition of negation – in particular the placement of negation words – depends to a large degree on the acquisition of finiteness (Becker 2005, Meisel 1997, Verhagen and Schimke 2009). In many contributions, negation has thus not been studied in its own right, but rather as a window into structural development in other domains (most prominently, in studies on the acquisition of syntax, as a measure of verb raising).

This overview focuses on core issues relating to the acquisition of meaning and form of negation in L1 and briefly compares the main findings to the evidence available from L2. Single-word negation, anaphoric negation, clausal negation, and constituent negation are covered, whereas the acquisition of negative polarity items, negative concord, and negative word formation are not. To date, the acquisition of negation in a large array of target languages has been studied. Table 1 provides a selective overview.

Table 1. Studies on L1 or L2 acquisition of negation in different target languages

Target language	Acquisition type	Sources
ASL	L1	Anderson and Reilly 1997
Cantonese	L1	Tam and Stokes 2001
Dutch	L1	Jordens 1987, van der Wal 1996
	L2	Jordens 1987, Verhagen 2009
English	L1	Bloom 1991, Cameron-Faulkner et al. 2007, Choi 1988, Clahsen 1988, Drozd 1995, Felix 1987, Harris and Wexler 1996, Klima and Bellugi 1966, Pea 1980
	L2	Cancino et al. 1978, Giuliano 2003, Milon 1974, Perdue et al. 2002, Silberstein 2001, Wode 1981
Finnish	L1	Bowerman 1973
French	L1	Choi 1988, Clark 1985, Meisel 1997, Verrips and Weissenborn 1992, Weissenborn et al. 1989
	L2	Giuliano 2003, Meisel 1997, Perdue et al. 2002, Schimke 2008, Véronique 2005
German	L1	Clahsen 1988, Drenhaus 2002, Felix 1987, Hummer et al. 1993, Meisel 1997, Schaner-Wolles 1995/96, Stromswold and Zimmermann 1998, Verrips and Weissenborn 1992, Weissenborn et al. 1989, Wode 1977
	L2	Becker 2005, Dietrich and Grommes 1998, Meisel 1997, Parodi 2000, Dimroth 2008, Schimke 2008
Hebrew	L1	Weissenborn et al. 1989
Hungarian	L1	Barbarczy 2006
Italian	L1	Volterra and Antinucci 1979
	L2	Bardel 1999, Bernini 2000
Japanese	L1	Clancy 1985, Ito 1981, McNeill and McNeill 1973, Sano 1998, Wakabayashi 1983
Korean	L1	Choi 1988, Hahn 1981
Latvian	L1	Rūķe-Draviņa 1972
Mandarin	L1	Lee 1982
Polish	L1	Smoczyńska 1885
Russian	L1	Snyder and Bar-Shalom 1998
Swedish	L1	Lange and Larsson 1973
	L2	Hyltenstam 1977
Tamil	L1	Ramadoss and Amritavalli 2007, Vaidyanathan 1991
Turkish	L1	Aksu-Koç and Slobin 1985

It is not possible to summarize the findings on the acquisition of negation in these different languages, but cross-linguistic variation will be discussed where it becomes important for more general issues, e.g. the question of whether it is easier to acquire different functions of negation if there is a different negation word for each of them, or the question of whether word order regularities in a given language influence the acquisition process.

Section 2 deals with developmental patterns in the L1 acquisition of negation and covers research on children's acquisition of negative meanings (2.1) and forms (2.2), thereby focusing on the acquisition of lexical items (2.2.1) and their position in the sentence in relation to their scope (2.2.2). Section 3 compares these findings to what is known about negation in L2 acquisition and presents similarities (3.1) and differences (3.2). For the reasons mentioned above it focuses on formal properties of negative utterances, in particular the acquisition of lexical items and their position and scope properties. Section 4 presents some conclusions.

2. Developmental patterns in L1

This section summarizes findings from the acquisition of meaning and form of negation, focusing on orders of acquisition and their implications for L1 development.

2.1. How children acquire the functions of negation

Young children start to express negation before they actually learn to speak, but their early negation gestures and words do not yet cover the entire array of negative meanings available in adult language. It is widely accepted that development starts with affective or volitional functions of negation like the ones that have already been described by Stern and Stern (1928). These authors concluded "Das erste *nein* des Kindes bedeutet nicht „nein, das ist nicht so", konstituiert also nicht ein negatives Urteil, sondern es bedeutet: „nein, es soll nicht so sein", „nein, das will ich nicht", „nein, das sollst du nicht tun", stellt also eine abwehrende Stellungnahme dar" (p. 267).[2] Stern and Stern also found that the development from here to the acquisition of truth-functional negation often takes several months.

---

[2] "Children's first *no* does not mean 'no, this is not the case' and thus does not constitute a negative judgment, but means 'no, it should not be like this', 'no, I don't want that', 'no, you should not do this', and therefore presents a rejecting statement."

Since this early distinction of the two extreme points in this development (from affective/volitional to truth-functional), researchers have applied a more fine-grained matrix. In particular, many studies have addressed the question of whether there is developmental continuity between the early affective functions of negation and the higher order logical or truth-functional negation (Pea 1980: 157). This has given rise to methodological discussions, since the sequential development of negative meanings does not necessarily equal the order of acquisition of negative expressions. As long as children do not make formal distinctions it is unclear if different meanings correspond to psychologically real semantic categories in their conceptual system (Cameron-Faulkner et al. 2007: 275).

As a consequence it is unclear how many categories should be distinguished and quite a few different taxonomies have been suggested. Table 2 shows the relatively fine-grained classification proposed by Choi (1988) as an example because it also includes negative meanings emerging relatively late.

Table 2. Categories for a classification of early negative meanings (following Choi 1988)

Category	Description	Example
NONEXISTENCE	child's expectation of the presence of an entity at a particular place or time was not met	*allgone X / X allgone*
FAILURE	negates occurrence of a specific event	*not work, not fit*
REJECTION	negation of object or action present in the context	*not eat*
PROHIBITION	negation of others' action present in the context	*no!*
DENIAL	negated proposition represented in the interactants verbal assertion	*is this a car? – no, that's a pony*
INABILITY	child negates a physical ability	*I can't*
EPISTEMIC NEGATION	child negates possession of knowledge	*I don't know*
NORMATIVE NEGATION	discrepancy between state of affairs and the child's expectations (norms) of the objects involved	*him (= horse) can't go on a boat*
INFERENTIAL NEGATION	involves inference about listener – no overt assertion precedes the child's negation	*I not broken this*

Given the difficulties in distinguishing these different types in particular in the early phases of acquisition, other researchers have collapsed some of the categories. A review of a variety of studies reveals that the following three broader categories are widely used (e.g. Pea 1980, Hummer et al. 1993).[3]

A. REJECTION/REFUSAL
B. DISAPPEARANCE/NONEXISTENCE/UNFULFILLED EXPECTATION
C. DENIAL

The following survey on the acquisition of negative meanings from early nonverbal behavior up to the one word stage is based on this tripartite distinction.

2.1.1. Functions of negation expressed up to the one-word stage

Cross-linguistic evidence suggests that REJECTION/REFUSAL is the first semantic category of negation children are able to express (Pea 1980). It comprises children's expressions of rejection of an external object or action as well as those of refusal to comply with a request. Before they start to speak young children can push an undesired object away or turn their head in order to express this function. Rejection is also the first type of negation that surfaces in speech: The first form used for this function in English is the word *no*, but gestural messages persist later (Guidetti 2005). As Horn (1989: 166f.) points out there are many extra-linguistic ways of expressing meanings like rejection or disappearance, and even when expressed in speech, these meanings do not require the presence of overt morpho-syntactic negative markers. For example, an earlier negative statement can be rejected by means of a positive utterance. Studies of the acquisition of negative meanings have widely ignored these possibilities and focused on overt negation in gesture and speech.

The early acquisition of rejection is probably due to its simplicity. Rejection typically applies to holistic actions or events and the topic of this type of negation has no need for internal representation, because it is always present in the context.

The second negative function that is typically attested in children's gestures and speech refers to expressions of DISAPPEARANCE or of NON-EXISTENCE. The latter type always involves the non-existence of an ex-

---

[3] A different order of acquisition has been found for negative meanings expressed in multi-word speech (e.g. Bloom 1970; see next section).

pected entity and is therefore often seen as related to negation occurring in the context of UNFULFILLED EXPECTATIONS (Pea 1980). The reason why this type of negation appears later than instances of rejection is seen in the fact that denying the presence of something requires an abstract cognitive representation of the relevant object (Pea 1980). This type of negation has therefore been studied in relation to the development of object permanence. Tomasello and Farrar (1984) found a clear relation between this cognitive ability and the emergence of what they called "absence-relational words" (1984: 486) such as *allgone*, but also *more, find, another*.

The last function of negation to appear developmentally is DENIAL, used to negate the truth of a statement, in the beginning mainly statements expressing facts in the world. Denial or truth-functional negation is acquired relatively late because it requires that children can simultaneously represent two mental models, one representing the true state of the world and one representing its false counterpart (Hummer et al. 1993).

Volterra and Antinucci (1979) focus on children's pragmatic development and assume that they have to learn how to negate "a corresponding positive presupposition attributed to the listener" (p. 283). At first the information which the child presupposes and then negates is always available in the ongoing event or the immediately preceding utterance. Later the child must reconstruct such presuppositions internally because they become less recoverable from situational cues in the context. Pea (1980: 31), however, argues convincingly that Volterra and Antinucci's (1979) emphasis on the role of the addressee's presupposed beliefs would overburden the inferencing capacities of pre-two-year-olds.

These issues cast doubts on the assumption that the first occurrences of denials (between 1;8 and 1;11 according to Hummer et al. 1993) really indicate that children have acquired higher order truth-functional negation (Pea 1980). Can we assume that this acquisition gets by without intermediate steps or is there a more continuous developmental progression from rejection and disappearance negation to the truth-functional negation of statements?

Gopnik and Melzoff (1985) assume that the expression of failure of plans can serve such a bridging function, because it occurs also in child monologues and is thus no longer bound to social action. Hummer et al. (1993) cite cross-linguistic support for a more continuous development that comes from the acquisition of languages using different lexical items for rejection and higher order negation. There is some evidence that words used for early pre-logical negation are later over-generalized for the expression of denials, where the target language would use another negation word (e.g. Clancy 1985 on Japanese, Vaidyanathan 1991 on Tamil).

A number of studies (Drozd 1995, Hummer et al. 1993) suggest in addition that what superficially looks like denial is actually often expressing a kind of metalinguistic negation (for an overview see Horn 1989: chapter 6). Hummer at al. (1993) study children's negative answers to yes/no questions involving a wrong comment (e.g. *Is this a cow?* while pointing to the picture of a dog). The authors suggest that children's denials in such a context need not be interpreted as instances of truth-functional negation, but could instead mean something like "your utterance violated the rules of the naming game" (Hummer et al. 1993: 617). Context analyses of spontaneously occurring denials likewise show that young children often use these in order to reject another person's use of language that is inappropriate in their eyes (Drozd 1995). Early denials might thus not be instances of truth-functional negation but rather they may be similar to the earlier functions expressing unfulfilled expectations with respect to a habitual norm of language use. Hummer et al. (1993) conclude that "there seems to be empirical evidence for the developmental emergence of the concept of falsity from a much more inclusive category of disagreement. (...) most children below five are not able to differentiate between false statements (...) and anomalous statements." (p. 617).

Even though there are still open questions concerning the exact cognitive prerequisites and the number of intermediate developmental steps one can conclude that new meanings emerge when progress in cognitive development allows more abstract topics to be negated. The different negative meanings are acquired in a relatively invariant sequence (Pea 1980), whereas individuals differ as to which lexical items they use to express the same meanings and in the meanings most prevalent in their language use.

2.1.2. Functions of negation expressed in multi-word speech

So far we have been concerned with acquisition orders of negative meanings that children express with gestures or negative lexical items used in isolation. Interestingly, slightly different orders of acquisition have been found in multi-word speech, where what is negated is expressed verbally and not inferred from the (extra-)linguistic context. There is converging evidence from several languages (Choi 1988 on English, Korean, and French, McNeill and McNeill 1973 on Japanese, Bloom 1970 on English) that NON-EXISTENCE (and not REJECTION) is the first semantic category of negation to be expressed in multi-word speech. Table 3 summarizes Bloom's (1970: 172–173) findings.

Table 3. Negative meanings in multi-word speech (following Bloom 1970)

Category	description	example
NONEXISTENCE	referent not manifest in context, where it was expected	*no pocket*
REJECTION	existing or imminent referent rejected or opposed by the child	*no dirty soup*
DENIAL	an actual or supposed predication was not the case	*adult: there's the truck child: no truck ('this isn't a truck')*

Choi (1988) also assumes that rejection is expressed later than nonexistence in multi-word speech. Children's early one-word utterances are sufficient for the rejection of objects or actions that are salient in the context. This is different for disappearance and nonexistence where children comment about objects that are not or no longer directly evident in the context. There is thus more pressure on further formal development, and the first word combinations involving negative items are thus often of the type *no X / X no* or *allgone X / X allgone*.

In isolated occurrences as well as in connected speech it is assumed that children's acquisition of new forms often lags behind their acquisition of new functions.[4] Bloom (1991) proposes that this might be due to properties of English as a target language where meaning differences are blurred because several types of negation can occur with the same negative form. As a consequence, researchers have hypothesized that target-like form-function mappings are acquired more rapidly if the different negative meanings are covered by different lexical items (Cameron-Faulkner et al. 2007: 255). The evidence from the acquisition of languages with such properties is mixed, however. McNeill and McNeill (1973) studied the acquisition of semantically contrasting negative items in Japanese. They found that children established the relevant semantic contrasts from early on, even though they did not always take all relevant semantic features into account. Rūķe-Draviņa (1972) on the other hand reports that children acquiring Latvian

---

[4] Compare also the discussion of overextended forms in Choi (1988: 530) and Pea (1980: 107).

used the available three different negation words in free variation despite their semantic contrast.

Direct comparisons involving unrelated target languages (e.g. Choi 1988) reveal a relatively large overlap between the developmental patterns. This suggests that the preconditions for the expression of different types of negation that are linked to young children's general cognitive development play a greater role for the acquisition of negative functions than the structure of the input language. Cross-linguistically children initially negate only events that they do not have to refer to linguistically because they are present in the immediate context. Only later do they learn to make negative comments about non-present objects or actions. The expression of denial requires even more abstract representation and thus appears even later. In a first step children seem to use denial in order to signal that their interlocutors' use of language in a concrete situation does not correspond to their expectations. Researchers agree that adult-like truth-functional negation is the last major semantic category that children acquire.

2.2. The formal expression of negation in child language

This section is devoted to the acquisition of formal properties of negation. It deals with the influence of the syntactic context on the choice of lexical items and with the position of the negators in children's utterances. Out of all the negative meanings discussed in the previous section, the relevant studies have mainly focused on the form of denials. An additional important distinction is the one between anaphoric and sentential (clausal) negation. Anaphoric negation relates to the content of an earlier utterance. It can occur in isolation or preceding an utterance that is following the negated one (*This is red. No, this is orange.*). Sentential negation applies its negative meaning to the utterance in which it occurs, instead of to an earlier one (*This is not red*).[5] Many of the target languages studied have a special lexical item for sentential negation, but it can sometimes not easily be distinguished from anaphoric negation in child language. This distinction is important because the question of whether children treat anaphoric and sentential negation differently (lexical items, positions) has played a crucial role in the debate on the acquisition of forms of negation, both with respect to lexical and with respect to syntactic development.

---

[5] The term 'sentential negation' does, however, not imply that we are dealing with full sentences. The negated utterances can be elliptical or (in particular in learner language) otherwise incomplete.

2.2.1. The acquisition of negative lexical items: anaphoric vs. sentential negation

Cross-linguistic evidence shows that children's first negation words typically correspond to the form of the adult anaphoric negator, presumably because this form often occurs in isolation or at least in sentence external position and is therefore more salient in the input. It is an open question, however, if the adult anaphoric negator is used for anaphoric negation only. Wode (1977) claims that there is a presumably universal developmental stage at which children use the anaphoric negative also for sentential negation.

More recent data speak against the universality of such a stage. French anaphoric *non*, for example, is never used for sentential negation: "children have a sentence external position for *non* and a sentence internal one for sentential negation with *pas*" (Weissenborn et al. 1989: 28; see also Choi 1988).

For other target languages the picture is not as clear. With respect to German, for example, there are contradicting claims concerning the occurrence of *nein* as a sentential negator (thus replacing the target-like *nicht*). Weissenborn et al. 1989 find some examples in their data (see example 1), but claim that these occurrences do not cluster in the transition phase between anaphoric and sentential negation, and are therefore unlikely to constitute a distinct acquisitional stage.

(1) *Nein ich putt mache*
    No   I   kaputt make   (Simone 2;2)

Deprez and Pierce (1993) on the other hand maintain that there is a phase during which children systematically overuse *nein* for sentential negation, sometimes even in utterance internal position; cf. example 2.

(2) *Ich nein schlafen*
    I   no   sleep   (from Felix 1987)

These authors assume that the phenomenon is not due to the overgeneralization of early anaphoric negation, but that learners tend to overuse the negative item with least phonetic complexity. Stromswold and Zimmermann (1998) point out that these contradicting findings are mainly due to the methodological difficulty of distinguishing sentential from anaphoric negation. They present a careful context analysis of the available transcripts and come to the conclusion that the vast majority of all potentially sentential negatives occurred with *nicht*. In German the picture is thus not as clear as in French, but there does not seem to be a stage in child language development at which *nein* would be systematically replacing *nicht* or at which both negators would occur in free variation.

The acquisition of English presents a different picture again. A classic study by Klima and Bellugi (1966) found that there was indeed a first stage at which anaphoric *no* and sentential *not* were used in free variation in contexts of sentential negation (see also Harris and Wexler 1996). In a later step *n't*-negators were added to the repertoire (starting with unanalysed *can't* and *don't*) and only after that children distinguished between *no* and *not* in an adult-like way.

Wode (1977: 100) assumes that the long period of confusion between *no* and *not* in English is due to the phonological resemblance between these negation words that makes it difficult for children to analyze and distinguish the relevant lexical items in the input. Whereas children learning other languages overcome this learning problem quite rapidly and start consistently using the target language's item for sentential negation, the acquisition of English shows a more gradual development.

Cameron-Faulkner et al. (2007) collected a dense longitudinal database of one English speaking child aged 2;3–3;4 and found that *no* and *not* did not actually occur interchangably as was assumed by Klima and Bellugi (1966). Although there was some overlap, the use of *no* for sentential negation (e.g. *no move, no reach*) was found to decrease at the age of 2;6 while the use of *not* in similar contexts (e.g. *not lose it, not play, not reach*) started to increase at around the same time and became the dominant negator by 3;0 (Cameron-Faulkner et al. 2007: 267). The distribution of negation in this big corpus shows that there is no stage of completely random distribution, but rather that children distinguish the two negators at least quantitatively. The authors suggest that the input frequency is responsible for *no* being the first negator used by the child. Once *no* is established as a familiar negator the child keeps relying on this established form and uses it also for multi-word negation instead of imitating existing multi-word combinations in the input.[6] This would correspond to a conservative learning strategy according to which new functions are first expressed by old items (Choi 1988), but one wonders why this is the case in some languages more than others (compare the evidence for French and German above). These findings concerning the acquisition of lexical items for sentential negation and the role that the earlier anaphoric negators play in this process are tightly linked to another important issue: the acquisition of a position for sentential negation.

---

[6] This presents an interesting challenge to the usage-based approach to language acquisition that Cameron-Faulkner et al. (2007) put forward.

2.2.2. The acquisition of word order and a position for negation

This section deals with the acquisition of word order in negated sentences. The focus of the research has again been on sentential negation, but here the term is not only used in contrast to anaphoric negation but also in contrast to constituent negation. Negation not only reverses the truth-value of a sentence, but additional means (e.g. word order, prosody, special lexical items) can be used to indicate which part of the negated sentence is incompatible with its positive counterpart (Klein, submitted). In some cases these additional means identify only one particular constituent as the incompatible part of the sentence. With the exception of some comprehension studies mentioned in the end of this section, such narrow scope cases have not been studied extensively in child language. Most studies have focused on so-called sentence negation instead, the most neutral case in which the entire sentence or at least the entire VP is in the scope of negation.

How do children integrate their newly acquired negation words into their utterances at the beginning of the multi-word stage? Longitudinal data from children learning English and German have been studied quite intensively because they exhibit a systematic use of non adult-like structures in the early stages of acquisition, in particular the so-called 'Optional Infinitive Stage' (see Jordens 2002; Wexler 1994, and the discussion below). Based on data from Bloom (1970), Klima and Bellugi (1966), Felix (1978) and Wode (1977), Felix (1987: 98) provides the following overview for the development of sentence negation in English and German.

Table 4. Stages in the development of sentence negation in English and German[7] [8]

English	German
Stage I (Neg + S)[7]	
no the sun shining	*nein hauen* (don't bang)
no Daddy hungry	*nein schaffe ich* (I can't manage)
no see truck	*nein spielen Katze* (I don't want to play with the cat)
Stage II (*no/nein* + VP)[8]	
Kathryn no like celery	*ich nein schlafen* (I don't sleep)
I no reach it	*das nein aua* (that doesn't hurt)
man no in there	*ich nein hat eins* (I don't have one)

---

[7] This corresponds to stage IIb in Wode (1977), who includes additional earlier phases in his model.

[8] See the discussion in the preceding Section concerning the use of *nein* for sentential negation.

Stage III (*not/nicht* + V; *don't* + V, V + *nicht*)	
Kathryn not go over there	*Eric nicht schlafen* (E. doesn't sleep)
I don't go sleep	*ich nicht essen mehr* (I don't eat more)
this one don't fits	*Henning brauch nicht Uni* (H. doesn't have to go to the university)

Whereas Felix's Stage II ('Neg + VP') is restricted to the negator *no* for English, other researchers maintain that early occurrences of forms like *can't* and *don't* in utterance internal position should equally be counted under stage II (see examples below). At this time, however, these are used as unanalysed holophrases, and their positive counterparts (*can, do*) do not yet occur.

(3) *I can't catch you*
 I don't know his name

When *don't, can't* and later *won't* and *didn't* are used as negatives at Stage III the positive versions *do, can*, and *will* also occur, which suggests that children can analyse *don't* as *do + not*. In fact, nonthematic verbs are now always correctly negated, but children still sometimes omit the auxiliary and the copula and then use the negator *not* to express negation as in the following examples.

(4) *I not hurt him*
 This not ice cream

Clark and Clark (1977) point out that the age at which children reach these developmental stages varies tremendously. They compared the development of the children Adam, Eve, and Sarah and found that Stage III was reached by Adam at 3;0, by Eve at 2;0, and by Sarah at 3;5.

With respect to the orders of acquisition summarized in Table 4, two major issues have given rise to further investigations. The first issue has to do with the evidence for sentence external negation at Stage I (Bloom 1991; Drozd 1995, Stromswold and Zimmerman 2000). The second issue relates to the reasons for the variability that is observed at Stage III (pre-verbal *not* vs. *don't* in English; pre-verbal vs. post-verbal *nicht* in German) and its relation to the acquisition of finiteness. We will take these issues up in turn.

2.2.2.1. Is there a development from external to internal positions of sentential negators?

According to a number of studies on the acquisition of English and German (see above) the first multi-word structure children use in order to express sentential negation is one in which the negator is placed in a sentence external position, preceding the negated sentence in the majority of cases. A variety of reasons for the occurrence of negation in this position have been put forward. McNeill and McNeill (1973) assume that such a sentence external position is a spell out of the basic ("deep structure") position for sentence negation. In English the negator has to be moved into its target-like sentence-internal surface position. No such movement is necessary in Japanese, where negation always appears in its sentence final basic position and children do not produce any deviant word orders.

Klima and Bellugi (1966) suggest that the reason for sentence external positions of negation is that children avoid interruptions in the core part of the sentence, while Wode (1977) assumes that the 'Neg + S' structure is due to the overgeneralization of a pattern that children encounter in anaphoric contexts in the adult input. Slobin (1985: 1239) proposes that children put negation in sentence initial position because they want to indicate "that the scope of negation should be the proposition, as indicated by the verb or the clause as a whole, rather than any particular nonverbal lexical item within the clause".

An explanation in a very similar vein is offered by Van Valin (1991). If these early negators have the whole proposition in their scope, the most natural place for them to occur is outside their scope domain. Importantly, Van Valin (1991: 22) assumes that external negation is likely to occur at a time in development when such propositions do not yet show an internal structure, but are rather unanalyzed holophrases.

These are plausible explanations, and some of them would also work hand in hand. The problem is, however, that a number of other studies have challenged the idea of a 'Neg + S' stage altogether. The most prominent opponent is probably Bloom (1991) who claims that "a stage of sentence external negation in early acquisition is a myth" (p. 144).

Bloom's main argument is that what looks like external negation is in the majority of cases an artifact of missing subjects. At the relevant stage, subjects are frequently omitted. If they are present, however, Bloom (1991) claims that negation tends to appear in sentence internal position (e.g. *Kathryn no shoe;* but see Déprez and Pierce (1993) for counter evidence).

Stromswold and Zimmerman (2000) present a context analysis that reveals that most of the putative instances of external sentential negation are compatible with an analysis as anaphoric negation in which the sentence initial negator has scope over the preceding rather than the following sentence. The following example (adapted from Stromswold and Zimmermann 2000) illustrates the difficulty (without an indication of a specific context) of assigning an anaphoric (A) or a non-anaphoric reading (B) to negation in early child speech.

(5) Andreas (2;1)

    Ann: *Thorsten holt einen, bleib du hier.* ('Thorsten went to get one, you stay here')

    And/A: *nee, ich guck.* (no I watch; 'No, I don't want to stay here, I'll (come and) watch')

    And/B: *nee ich guck.* (no I watch; 'I don't watch')

Given these ambiguities, Van Valin (1991) proposes to study the acquisition of negation in SOV languages. In such languages, sentence external negation, if present, would be expected in a sentence final position whereas anaphoric negation would be placed in sentence initial position. Van Valin cites examples of sentence external negation from Turkish.

(6) Turkish (Aksu-Koç and Slobin 1985)

Yap-ıcağ-ım ıth (*do*-FUT-1sg NEG; 'I won't do it.') where the adult form would be

Yap-mı-yacağ-im (*do*-NEG-FUT-1sg)

It is hard to judge from the available evidence, however, if these examples are isolated occurrences or reflect a more systematic pattern of use.

Drozd (1995) studies occurrences of pre-sentential negation in English that a context analysis identifies as clearly non-anaphoric and comes to the conclusion that in most cases, these do not express ordinary denials. Building on a hypothesis by Horn (1989: 462) he assumes that negation in these cases is metalinguistic (as opposed to truth-functional) because children use it in order to object to a previous utterance. The meaning of these uses could thus be glossed as *Don't say X (to me)*. An utterance like the one in (7) below would thus not mean 'It is not true that Nathaniel is a king', but rather 'I object to your saying *Nathaniel's a king*'" (Drozd 1995: 588).

(7) *No Nathaniel a king*

Additional evidence for the metalinguistic function comes from the observation that, out of the three early negative meanings distinguished in Section 2.1 (i.e. rejection, nonexistence, and denial) only rejection and denial occur with pre-sentential negation. Nonexistence statements like *No there's a pony* do not occur and would indeed not be compatible with a metalinguistic reading of the type explained above (Drozd 1995: 589). It is also shown that most utterances with pre-sentential negation occur in contexts in which children are echoing utterances of their adult interlocutors as in (8) below.

(8) Mother: *That's a cake. We're going to have it for dinner.*
     Peter (2;3): *No that's a cake.* (= 'No way that's a cake!')

Based on his finding that the vast majority of pre-sentential negations with *no* are echoic and consistent with an exclamatory paraphrase like *no way*, Drozd concludes that there is indeed no general stage of external sentence negation for the negator *no*. Instead, the occurrences in question are instances of 'metalinguistic exclamatory sentence negation'.[9]

This is particularly interesting because it speaks to the issues of continuity between the so-called pre-logical functions of negation (rejection and nonexistence) on the one hand and truth-functional denial on the other that was raised in Section 2.1. Considering the developmental gap between these two (but not restricting this claim to negation in a particular position) Hummer et al. (1993) had proposed like Horn (1989) and Drozd (1995) that what looks like truth-functional denial might actually often have a metalinguistic meaning expressing that some preceding statement has an unwarranted content or form rather than that it is false.

Taken together it seems as if the role that sentence external negation plays in L1 development varies cross-linguistically and probably also interindividually. Furthermore it is possible that its frequency has been over estimated because of the difficulty of distinguishing it from anaphoric negation. Where it occurs it often involves a different negator and it might also differ in function from sentence internal negation in that it is more metalinguistic. Concerning the latter point, however, it is doubtful that all of children's early sentence internal negations are of the truth-functional kind. They might in fact also express the rejection of an earlier statement on grounds of disagreement with any of its properties rather than only falsity.

---

[9] Utterances with initial *not* were less often echoic and compatible with exclamatory paraphrases. This is another argument against the claim (Klima and Bellugi 1966) that *no* and *not* are used in free variation.

Sentence internal negation would then differ from sentence external negation in that it is less echoic, but not necessarily in that it is less metalinguistic.

Wherever sentence external negation occurs, the different reasons that have been put forward (see above) in addition to assigning it a different function certainly go a long way in explaining it. But those accounts that are based on the assumption that children producing sentence external negation are actually using surface position in order to signal the sentence negator's wide scope (Slobin 1985; Van Valin 1991) must then also explain on which grounds children give this system up and acquire the target language's internal position for sentence negation. Van Valin (1991: 26) assumes that this developmental shift towards sentence internal negation occurs when children break up their holophrastic representation of a clause into a nucleus and a periphery and understand that it is the nucleus that is the reference point for operators like negation. A related way of saying this is that these early holophrastic units are assigned an internal information structure, whereupon negation tends to be placed between the topic and the comment of such utterances (Jordens and Dimroth 2006).

Once children consistently use sentence internal negation this still does not mean that they always put the negator in an adult-like position. Table 4 (above) illustrates that at Stage III, children produce pre-verbal as well as post-verbal internal negation.

2.2.2.2. What causes the variable word orders at Stage III?

Let us first consider the variability between *nicht*-V and V-*nicht* observed in German. It was noticed from early on (Wode 1977) that verbs in the *nicht*-V pattern mainly occurred in non-finite form (infinitives, participles) whereas verbs in the V-*nicht* pattern were mainly finite. This pattern mirrors the distribution of finite and non-finite verbs in the input. German declarative main clauses have two positions for verbs: whereas non-finite forms occur in sentence final position, finite verbs raise to the second position (V2), i.e. to the left of the negator that is typically placed in a pre-non-finite position (9a). If there is only one (finite) main verb, it is raised to V2 and the negator stays behind. In these cases its surface position can be sentence-final (9b).

(9) a. *Paul hat nicht gelacht.* (Paul has-3SG not laughed; 'Paul didn't laugh.')
    b. *Paul lacht nicht.* (Paul laugh-3SG not; 'Paul doesn't laugh')

Clahsen (1988) analysed occurrences of the type V-*nicht* as resulting from a morphological procedure in which children interpret the negator as a kind of affix to the finite verb. This would however imply that there is a stage at which no other constituents can intervene between the finite verb and the negator, which other researchers (e.g. Weissenborn et al. 1989) question on the basis of examples like the following.

(10) Simone (2;1): *das macht der maxe nicht* (this does Max not; ‚Max doesn't do this')

A finiteness-related distribution was not only linked to a bias in the adult input but also to children's own grammatical representations that reflect a strong dependency between verb raising and morphological finiteness (V-*nicht* with finite verb forms and *nicht*-V with non-finite verb forms). This distribution is either assumed to be the consequence of an UG constrained dependency that only young children have access to (e.g. Meisel 1997) or it is seen as the end point of an acquisition process in which finite nonthematic verbs like the copula or auxiliaries, play an important role (e.g. Jordens and Dimroth 2006).

In languages like Dutch, English, and German, such finite nonthematic verbs have been analysed as lexically empty carriers of assertion (Klein 2006). The copula and auxiliary verbs are the first unambiguous assertion markers attested in child language (Jordens and Dimroth 2006), but, importantly, they are acquired later than negation. Before the acquisition of finite nonthematic verbs, children can make (positive) assertions but they do not mark this with the target language's grammatical means. The positive counter-part of 11a (from Table 4) would thus be 11b (with a thematic verb that is not marked for finiteness).

(11) a. *Kathryn not go over there*
 b. *Kathryn go over there*

When auxiliaries first appear in child language they are always used in a finite form and in a finite position (thus preceding negation). This gives rise to the target English word order (*I don't go sleep*) and a similar word order with German or Dutch auxiliaries. In these languages, however, negation with thematic verbs does not necessitate the insertion of an auxiliary. Finite thematic verbs raise to the V2 position and leave negation behind. Before this regularity is acquired, negation follows nonthematic verbs, but precedes or follows thematic verbs. The variation attested at Stage III (see Table 4) is then the result of early, probably root-learned occurrences reflecting the distribution of patterns in the input that children partly reproduce before morphological finiteness marking becomes productive (Wode 1977: 94, Jordens 2002).

Importantly, this variation is not due to problems with the placement of negation but rather reflects different stages in children's acquisition of finiteness. Development goes from pre-verbal negation in non-finite utterances to post-verbal negation with nonthematic verbs, and, in languages where this is grammatical, post-verbal negation with finite thematic verbs.

2.2.2.3. Further development

When children have finally mastered the target position for sentential negation there is still more to learn. As pointed out in the introduction to this section, sentence negation does not only reverse the truth value of the sentence containing it, its position is often also used to indicate where the current sentence differs from its true counterpart (i.e. scope marking). Whereas in many languages, there are default positions for negation that are compatible with wide and narrow scope readings (scope must then be signaled by additional means, for instance intonation), there are also positions dedicated to the expression of narrow scope.

Production studies on the acquisition of narrow scope marking are scarce, however. What has been studied more intensively is children's interpretation of relative scope in sentences containing negation and another quantifier. In such cases, both operators can typically be interpreted in their linear order (in which case the first operator has scope over the second) or vice versa (so-called scope reversal). Musolino, Crain, and Thornton (2000) and Lidz and Musolino (2002) showed that children have a clear preference for translating surface order into scope relations when interpreting sentences like (12).

(12) *Every duck didn't cross the river*

Following their preference for linear interpretation children thus take (12) to mean that for every duck it is true that it did not cross the river, whereas adults prefer a scope reversal interpretation in these cases (= Not every duck crossed the river).

English sentences like (12) do not seem to be very natural.[10] This is different from a phenomenon in Dutch and German, where indefinite NPs are regularly 'scrambled' over negation. When trying to understand these con-

---

[10] Gennari and MacDonald 2006 found that children are biased by the existence of more natural adult-language realizations.

structions, children do not rely on the surface order but seem to prefer a non-adult-like inverse scope interpretation (Krämer 1998).

(13) a. *De jongen heeft een vis niet gevangen.* (the boy has a fish not caught)
 b. *De jongen heeft geen vis gevangen* (the boy has no fish caught)

Krämer (1998) finds that children between 4 and 7 do not make a difference between (13a) and (13b). When presented with a picture story in which a boy catches all but one fish both sentences are rejected as incorrect descriptions, whereas adults correctly accept (13a). Unsworth et al. (2008) show however that children can be pushed by the context to accept (13a). When it is made clear that it is the boy's task to catch all the fish and children are then explicitly asked if he really did, they readily accept (13a) as a true description. Unsworth et al. (2008) therefore conclude that children must have access to the adult-like readings as well.

3. Developmental patterns in l2

The acquisition of negation is one of the best studied domains in untutored L2 acquisition by older children and adult learners. Investigations of negation at different stages of L2 development have concentrated on Dutch, English, French, German, Italian, and Swedish as target languages (see Table 1). Starting from the assumption that L2 learners are already competent speakers of their L1, no special attention has been devoted to the emergence of negative functions like rejection and disappearance that were attested early children's speech. Instead, the focus has been on denials and the acquisition of the relevant formal means for their expression on the sentence level.

Evidence on the L2 acquisition of negation has often not been gathered in its own right, but rather in order to test hypotheses concerning the relation between the acquisition of finite verbal morphology and finite syntax. In many of the target languages finite verbs raise over negation to a position higher up in the syntactic structure of the relevant sentences. Negative sentences are thus a classic means to investigate the position of verbs in learner language, which is often ambiguous in simple utterances, in particular if L2 learners start out with an SVO word order (cf. Haberzettl 2005).

The findings reported in the subsequent sections are mainly concerned with the acquisition of different lexical items for the expression of negation and their integration in the developing sentential structures, partly as a function of wide vs. narrow scope in different information structures. The influence of the structure of negation in the learners' L1 (transfer) cannot be systematically treated here for reasons of space. In particular in the early stages of L2 acquisition patterns from L1 seem to be less influential than language internal principles for the development of utterance structure (Perdue 1993; Stauble 1984). The focus is on a comparison to the relevant findings from L1 acquisition. Similarities are summarized in 3.1, differences in 3.2.

3.1. Where L1 acquisition resembles L1 acquisition

For the L2 acquisition of languages with post-verbal (rather: post-finite) negation there is converging evidence for orders of acquisition that are similar to those observed in child language (see Section 2.2.2). The developmental patterns summarized in Table 5 below have been found in the ESF project (Perdue 1993), a longitudinal and cross-linguistic study on L2 acquisition by untutored adult learners. The orders of acquisition for sentence negation in English (Giuliano 2003; Silberstein 2001), French (Giuliano 2003; Véronique 2005), and German (Becker 2005; Dietrich and Grommes 1998) are largely confirmed by other studies (e.g. Cancino et al. 1978; Clahsen et al. 1983; Meisel 1997), and also for other target languages with post-finite negation (e.g. Swedish, see Hyltenstam 1977).

Table 5. Four major steps in the L2 acquisition of sentence negation in English, French, and German [11]

Phase	Structure	English	French	German
I	Neg-X	for me no very concentration	[nepade eprimeri][11] ((this is) no print shop)	nix andere kind (no other child)
II	Neg-V	I prove two time but no speak English	pas [parlar] français (not speak French)	ich nix komme (I not come)
III	$V_{nonthem.}$-Neg	I don't see very well	là bas [ne pe pa sortir] (over there neg can not leave; 'X was not allowed to leave the place')	ich kann nicht verkauf (I can not sell)
IV	V-Neg	—	la personne [ne travaj pa] (the person neg work not; 'the person doesn't work')	er arbeitet nicht gut (he works not well; 'he doesn't work well')

---

[11] Target deviant learner realizations are transcribed in square brackets. See Perdue (1993) for details

According to Perdue et al. (2002), Stage I reflects a nominal utterance organisation. At this stage negation can directly precede all sorts of (mainly nominal) lexical items. From Stage II onwards, verbs and their argument structure play an important role for utterance organisation (this stage corresponds to the 'Basic Variety'; Klein and Perdue 1997). In learners of English, French, and German, negation precedes the VP. Importantly, at this stage verbs are not yet productively marked for finiteness. Stage III corresponds to the first step in the development of a finite utterance organisation, in which non-thematic verbs (copula, auxiliaries, modal verbs) are the first verbal items productively used in finite forms and positions. In the target languages presented in Table 5 these items appear to the left of the negator in surface structure without exception. Syntactic and semantic reasons for this special behavior of non-thematic verbs have been put forward (see Becker 2005, Giuliano 2003, Jordens and Dimroth 2006, Parodi 2000). The gist of these approaches is that the category of finiteness is easier to grasp for L2 learners when it is carried by non-thematic verbs that do not have much lexical content.

At Stage IV learners have understood how to mark utterances with thematic verbs for finiteness in cases where no such independent carrier of finiteness is present. In French and German this means that from Stage IV onwards, negation occurs to the right of thematic verbs as illustrated by the examples in Table 5. As Becker (2005: 29) points out, however, the acquisition of finite verb morphology is a long process and some variation in the form and position of finite verbs is to be expected.

Overall it can be concluded that the way in which negation is integrated depends on the structure of the underlying utterances – in particular on non-finite as opposed to finite utterance organisation. In 2.2.2 we saw a similar tendency in the development of negation in L1 acquisition. We can conclude that in L2 as well as L1 acquisition, the expression of negation precedes the grammatical marking of assertion through finiteness; see Klein 2006).

Concerning the later acquisition of finiteness, it is relatively undisputed that non-thematic verbs are the first verbs used productively in finite form and position (i.e. to the left of negation in verb raising languages). There is, however, an ongoing debate concerning the relation between morphological and syntactic features of the finiteness of thematic verbs in L1 vs. L2 acquisition. The claim is that there is a direct relation between morphological finiteness and verb raising only in L1 acquisition whereas there is more variability in L2 acquisition, i.e. both finite and nonfinite verbs can show up to both sides of negation (Meisel 1997). Verhagen and Schimke (2009) assemble the available evidence for L1 and L2 acquisition of French and German and demonstrate that there is a certain degree of variability in both L2 and L1 data, so that the differences between both types of acquisition in this domain are probably smaller than previously assumed.

3.2. Where L2 acquisition differs from L1 acquisition

Careful analyses of data from the first phases of L2 acquisition (e.g. Becker 2005; Bernini 2000; Silberstein 2001) reveal that L2 learners differ from L1 learners in the communicative potential of even their earliest productions of negative utterances. Whereas communicative needs and communicative means develop largely on a par in L1 acquisition (cf. the stepwise acquisition of negative meanings discussed in Section 2.1), this is clearly not the case for L2 acquisition. There is no need for L2 learners to go through the process of acquisition of negative functions a second time.

Utterances produced by older children or adults learning an L2 are more complex from early on. Whereas it has been proposed for L1 acquisition that young children tend to start from a holistic representation for what is negated (Jordens 2002, Van Valin 1991), it is less clear that there is such a stage in L2 acquisition. Studies relying on detailed contextual analyses of beginning L2 speech (Andorno 2008, Perdue 1996) found that early L2 utterances have an internally partitioned structure that learners can exploit for the expression of negation. From the ‚Neg-X' Stage onwards (Table 5) learners can put elements that do not belong to the scope of negation into sentence initial (topic) position (Becker 2005, Silberstein 2001). As illustrated in (14) for L2 English, such topics can be relatively complex (‚+' indicates a short break).[12]

(14) a. *in london + no very sun*
b. *me + for holiday + no september*

Beginning L2 learners not only mark that topical elements do not belong to the scope of negation, they also block scope extension to the right of the negator by stressing the affected element and de-stressing the following part of the utterance (stress indicated through capitals).

---

[12] All L2 examples in this Section are from the ESF database hosted at the Max-Planck-Institute for Psycholinguistics in Nijmegen; (14) – (16) are from the learner Santo (L1 Italian, L2 English), (17) is from Angelina (L1 Italian, L2 German).

(15) M: *so you're a very good cook then*
      S: *no VERY good*

Another important device for the marking of information structure that occurs in the beginning phases of L2 acquisition is the use of so-called I-Topics. Utterances with I-Topics have a special intonation contour and evoke a contrastive reading (Büring 1995) that can be used by learners in order to express that some topical constituent is *within* the scope of negation, in contrast to another (implicit or explicit) topic.

(16) Interviewer: *do you have your driving license?*
      Learner: *original copy no* [Silberstein 2001]

(17) Learner: *mein mann (h)aber de auto, ich nix* (my husband have the car, I not) [Becker 2005]

The main differences between older L2 learners and young children acquiring negation in their L1 lies in the availability of the functions of negation at the outset of acquisition. Even though the formal repertoire at the beginning of the acquisition process is equally restricted in L1 and L2 acquisition, clever usage of the different utterance patterns and their information structure in context allows L2 learners to express negation with different scope properties even before they have acquired the relevant target language means.

4. Conclusion

The acquisition of negation has been studied intensively in both L1 and L2 development. Wherever researchers have undertaken direct comparisons of both types of acquisition processes, they have come to very different conclusions with respect to overall similarities and differences. Cook (1993: 43), for example, raises the issue of "the specific developmental structures 'Neg + X' and 'Subject Neg VP', which universally occur in studies of L1 acquisition across languages as diverse as Latvian and Hungarian, and in studies of L2 acquisition of English and German as well as in pidgin and Creole languages". He wonders why it is that learners produce such structures in the absence of models in the input and comes to the conclusion that "There is evidence for a common grammar at particular stages of language development, more or less regardless of L1 or L2." (Cook 1993: 43; see e.g. Jordens 1987 for a similar claim). Meisel (1997: 228), on the other hand, points out that L2 learners, as opposed to L1 learners "do not refer to the [± finite] distinction when acquiring the placement of the negative element" and consequently views "...first and second language acquisition as fundamentally different in nature."

Clearly, however, these claims apply to different phases in development. Cook's (1993) statement of similarity refers to the use of negation before the acquisition of assertion marking through finiteness. Meisel (1997) investigates the behavior of negation during the acquisition of finiteness and finds that children are more consistent than adult L2 learners in raising verbs with finite inflections over negation and leaving non-finite verbs in an unraised position within VP. Far-reaching claims are based on these observations. Meisel assumes that due to maturational constraints only children have access to procedures of language acquisition that conform to UG (e.g. concerning the relation between feature strength and verb raising) whereas adults have to rely on domain-general problem solving faculties instead (see also Bley-Vroman 1990). There is, however, an ongoing debate concerning adult L2 learners' access to UG (e.g. Prévost and White 2000) and concerning the empirical evidence underlying claims about the placement of verbs relative to negation in L1 and L2 (Verhagen and Schimke 2009). Some researchers therefore assume that similarities in utterance structure outweigh differences also at these later stages (e.g. Dimroth et al. 2003).

What we can maintain at this point is that both L1 and L2 learners explicitly express negation before they acquire finiteness as the target language's grammatical means of assertion marking. While, at this time, L1 learners are still struggling with the acquisition of negative meanings and with the reanalysis of putatively holistic representations of the information to be negated, L2 learners can exploit the communicative potential of relatively simple structures more successfully.

References

Aksu-Koç, Ayhan and Dan I. Slobin
    1985   The acquisition of Turkish. In Dan I. Slobin (ed.), 839–878.
Anderson, Diane E. and Judy S. Reilly
    1997   The puzzle of negation: How children move from communicative to grammatical negation in ASL. *Applied Psycholinguistics* 18: 411–429.
Andorno, Cecilia
    2008   Entre énoncé et interaction: le role des particules d'affirmation et negation dans les lectes d'apprenants. *Acquisition et Interaction en Langue Etrangère* 26: 173–190.
Barbarczy, Anna
    2006   Negation and word order in Hungarian child language. *Lingua* 116: 377–392.

Bardel, Camilla
1999 Negation and information structure in the Italian L2 of a Swedish learner. *Acquisition et Interaction en Langue Etrangère*, Special Issue Nr 2: 173–188.

Becker, Angelika
2005 The semantic knowledge base for the acquisition of negation and the acquisition of finiteness. In *The structure of learner varieties*, Henriette Hendriks (ed.), 263–314. Berlin/New York: Mouton de Gruyter.

Bernini, Giuliano
2000 Negative items and negation strategies in non-native Italian. *Studies in Second Language Acquisition,* 22, 399–440.

Bley-Vroman, Robert
1990 The logical problem of foreign language learning. *Linguistic Analysis* 20: 3–49.

Bloom, Lois
1970 *One word at a time.* The Hague: Mouton.

Bloom, Lois
1991 *Language Development from Two to Three.* Cambridge: Cambridge University Press.

Bowerman, Melissa
1973 *Early syntactic development: a cross-linguistic study with special reference to Finnish.* Cambridge: Cambridge University Press.

Büring, Daniel
1995 The 59th Street Bridge Accent. On the meaning of topic and focus. PhD dissertation, Tübingen University.

Cameron-Faulkner, Thea, Elena Lieven and Anna Theakston
2007 What part of no do children not understand? A usage-based account of multiword negation. *Journal of Child Language* 33: 251–282.

Cancino, Herlindo, Ellen Rosansky and John Schumann
1978 The acquisition of English negative and interrogatives by native Spanish speakers. In *Second Language Acquisition*, Evelyn Hatch (ed.), 207–230. Rowley MA: Newbury House.

Choi, Soonja
1988 The semantic development of negation: a cross-linguistic longitudinal study. *Journal of Child Language* 15: 517–531.

Clahsen, Harald
1988 Critical phases of grammar development: a study of the acquisition of negation in children and adults. In *Language development*, Peter Jordens and Josine Lalleman (eds.), 123–148. Dordrecht: Foris.

Clahsen, Harald, Jürgen Meisel and Manfred Pienemann
1983 *Deutsch als Zweitsprache. Der Spracherwerb ausländischer Arbeiter.* Tübingen: Narr.

Clancy, Patricia M.
   1985   The acquisition of Japanese. In Dan I. Slobin (ed.), 373–524.

Clark, Herbert H. and Eve V. Clark
   1977   *Psychology and Language: An introduction to psycholinguistics.* New York: Harcourt Brace Jovanovich.

Clark, Eve V.
   1985   The acquisition of Romance, with special reference to French. In Dan I. Slobin (ed.), 687–782.

Cook, Vivian
   1993   *Linguistics and second language acquisition.* New York: St. Martin's Press.

Déprez, Viviane and Amy Pierce
   1993   Negation and functional projections in early grammar. *Linguistic Inquiry* 24: 25–67.

Dietrich, Rainer and Patrick Grommes
   1998   'nicht'. Reflexe seiner Bedeutung und Syntax im Zweitspracherwerb. In *Eine zweite Sprache lernen. Empirische Untersuchungen zum Zweitspracherwerb*, Heide Wegener (ed.), 173–202. Tübingen: Narr.

Dimroth, Christine
   2008   Age effects on the process of L2 acquisition? Evidence from the acquisition of negation and finiteness L2 German. *Language Learning* 58: 117–150.

Dimroth, Christine
   2009   Stepping stones and Stumbling Blocks. Why negation accelerates and additive particles delay the acquisition of finiteness in German. In *Functional Elements: Variation in Learner Systems*, Christine Dimroth and Peter Jordens (eds.), 137–170. Berlin / New York: Mouton de Gruyter.

Dimroth, Christine, Petra Gretsch, Peter Jordens, Clive Perdue and Marianne Starren
   2003   Finiteness in Germanic languages. a stage-model for first and second language development. In *Information Structure and the Dynamics of Language Acquisition*, Christine Dimroth and Marianne Starren (eds.), 65–94. Amsterdam: Benjamins.

Drenhaus, Heiner
   2002   On the acquisition of German *nicht* 'not' as sentential negation. *The proceedings of the 26nd Annual Boston University Conference On Language Development*, 166–174. Somerville: Cascadilla Press.

Drozd, Kenneth F.
   1995   Child English pre-sentential negation as metalinguistic exclamatory sentence negation. *Journal of Child Language* 22: 538–610.

Felix, Sascha
   1978   *Linguistische Untersuchungen zum natürlichen Zweitspracherwerb.* München: Fink.

Felix, Sascha
1987 *Cognition and Language Growth*. Dordrecht: Foris.
Gennari, Silvia P. and Maryellen C. MacDonald
2006 Acquisition of Negation and Quantification: Insights From Adult Production and Comprehension. *Language Acquisition* 13: 125–168.
Giuliano, Patricia
2003 Negation and relational predicates in French and English as second languages. In *Information Structure and the Dynamics of Language Acquisition*, Christine Dimroth and Marianne Starren (eds.), *Information Structure and the Dynamics of Language Acquisition*, 119–158. Amsterdam: Benjamins.
Gopnik, Alison and Andrew N. Meltzoff
1985 From people, to plans, to objects: changes in the meaning of early words and their relation to cognitive development. *Journal of Pragmatics* 9: 495–512.
Guidetti, Michèle
2005 Yes or no? How young French children combine gestures and speech to agree and refuse. *Journal of Child Language* 32: 911–924.
Haberzettl, Stefanie
2005 *Der Erwerb der Verbstellungsregeln in der Zweitsprache Deutsch durch Kinder mit russischer und türkischer Muttersprache*. Tübingen: Niemeyer.
Hahn, Kyung-Ja P.
1981 The development of negation in one Korean child. PhD dissertation, University of Hawaii.
Harris, Tony and Ken Wexler
1996 The optional-infinitive stage in child English: evidence from negation. In *Generative perspectives on language acquisition*, Clahsen, Harald (ed.), 1–42. Amsterdam: Benjamins.
Horn, Laurence R.
1989 *A natural history of negation*. Chicago: Chicago University Press.
Hummer, Peter, Heinz Wimmer and Gertraud Antes
1993 On the origins of denial negation. *Journal of child language* 20: 607–618.
Hyltenstam, Kenneth
1977 Implicational Patterns in Interlanguage Syntax Variation. *Language Learning* 27: 383–411.
Ito, Katsutoshi
1981 Two aspects of negation in child language. In *Child language – an international perspective*, Philip S. Dale and David Ingram (eds.), 105–114. Baltimore MD: University Park Press.
Jordens, Peter
1987 Neuere theoretische Ansätze in der Zweitspracherwerbsforschung. *Studium Linguistik* 22: 31–65.

Jordens, Peter
  2002   Finiteness in early child Dutch. *Linguistics* 40: 687–765.
Jordens, Peter and Dimroth, Christine
  2006   Finiteness in children an adults learning Dutch. In Natalia Gagarina and Insa Gülzow (eds.), *The Acquisition of Verbs and their Grammar* (pp. 173–200). Dordrecht: Springer.
Klein, Wolfgang
  2006   On finiteness. In *Semantics in acquisition*, Veerle van Geenhoven (ed.), 245–272. Dordrecht: Springer.
Klein, Wolfgang
  submitted A simple analysis of sentential negation in German. *Language*.
Klein, Wolfgang and Clive Perdue
  1997   The Basic Variety. Or: Couldn't natural languages be much simpler? *Second Language Research* 13: 301–347.
Klima, Edward S. and Ursula Bellugi
  1966   Syntactic regularities in the Speech of Children. In *Psycholinguistic Papers*, John Lyons and Roger J. Wales (eds.), 183–208. Edinburgh: Edinburgh University Press.
Krämer, Irene
  1998   Children's interpretations of indefinite object noun phrases: Evidence from the scope of negation. In *Linguistics in the Netherlands*, Renée van Bezooijen and René Kager (eds.), 163–174. Amsterdam: John Benjamins.
Lange, Sven and Kenneth Larsson
  1973   Syntactical development of a Swedish girl Embla, between 20 and 42 months of age. Project Child Language Syntax, Report No. 1., Stockholm University: Institutionen för Nordiska Sprak.
Lee, Thomas H.-T.
  1982   The development of negation in mandarin-speaking child. *Language Learning and Communication* 2: 269–281.
Lidz, Jeffrey and Julien Musolino
  2002   Children's Command of Quantification. *Cognition* 84: 113–154.
McNeill, David and Nobuko B. McNeill
  1973   What does a child mean when he says 'no'? In *Studies in child language development*, Charles Ferguson and Dan I. Slobin (eds.), 619–627. New York: Holt, Rinehart and Winston.
Meisel, Jürgen
  1997   The acquisition of the syntax of negation in French and German: contrasting first and second language development. *Second Language Research* 13: 227–263.
Milon, John P.
  1974   The development of *negation* in English by a second-language learner. *TESOL Quarterly* 8: 137–143.

Musolino, Julien, Stephen Crain and Rosalind Thornton
    2000   Navigating negative Quantificational Space. *Linguistics* 38: 1–32.
Parodi, Teresa
    2000   Finiteness and verb placement in second language acquisition. *Second Language Research* 16: 355–381.
Pea, Roy D.
    1980   The Development of Negation in Early Child Language. In *The Social Foundations of Language and Thought: Essays in Honor Of Jerome S. Bruner*, David R. Olson (ed.), 156–186. New York: W. W. Norton.
Penner, Zvi, Rosemarie Tracy and Jürgen Weissenborn
    2000   Where Scrambling begins: Triggering Object Scrambling at the Early Stage in German and Bernese Swiss German. In *The Acquisition of Scrambling and Cliticization*, Susan M. Powers and Cornelia Hamann (eds.), 127–164. Dordrecht: Kluwer.
Perdue, Clive (ed.)
    1993   *Adult language acquisition: cross-linguistic perspectives*. Cambridge: Cambridge University Press.
Perdue, Clive
    1996   Pre-basic varieties. The first stages of second language acquisition. *Toegepaste Taalwetenschap in Artikelen,* 55, 135–150.
Perdue, Clive, Sandra Benazzo and Patrizia Giuliano
    2002   When finiteness gets marked. The relation between morpho-syntactic development and use of scopal items in adult language acquisition. *Linguistics* 40: 849–890.
Prévost, Philippe and Lydia White
    2000   Missing surface inflection or impairment in second language acquisition? Evidence from tense and agreement. *Second Language Research* 16: 103–33.
Ramadoss, Deepti and R. Amritavalli
    2007   The acquisition of functional categories in Tamil with special reference to negation. *Papers from the consortium workshops on linguistic theory* 1: 67–84.
Rūķe-Draviņa, Velta
    1972   The emergence of affirmation and negation in child language: some universal and language-restricted characteristics. In *Colloquium Paedolinguisticum*, Karel Ohnesorg (ed.), 221–241). The Hague: Mouton.
Sano, Tetsuya
    1998   Psycholinguistic investigation of the acquisition of negation in Japanese. *Japanese/Korean Linguistics* 8: 131–144.
Schaner-Wolles, Chris
    1995/96 The acquisition of negation in a verb second language: From 'anything goes' to 'rien ne va plus'. *Wiener Linguistische Gazette* 53: 87–119.

Schimke, Sarah
   2008   The acquisition of finiteness in Turkish learners of German and Turkish learners of French. PhD Thesis, Max Planck Institute for Psycholinguistics, Nijmegen, The Netherlands.

Silberstein, Dagmar
   2001   Facteurs interlingues et specifiques dans l'acquisition non-guidée de la négation en anglais L2. *Acquisition et Interaction en Langue Etrangère* 14: 25–58.

Slobin, Dan I.
   1985   Crosslinguistic evidence fort the Language-Making Capacity. In Dan I. Slobin (ed.), 1157–1249.

Slobin, Dan I. (ed.)
   1985   *The Crosslinguistic Study of Language Acquisition*. Hillsdale, NJ: Lawrence Erlbaum.

Smoczyńska, Magdalena
   1985   The acquisition of Polish. In Dan I. Slobin (ed.), 595–686.

Snyder, William and Eva Bar-Shalom
   1998   Word order, finiteness, and negation in early child Russian. *The proceedings of the 22nd Annual Boston University Conference On Language Development*, 717–725. Somerville: Cascadilla Press.

Stauble, Anne-Marie
   1984   A Comparison of a Spanish-English and Japanese-English Second Language Continuum: Negation and Verb Morphology. In *Second Languages: A Cross-linguistic Perspective*, Roger Andersen (ed.), 323–354. Rowley, MA: Newbury House.

Stern, Clara and William Stern
   1928/75 *Die Kindersprache. Eine psychologische und sprachtheoretische Untersuchung*. Darmstadt: Wissenschaftliche Buchgesellschaft.

Stromswold, Karin and Kai Zimmermann
   1998   Acquisition of negation in German. *The proceedings of the 29th annual Child Language Research Forum*: 231–240.

Stromswold, Karin and Kai Zimmermann
   2000   Acquisition of *Nein* and *Nicht* and the VP-Internal Subject Stage in German. *Language Acquisition* 8: 101–127.

Tam, Clara W.-Y. and Stephanie F. Stokes
   2001   Form and function of negation in early developmental Cantonese. *Journal of Child Language* 28: 371–391.

Tomasello, Michael and Michael J. Farrar
   1984   Cognitive bases of lexical development: object permanence and relational words. *Journal of Child Language* 11: 477–493.

Unsworth, Sharon, Andrea Gualmini and Christina Helder
   2008   Children's interpretation of indefinites in sentences containing negation: A re-assessment of the cross-linguistic picture. *Language Acquisition* 15: 315–328.

Vaidyanathan, Raghunathan
  1991   Development of forms and functions of negation in the early stages of language acquisition: a study in Tamil. *Journal of Child Language* 18: 51–66.
van der Wal, Sjoukje
  1996   *Negative Polarity Items & Negation: Tandem Acquisition.* Groningen Dissertations in Linguistics 17.
Van Valin, Robert
  1991   Functionalist linguistic theory and language acquisition. *First Language* 11: 7–40.
Verhagen, Josje
  2009   The acquisition of finiteness in Dutch as a second language. PhD Thesis, VU University, Amsterdam.
Verhagen, Josje and Sarah Schimke
  2009   Differences or fundamental differences. *Zeitschrift für Sprachwissenschaft* 28: 97–106.
Véronique, Daniel
  2005   Syntactic and Semantic Issues in the acquisition of negation in French. In Jean-Marc Dewaele (ed.), *Focus on French as a Foreign Language* (pp. 114–134). Clevedon: Multilingual Matters.
Verrips, Maike and Jürgen Weissenborn
  1992   Routs to Verb Placement in Early Child German and French: The Independence of Finiteness and Agreement. In Jürgen Meisel (ed.), *The Acquisition of Verb Placement* (pp. 283–331). Dordrecht: Kluwer.
Volterra, Virginia and Francesco Antinucci
  1979   Negation in child language: a pragmatic study. In *Developmental pragmatics*, Elinor Ochs and Bambi Schieffelin (eds.), 281–303). New York: Academic Press.
Wakabayashi, Setsuko.
  1983   Early stages in the acquisition of Japanese negation. *First Language* 4: 154–155.
Weissenborn, Jürgen, Maike Verrips and Ruth Berman
  1989   Negation as a window to the structure of early child language. Ms., Max-Planck-Institute for Psycholinguistics, Nijmegen, The Netherlands.
Wexler, Kenneth
  1994   Optional infinitives, head movement and the economy of derivations. In *Verb Movement*, David Lightfoot and Norbert Hornstein (eds.), 305–350). Cambridge: Cambridge University Press.
Wode, Henning
  1977   Four early stages in the development of LI negation. *Journal of Child Language* 4: 87–102.
Wode, Henning
  1981   *Learning a second language.* Tübingen: Narr.

On the diachrony of negation[1]

Johan van der Auwera

1. Introduction

This study deals with the genesis of negators. In the first part I discuss the negators that are involved in what Miestamo (2005) calls 'standard negation'. In the second part I discuss three types of non-standard negation. In standard negation, the scope of the negation is the entire clause, the clause is a declarative, its main predicate is a verb, and the negative strategy is a general (productive) one. A negation that lacks any of these properties is 'non-standard'. In practice, the distinction is not that easy to draw, but I can refer the reader to Miestamo (2005) itself and here just illustrate the two types, i.e., standard negation in (1) and non-standard negation in (2).

(1)   John *does not* like Fred.

(2)   a. This is *im*possible.
      b. *Doncha* come no more!
      c. Fred is *no* stranger to us.
      d. I like *nothing*.
      e. I *can't* be bothered.

In (2a) the scope is not clausal, in (2b) the clause is not declarative, in (2c) the predicate is not verbal, and (2d) and (2e) do not use the general strategy with *do* periphrasis.[2]

---

[1] I gratefully acknowledge the financial support of (i) the Belgian Federal Government for grant P6/44 ('Grammaticalization and (Inter)Subjectification') and (ii) the Research Foundation Flanders for grant G.0152.09N ('Jespersen cycles'). The paper was written in sync with van der Auwera (2009), which gives a more complete picture of the 'Jespersen cycles' discussed in section 2.1. The paper also relates to van der Auwera (2006), the basis for the discussion in section 3.1. Special thanks are due to Maud Devos (Tervuren), Volker Gast (Berlin), Eitan Grossman (Jerusalem), Larry Horn (Yale), Pierre Larrivée (Birmingham) and the Leverhulme Trust, William McGregor (Aarhus), and Ljuba Veselinova (Stockholm).

[2] (2d) and (2e) are, of course, different. (2d) has a construal that uses standard negation, but (2e) does not have this: *I don't like anything* vs. **I don't can be bothered.*

Though standard and non-standard negation will be discussed in separate sections, the two types of negation are by no means unrelated. For one thing, it seems that everything that is relevant for the diachrony of standard negation is also relevant for the diachrony of non-standard negation, even though the latter typically involves additional complications. And for another thing, non-standard negators may turn into standard ones or into other types of non-standard negators. These links between the standard and non-standard negators will be discussed too.

I will restrict the attention to processes of change internal to a language and not discuss borrowing and calquing nor cases for which one can argue that both internal and external factors are at play.[3]

At a very abstract level, a negator, whether a standard one like *not* or a negative pronoun like *nobody*, may come about in three ways. First, some element X that is not itself a negator but usually has a related meaning, like 'to lack', develops into a negator. Second, some element that is not itself a negator but that frequently collocates with a negator, undergoes a 'contamination' (or 'contagion', Bréal 1897: 221–226) and thus develops a negative meaning on its own. An older negator NEG1 thus gets joined and possibly even replaced by a new one NEG2. In French *pas* was a collocate of the negator *ne,* and *pas* can now provide the negative meaning all by itself. Third, some element is not itself a negator, but frequently collocates with one, and they become one word ('univerbation'). Here too we get a new negator NEG2, but this NEG2 comprises the NEG1 marker. The English pronoun *nobody* is a simple example. The three types are schematized in (3)[4]. I abstract from word order.

(3)   a.   X   →   NEG
       b.   NEG1 X   →   NEG1 NEG2   →   Ø NEG2
       c.   NEG1 X   →   [NEG1-X]$_{NEG2}$

In what follows I will provide illustrations and hypotheses for each of three processes.

---

[3] Contact influence has been taken most seriously for the workings of the 'Jespersen cycles', to be discussed in 2.1. The areal perspective has been around since Bernini and Ramat (1992/1996), esp. for Europe, but the southern Mediterranean region has been looked at from this point of view as well (Lucas 2007, 2008).

[4] The schematization in (3a) merely says that some element X becomes a negator. It does not show that there must be some relation between the original meaning and that of negation. As (33) will show, the relation can be quite indirect.

The first process will be central in the section on prohibitive negation (3.1), which is a subtype of non-standard negation. For standard negation, the process is poorly documented (see e.g. the scarcity of data on the origin of standard negators in Heine and Kuteva (2002) or the skepticism voiced by Dryer (1988: 112–113) on a proposal by Givón (1984: 337–338) to relate the position of negators to that of their verbal sources). Yet what is well documented for standard negation and intensely studied, from a formal point of view as well (e.g. van Gelderen 2008), is the second type of process, that of genesis by contamination. It goes under the name of 'Jespersen cycle' and will be the central topic in section 2. The third type, genesis by univerbation, will be central in the discussion of two non-standard types, non-verbal and existential negation in section 3.2, and negative pronouns and negative adverbs in section 3.3.

2. Standard negation

Consider the sentence in (1) again, repeated as (4b) and accompanied by its positive counterpart.

(4) a. John likes Fred.
    b. John *does not* like Fred.

(4b) says that it is not the case that John likes Fred. The clausal negator is *not*, and in section 2.1 I will discuss how markers like *not* came about. The process is often called a 'Jespersen cycle', and I will follow this use. Note that when comparing (4b) to (4a), one can see that they differ in more than just the 'presence vs. absence of *not*': (4b) also contains the auxiliary *do*. This asymmetry will involve us in section 2.2.

2.1. The Jespersen cycles

2.1.1. Jespersen (1917) and Meillet (1912)

The term 'Jespersen cycle' (also 'Jespersen's cycle' or just 'Negative cycle') has been associated with the process of negator renewal such as it is found in French. French inherited the Latin preverbal clausal negator *non*, which became *ne* in French. It then developed a more complex pattern in which *ne* was accompanied by *pas*, originally meaning 'step'. In colloquial registers *ne* then fell into disuse, thus yielding a pattern with just *pas*. This pattern again employs a single negator, which can in principle enter a new cycle. This scenario is represented in (5).

(5) Stage 1 ⟶ Stage 2 ⟶ Stage 3
     ne   ne ... pas   pas

The sketch in (5) does not by itself say all that much. I will mention three issues that the representation in (5) leaves open. First, (5) does not say why languages (or rather speakers) start this kind of process. On this issue, a consensus is growing that the Jespersen cycle has its origin in the fact that languages probably always have ways to emphasize the negation. So instead of the *not* in (6a), speakers can make the denial more emphatic and say *not ... at all*.

(6) a. John *does not* like Fred.
    b. John *does not* like Fred *at all*.

If speakers overuse the emphatic strategy, the latter may lose its emphatic effect and become as neutral as the simple strategy. From then on the language has two neutral negative strategies, and there is the option – the very option schematized in (5) – that the older construction loses out, both as a negator on its own (i.e., *ne*) and as part of a complex strategy (i.e., *ne...pas*). In this way, the Jespersen cycle is a normal instance of grammaticalization, which indeed often originates in the availability of two patterns, a neutral and a more expressive one, with the latter bleaching and becoming neutral too, and consecutively replacing the earlier neutral pattern. This is actually not the scenario that Jespersen (1917) had in mind, the man after whom Dahl (1979) named the cycle and whose work proved very influential. In Jespersen's view it was not the mere availability of an emphatic strategy and its potential for inflationary overuse that is the origin but rather the phonetic weakness of the old negator and the need for a phonetically stronger exponent. Yet the grammaticalization perspective does have authoritative credentials: it is found in Meillet (1912), which antedated Jespersen (1917)[5]. Interestingly, Meillet (1912) also stressed the cyclical nature of the process:

> Les langues suivent ainsi une sorte de développement en spirale: elles ajoutent des mots accessoires pour obtenir une expression intense: ces mots s'affaiblissent, se dégradent et tombent au niveau de simples outils grammaticaux; on ajoute de nouveaux mots ou des mots différents en vue de l'expression; l'affaiblissement recommence et ainsi sans fin. (Meillet 1912: 394 [1926: 139–140])

---

[5] See van der Auwera (2009) for an overview of a century's worth of scholarship on the Jespersen cycle. Pride of place is given to Gardiner (1905), the Egyptologist who describes the genesis of negation in Egyptian and links it up with the French and English scenarios.

A second issue that a representation like (5) says nothing about is why an element like *pas* proved to have a strengthening effect. In the case of *pas* one assumes that the use of the 'step' meaning with negative movement verbs acquired a minimizing effect, taking it from literally 'not a step' to 'not even a step', i.e., 'not even the minimal progression of a step', which then further generalized to 'not at all', making it usable with verbs other than those of movement. For the specific case of *pas*, there is no direct textual evidence for this hypothesis though: the earliest French texts are too late, and the last Latin ones too early (Buridant 2000: 708), but in French and in many other languages, there are directly attested developments for other minimizers, such as French *mie* and *point*, which were the main competitors of *pas* and which developed from 'crumb' and 'point', respectively ('not a crumb/point' → 'not even a crumb/point' → 'not at all').[6]

Note that the minimizing effect is not dependent on negation: it suffices to have a so-called 'negative polarity' context, a characterization encompassing negation proper but also contexts like conditionals and questions (see Hoeksema, this volume). Thus, just like English *at all,* the minimizer is not restricted to negation (as in (6b)), but is found in conditionals and questions (see (6c–d)); *pas*, *mie* and *crumb* are either assumed (in the case of *pas*) or attested (in the case of *mie* and *point*) with the 'at all' sense in these contexts too (Eckardt 2003).

(6) c. If you see him *at all*, tell him that Fred is home now.
    d. Have you seen him *at all*?

Note also that not only does a minimizing use like 'even a step' not require emphatic negation, emphatic negation does not require a noun referring to a minimal entity either. In Dutch, for instance, and equally so in German and English, the strengthener *niet* does not go back to a construction referring to a minimal quantity, but to a construction meaning 'nothing', which refers to a quantity below minimality. In the Penutian language Wintu, the strengthener was a negative existential verb, a construction for which Croft (1991: 10, 13–14) proposes a cycle of its own, and in the Vanuatu language Motlav, the strengthener was a partitive marker, a construction getting its emphasis through the interpretation that the event does not (even) involve part of something.

---

[6] Horn (1989: 452, 2001: 188) points out that the crucial role of minimizers was described as early as Pott (1833: 410).

(7) Dutch (Middle Dutch as well as present-day West-Flemish and East-Flemish)
Ik *en* heb hem *niet* gezien
Ik NEG have him NEG seen
'I haven't seen him.'

(8) Wintu (Croft 1991: 10, based on Pitkin 1984: 198)
ʔelew-be:sken   haraa:-wer-mina
NEG.EX-you.IMPF go-FUT-NEG
'You were not supposed to go.'

(9) Motlav (François 2003: 313)
Et igni-k *te.*
NEG wife-my NEG
'This is not my wife.'

A third issue left unclear in the representation in (5) and on which (5) is actually misleading is the possible contemporaneity of the three stages. This is a complex matter. First, nobody doubts that there are transition periods between each of the three stages and in these the old pattern is in competition with the new one. This effectively turns the three-stage model into a five-stage model, for there are of course two transition stages.

(10) Stage 1 ⟶ Stage 2 ⟶ Stage 3 ⟶ Stage 4 ⟶ Stage 5

ne   *ne*   *ne ... pas*   *ne... pas*   *pas*
                  ne ... pas   *pas*

Second, even (10) is still a simplification, for one thing, because one would expect that a transition period will show stages of a decreasing use of the old construction in tandem with an increasing use of the new one. And for another thing, there could also be a comporaneity of each of the three 'basic' stages. An older pattern may, but need not, remain an option, either as a general option or as one that is restricted in some way, when even the successor of its successor has appeared. French is a case in point. The older *ne* strategy is still acceptable in some contexts (and registers), as with the verb *pouvoir* (Muller 1991: 229) (and other relic forms, e.g. *Je ne saurais vous dire...* 'I couldn't tell you' cf. English *I kid you not, She loves me not*), the standard *ne ... pas* is the normal option, and *pas* by itself is the progressive one.

(11) French
  a. Il *ne* peut   venir ce soir.
  b. Il *ne* peut *pas* venir ce soir.
  c. Il   peut *pas* venir ce soir.
     he NEG can NEG come this evening
     'He can't come tonight.'

This contemporaneity is represented in (12).[7]

(12) Stage 1 ⟶ Stage 2 ⟶ Stage 3 ⟶ Stage 4 ⟶ Stage 5

	ne	*ne*	*ne... pas*	
ne	*ne ... pas*	*ne ... pas*	*pas*	*pas*
		pas		

When two or more variants are available at any one phase, the choice could be steered by a multiplicity of factors,

(i) register: e.g. in (11), which contrasts the archaic option (11a), the written and spoken standard option (11b), and the spoken progressive one (11c);

(ii) region: e.g. Old French *ne ... mie* seems to have been typical for the North and the East, where *ne ... pas* characterized the centre and the West (Kawaguchi in print)

(iii) discourse-pragmatics: e.g. it has been claimed by Hansen (2009), in the wake of Schwenter (2006), that Old and Middle French *ne ... pas* and *ne ... mie* differed from the simple *ne* construction in that the former were for a long time only used for discourse-old information.

2.1.2. Variation at the 2nd stage

The scenarios discussed or alluded to in 2.1.1 have in common that an originally simplex negator is joined by a second negator, which (i) is formally distinct from the first and (ii) has its origin in an emphatic negative con-

---

[*] An account that strongly and persuasively emphasizes comtemporaneity is Martineau and Mougeon (2003). Note that Stage 3 of (12) could also feed into a stage with a choice between with only *ne* and only *pas*. In other words, the doubling strategy would disappear before the old simplex strategy. I do not know of any attested case, but this kind of development should be possible.

struction (at least in the interpretation offered by Meillet 1912). Neither property is necessary, however.[8]

One point of variation concerns the nature of the second negator.[9] In French and Dutch the second negators *niet* and *pas* differ from the first negators *ne* and *en* respectively. In Brazilian Portuguese (13), however, the second negator seems to be a copy of the first.

(13) Brazilian Portuguese (Schwegler 1991: 209)

Eu *não* quero *não*.
I   NEG   want   NEG
'I don't want to.'

This kind of negator doubling is attested in other Romance languages: Schwegler (1990, 1996) discusses Spanish, and Ramat (2006) and Floricic and Molinu (2008) Italian. For Germanic the phenomenon did not escape the attention of Jespersen (1917: 72), who focused on Swedish; Afrikaans is probably the best known Germanic illustration (see Roberge 200, Biberauer 2008), but there is Dutch as well (Pauwels 1958). Jespersen (1917) did not relate this phenomenon to what would later be called the 'Jespersen cycle'. Jespersen called it 'resumptive negation', took it to originate outside of the clause (see below), and considered it a technique for emphasis, a technique that has the result "that the negative effect is heightened" (Jespersen 1917: 72). This view is actually in full accord with the current understanding of the origin of the Jespersen cycle, but not with Jespersen's own analysis. This takes us to the second point of variation.

For Jespersen the doubling strategy as in French and Dutch did not originate as a bleaching of an emphatic one, finding itself in competition with a healthy simplex strategy. For Jespersen, the simplex strategy was not healthy, it was phonetically weak, and it is this weakness that called for a clearer exponent. Jespersen (1917: 4) does also use the term 'strengthening' ("the original negative adverb is first weakened, then found insufficient and

---

[8] Alternatively, one does make these properties criterial and then the variations discussed in this section do not count as 'Jespersen cycles' but as 'related cycles'. This is a position found in De Cuypere *et al.* (2007) and De Cuypere (2008).

[9] Another point is whether or not the second negator, if emphatic at the first stage, has to lose its emphatic nature at the doubling stage and may thus remain emphatic when it becomes the only exponent of negation. It has been shown that emphasis may indeed linger on (see Horn 2001, Postal 2005: 159–172 and Hoeksema 2009).

therefore strengthened"), but it would not be a matter of strengthening for emphasis, but simply for clarity. As mentioned in 2.1.1. this view has been given up for at least the classical cases, involving French *pas* and Dutch *niet*, but that does not mean that a clarity-based account could not hold for the other cases.

Let us now look at (13) again. What would seem to be relevant in understanding this kind of doubling is whether or the second negator can be shown or assumed to result from the integration of a homomorphic negative answer particle, i.e., from the marker meaning 'No!'. For Brazilian Portuguese, this case has been made by Schwegler (1991: 209). The ancestor of (13) is claimed to be (14), a full clause, which is tagged with the particle 'No!', following the clause with a separate intonation contour.

(14) Brazilian Portuguese (Schwegler 1991: 209)
 Eu *não* quero, *não*!
 I NEG want no
 'I don't want to, absolutely not!'

The meaning is emphatic and for this reason Schwegler uses the gloss 'absolutely not'. What happens in between (13) and (14) is that the negative tag is integrated into the clause or, as Jespersen (1917: 72) puts it, 'the supplementary negative may be felt as belonging with the sentence, which accordingly comes to contain two negatives'.[10] Calling the second negator a copy of the clausal negator is therefore a bit misleading, at least diachronically. The second negator is diachronically not 'the same' as the clausal negator; but rather the negative answer particle, which, of course, may ultimately have the same source as the clausal negator.

But clause-final negator copying is also possible when the negator and the negative answer particle are not homomorphic. This is the case in Dutch, in which the clause-final negator copy is currently only a marginal option for Belgian Brabantic dialects, and also in Afrikaans, where it is a property of the standard language. (15) illustrates Belgian Brabantic.

---

[10] As befits a Jespersen cycle, one would also expect a stage in which only the new negator, i.e., the clause-final one, is used, and Brazilian Portuguese does indeed bear out this expectation (Schwegler 1991: 206).

(15) Belgian Brabantic Dutch
Hij wil geen soep niet meer eten *niet.*
he wants no soup NEG more eat NEG
'He doesn't want to eat any more soup.'

The point relevant now is that in Dutch, Brabantic or not, the negative answer particle is *nee(n)* rather than *niet* [11].

(16) Dutch
Ik ga niet, *nee*!
I go NEG no
'I am not going, no!'

So Brabantic Dutch (16), as opposed to Brazilian Portuguese (13), must truly involve the repetition of the negator, and we see Pauwels (1974) comparing the repetition of this negator with a repetition of a preposition, again in Belgian Brabantic Dutch.

(17) Belgian Brabantic Dutch (Pauwels 1974: 76)
Ik kan nie *aan* het plafond *aan.*
I can NEG at the ceiling at
'I can't reach the ceiling.'

The preposition doubling of (17) is not emphatic nor is it imaginable that it ever was: *aan ... aan* only means 'at' and not something like 'really at'. This kind of non-emphatic redundancy may have been at the origin of *niet ... niet* too.

The table in (18) summarizes the discussion about the nature of the second negator.

(18)

The second negator	has an emphatic origin	has no emphatic origin
is different from the first negator	French *pas*	??
is identical to the first negator	Brazilian Portuguese *não*	Brabantic Dutch *niet*

---

[11] The point is not uncontroversial (cp. the reaction of Roberge 2000: 146–147 to Bernini and Ramat 1996: 78): *niet* is in fact attested as an answer particle, but this is very rare (Joop van der Horst, p.c.) and thus, I assume, beyond the threshold frequency necessary for routinization and subsequent reanalysis.

The table in (18) also raises the question whether a second negator could be different from the first one and yet not have an emphatic origin. A positive answer exists at least since Kroskrity's (1984) analysis of Arizona Tewa. In this language negation is expressed by two morphemes embracing the verbal forms, *we-* and *-dí*.

(19) Arizona Tewa (Kroskrity 1984: 95)
      Sen   kʷiyó   *we*-mán-mun-*dí*.
      man   woman   NEG-3>3.ACT-see-NEG
      'The man did not see the woman.'

The second negator *-dí* also functions as a marker of subordination and Kroskrity (1984) argues that this is the origin of the second negator: originally, negative statements were subordinate[12] and through time the subordinate marker was reanalyzed as an exponent of negation. Kroskrity (1984)'s analysis was endorsed by Honda (1996: 41–44) and meanwhile other positive answers have been given, either implicit ones (Early 1994a: 76; see 3.1 below) or explicit ones (Lucas 2007: 427, De Cuypere et al. 2007: 309–312, De Cuypere 2008: 238–245), so I believe that there is no doubt that the top right hand corner of the table in (18) can be filled.

2.1.3. *Variation at the third stage*

The classical and less classical Jespersen cycles all involve a development from a construction with a single exponent of negation to a construction involving two exponents. The third stage is one which has a single exponent again, with the new negator. But this is not the only possibility. First, there is no reason why a construction with two exponents could not attract yet a third exponent. This issue was already raised in 1923, without reference to Jespersen (1917) or Meillet (1912) (or any other general discussion) and the one who raised it was the Flemish dialectologist Blancquaert. Blancquaert (1923: 68) asked whether the Dutch *en ... niet* pattern was combinable with a clause-final copy of *niet*. In 1958, the Flemish dialectologist Pauwels, already mentioned for a hypothesis about the origin of clause-final *niet*, provided a positive answer. Pauwels (1958: 454) claimed that in at least some

---

[12] That language may code negative sentences as subordinate is well-known (see Miestamo 2005: 173–175 and section 2.2 below).

Belgian Brabantic Dutch of his time (mid twentieth century), the combination was indeed possible, at least in subordinate clauses.[13]

(20) Brabantic Belgian Dutch
Pas op dat ge *niet en* valt *nie.*
fit on that you NEG NEG fall NEG
'Take care that you don't fall.'

Since then further cases have been attested; see Early (1994a,b) on the Vanuatu language Lewo, Parry (1997) on North Italian dialects and Mukash Kalel (1982: 323) on the Bantu language Kanyok. In terms of emphasis, it seems that both emphatic and non-emphatic tripling occurs. (20) is non-emphatic; for an emphatic scenario, see the analysis of Lewo in section 3.1 below.

Second, at the third stage the old exponent of negation may indeed disappear with its negative meaning, but why couldn't it remain in the language, but with another meaning? Thus Wallage (2008) argues that Middle English *ne* survived as a marker of negative polarity at a time when it was no longer a true negator (see also Breitbarth 2009), and for Dutch the case of the non-negative survival of an old negator has been argued for on two occasions: Both Overdiep's (1933: 23) and Neuckermans' (2008: 21) analyses show that Dutch *en* may undo its negative and also its negative polarity meaning and survive only as a marker of subordination.

(21) Middle English (Wallage 2008: 666)
No man douteth that he *ne* is strong in whom he seeth strengthe
'No one doubts that that person is strong in whom he sees strength.'

(22) Belgian Brabantic Dutch (Neuckermans 2008: 176)
Ze pakte eu portefueille waar dase eu sleutel in *en* doet.
She took her wallet where that.she her key in *en* does
'She took the wallet of hers in which she puts her key.'

---

[13] That the construction would be restricted to subordinate clauses is related to the fact that the Dutch doubling pattern *en ... niet* is holding out longest in subordinate clauses. The observation is at least as old as Overdiep (1937: 423) and has been confirmed in Dutch dialectological work up to Barbiers *et al.* (2009). See also the point about example (22) below.

2.1.4. Conclusion

(23) summarizes much of the preceding discussion in the form of a three-stage Jespersen model. At the first stage we may have a marker X that will later be interpreted as a negator and this NEG1 X construction is either emphatic (as with French *ne ... pas*) or not (as with Arizona Tewa *we- ...-dí*).[14] The first stage may also lack the X marker: I represent its absence by 'Ø'. Here emphasis is not a relevant parameter. At the second stage, the doubling constructions may be emphatic or non-emphatic. Emphatic ones result from either emphatic NEG1 X (as in French *ne ... pas*) or from NEG1 Ø (as with Brazilian Portuguese *não ... não*). Non-emphatic ones have three origins: (i) emphatic NEG1 NEG2 (as with French *ne ... pas*), NEG1 Ø (as with Brabantic *niet ... niet*), and non-emphatic NEG1 X (as with Arizona Tewa *we- ...-dí*). At the third stage, a non-emphatic NEG1 NEG2 may turn into the simpler Ø NEG2 or into the more complex NEG1 NEG2 NEG3, both of which are non-emphatic. Note that (23) does not show that an old negator may have an afterlife outside of the negative domain, either as a negative polarity item (see Middle English (21)) or as a marker of subordination (see Belgian Dutch (20)).[15]

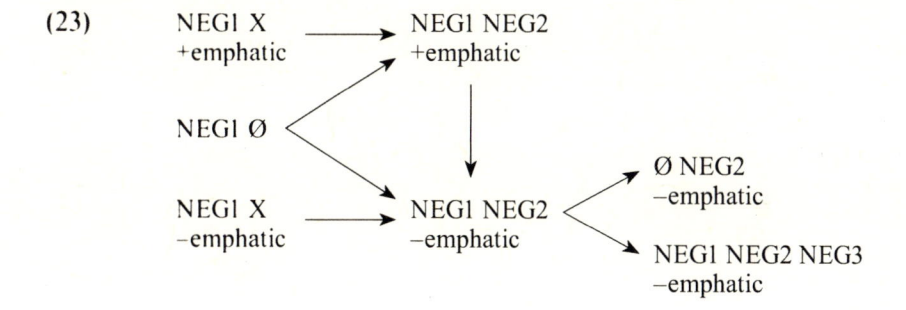

(23)   NEG1 X   ⟶   NEG1 NEG2
       +emphatic   +emphatic

       NEG1 Ø

       NEG1 X   ⟶   NEG1 NEG2   ⟶   Ø NEG2
       −emphatic   −emphatic   −emphatic

                                                       NEG1 NEG2 NEG3
                                                       −emphatic

---

[14] The notations abstract from word order properties. So a construction counts as "NEG1 X" both when the negator precedes and when it follows. This is not to deny, however, that word order properties play a role, but this topic falls outside the scope of this discussion.

[15] As already hinted at in note 1, the representation in (23) does not exhaust the variation found in Jespersen cycles. For a more general account, see van der Auwera (2009).

2.2. Negation's asymmetry

The non-emphatic element *di* in the Arizona Tewa construction illustrated in (19) was a subordination marker. The reason why this subordination marker could be reanalyzed as a negator is that in this language negative sentences were either always or at least frequently subordinate. The subordinate marker was thus a collocate of the negator. Something similar is found in English (1), repeated as (24b). In one sense, it is *not* that is responsible for the negative meaning, yet in modern English, the mere addition of *not* to a positive declarative does not generally yield a grammatical sentence.

(24) a. John likes Fred.
     b. John does *not* like Fred.
     c. *John *not* likes Fred.
     d. *John likes *not* Fred.

So for English, the *do* auxiliary is a collocate of the sentential negator.

The fact that negative declaratives often differ from their positive counterparts in more than just the absence vs. presence of a negator has been called 'asymmetry' by Miestamo (2005). He distinguishes subtypes along two cross-cutting dimensions. The first dimension concerns the question whether the entity that is asymmetric is an individual construction or a paradigm (i.e., a set of constructions). Thus the English construction in (24b) is asymmetric, but one would not want to say that the paradigm of all negative declarative sentence types distinguished in terms of e.g. tense or aspect is asymmetric. The situation is different in the Australian Yiwaidjan language Maung. Consider (25).

(25) Maung (Miestamo 2005: 97, based on Capell and Hinch 1970: 67)
    a. ŋi-udba
       1SG.3-put
       'I put'
    b. ni-udba-ji
       1SG>3-put-IRR.NPST
       'I can put'
    c. *marig* ni-udba-ji
       NEG   1SG>3-put-IRR.NPST
       'I do not / cannot put'

The negative declarative in (25c) is 'constructionally symmetric' with one of the two positive declaratives, viz. the irrealis one in (25b), but there is also paradigmatic asymmetry, in the sense that whereas positive declaratives may be realis (as in (25a)) or irrealis (as in (25b)), the negative counterparts must be irrealis.

The second dimension of asymmetry concerns the grammatical status of the collocate that causes the asymmetry, whether constructional or paradigmatic. Miestamo (2005) distinguishes four types. Arizona Tewa and, arguably, English illustrate the first type: the negative declaratives manifest a decrease in finiteness: the lexical verb appears in either a subordinate form (earlier Arizona Tewa) or an infinitival or bare stem form (English). Maung illustrates a second type: negatives are obligatorily irrealis. In a third type the negative contains a marker that marks emphasis in positive declaratives. English, already brought in to illustrate the first subtype, may be taken to illustrate this subtype too: *do* is of course possible in positive declaratives, and then it marks emphasis.[16]

(25) d. John does like Fred.

In a fourth type, the difference typically concerns the neutralization of verbal categories like tense, aspect, person or number (but there are other subtypes too and hence the fourth type is a somewhat of a wastebasket category). This type is illustrated with Burmese (26).

(26) Burmese (Miestamo 2005: 123, based on Cornyn 1944: 12–13)
    a. θwâ-dé
       go-ACT
       'He goes/went.'
    b. θwâ-mé
       go-POT
       'He will go.'
    c. θwâ-bí
       go-PERF
       'He has gone.'

---

[16] Periphrastic *do* obviously occurs in polar questions also, but this is not relevant from Miestamo's perspective. It also occurs in positive elliptical formations: *John does not like Fred but I do*.

d. *ma-θwâ-bû*
NEG-go-NEG
'He does/did/will not go, has not gone'

This typology of asymmetry obviously poses a diachronic question: how does asymmetry arise? Miestamo's (2005) answer is that for each of the asymmetries languages will have grammaticalized a typical semantic or pragmatic property of negation. Thus negation, Miestamo claims, is typically stative (basically, when something is not happening, nothing changes) and languages may grammaticalize this tendency with negative constructions showing reduced finiteness. Negative states of affairs are non-real in a strong sense, but there are other non-real states of affairs like futures, pasts or states of affairs scoped by a capacity (as illustrated in (25b)) and languages may reflect this similarity. Negative declaratives are often emphatic: the Jespersen cycles are a testimony to this effect too, and languages may reflect this in ways not yet covered by the Jespersen cycles. And finally, negative declaratives typically function in a discourse in which the positive counterpart is either explicit or implicit and it is thus not necessary to repeat the semantics of tense and aspect, for instance. The neutralization of tense and aspect illustrated with Burmese (26) makes sense as a grammaticalization of this tendency.

To sum up: an important dimension of negators is their collocational patterning. Miestamo (2005) has shown that the genesis of this patterning can be interpreted as a grammaticalization of a few typical semantic or pragmatic properties of negation.

A final point: Jespersenian strengthening can be seen as collocational patterning too. The minimizing 'not even a step' can be seen as a collocate. But while Miestamo stops at explaining the genesis of collocational asymmetry, the goal of Jespersenian cycles is explaining how a collocate of a negator can become a negator.

3. Non-standard negation

In this section I turn to non-standard negation. The discussion treats prohibitive negation, then non-verbal and existential negation, and finally some aspects of the formation of negative pronoun and adverbs.

3.1. Prohibitive negation

The generalizations described for standard negation are relevant for prohibitive negation too: (i) for the relevance of the Jespersen scenarios, see the discussion of Lewo (28) below, and (ii) for the relevance of the generalizations about asymmetry, see Miestamo and van der Auwera (2007). But prohibitive negators also present a story of their own, as is already suggested by the fact that prohibitive negators are often different from declarative ones. In a large (though not fully balanced) sample of 495 languages, van der Auwera *et al.* (2005) have argued that up to two thirds of the world's languages may have special prohibitive negators. Many special prohibitive negators derive from what could be called 'indirect prohibitive speech acts'. There are many subtypes. Consider first the speech acts in (27).

(27) a. Abstain from dancing.
b. Stop dancing.

(27) illustrates prohibitions, but only indirect ones, at least from the point of grammar, for the prohibitive sense arises from the lexical meaning of the verbs *abstain* and *stop*. (27a) is a 'prospective' imperative not to start an activity, and (27b) a 'retrospective' one to stop an activity (the 'retrospective' – 'prospective' terminology is due to De Clerck 2006). What is relevant for the present discussion is that both the lexical items or phrases for abstaining and stopping may grammaticalize into dedicated prohibitive negators. Thus we find 'abstain' as the hypothesized origin of prohibitive markers in the Australian language Kuku-Yalanji (Patz 2002: 194) and in the Vanuatu language Lewo (Early 1994a, 1994b). The case of Lewo is particularly interesting, because it brings in the Jespersen cycles again and in two different ways.

Lewo has three prohibitive strategies. The first one just uses the discontinuous irrealis strategy *ve ... re*, which is also found in declaratives, and of which the *re* element has a Jespersenian partitive origin, like the *te* element of the related language Motlav (see (9)). The second one strategy is periphrastic, with the verb *toko* 'abstain' in combination with a nominalization.

(28) Lewo (Early 1994a: 76)
a. *Ve* a-kan *re*!
NEG.IRR 2SG-eat NEG
'Don't eat it!'
b. Na-kan-ena *toko*!
NOM-eat-NOM desist
'Don't eat it!' or 'Abstain from eating it!'

toko is clearly a verbal element and as such it takes a nominalized complement. But there is also a *toko* element which shows up in a third prohibitive construction, the one that has developed most recently (Early 1994a: 77). In this strategy we find both *toko* and the structure with the normal negative markers.

(28) Lewo (Early 1994a: 76)
    c. *Ve   a-kan   re   toko!*
       NEG. IRR 2SG-eat NEG PROH
       'Don't eat it!'

In (28c) *toko* does not combine with a nominalization. From that point of view, *toko* is not a complement taking verb (or auxiliary) anymore. It would not make semantic sense either, for then (28c) would have to mean 'Desist from not eating!', i.e., 'Eat!'. Early (1994a: 77) also makes clear that there is no intonation break between *toko* and the preceding phrase. So one should not gloss the sentence as 'Don't eat it, don't!', a strategy with Jespersenian clause final negator emphasis. Yet I think – and here I go beyond Early's (1994a) account (and also Early 1994b) – this does make sense as the source of the new strategy. As opposed to the *toko* of (28b) then, which is simply the verb 'abstain', the *toko* of (28c) is a prohibitive particle. And – an interesting aside – since (28c) now has three negative particles, we can also use this construction as evidence for Jespersenian tripling (see section 2.1.3).

The Lewo case illustrates the grammaticalization of a prospective indirect prohibitive. The 'retrospective' scenario of telling a hearer to stop an ongoing activity can also grammaticalize and in their survey of the sources of grammatical markers, this is in fact the only source that Heine and Kuteva (2002: 335) list for prohibitives.[17] (29) is one of their examples.

(29) Seychelles Creole   (Heine and Kuteva 2002: 284,
    Arret vol   sitrô!   based on Corne 1977: 184)
    PROH steal lemon
    'Don't steal lemons!

---

[17] This is line with the hypothesis that the restrospective prohibitive is a 'more prominent' use than the prospective one (Birjulin & Xrakovskij 2001: 13), but it is not clear what 'prominence' amounts to. It does not imply a higher frequency, since at least in English and Dutch prospective prohibitives are much more frequent than retrospective ones (Declerck 279–284, 286; Van Olmen & van der Auwera 2008).

Arret has a verbal origin (= 'stop'), and it remains to be seen to what extent the conventionalized prohibitive is in fact still the verb. For Lewo (28c), at least, there is clear evidence that the erstwhile verbal status has been lost.

(30) illustrates some further prohibitive source constructions.

(30) a. You must not dance.
 b. Dancing is not possible.
 c. Dancing is not wanted.
 d. Dancing is taboo.
 e. Dancing is not good.

Each of these constructions may grammaticalize into prohibitive negators. Thus we get a derivation from the negative in combination with a necessity modal in Afrikaans *moenie* (from *moe* 'must' and *nie* 'not') and the Serbian/Croatian prohibitive negator *nemoj* derives from the negative *ne* and the imperative of the verb *moći* 'may' (Greenberg 1996: 164; Hansen 2004). The Mandarin prohibitive *bié* would be a shortened form of *buyào*, deriving from the ordinary negative *bú* followed by the necessity modal *yào* (although there is an alternative etymology relating *bié* to the meaning 'other'; Norman 1988: 127, Li 2004: 268–270). This *yào* element also means 'want', hence *buyào* could also be seen as deriving from 'not want', a path well known from Latin *noli*. 'Taboo' is the prohibitive source in the Oceanic language Tongan (Broschart 1999: 109), and 'not good' gives the Cantonese prohibitive *mhóu* (S. Matthews, p.c.).

(30) is by no means a complete list. Whenever we find a synchronic polyfunctionality for a prohibitive marker, a diachronic scenario is bound to relate the various meanings. It is not always clear, though, what the directionality is. Thus we see languages having elements serving for prohibitive and 'lest' (negative purpose) (e.g. the Gur language Pasaale Sisaale McGill, Fembeti & Toupin 1999: 139), but it is not a priori clear that the negative purpose comes first. Interestingly, also, both negative irrealis constructions, as in (31a), and negative realis ones, as in (31b), can develop into prohibitive markers.

(31) a. (That) you would not be dancing.
 b. You are not dancing.

Spanish and Latin are classical illustrations of irrealis (subjunctive) prohibitives, and the Bantu language Kela is another. (32) illustrates the imperative and the prohibitive singular. The former ends in *-a* and the latter carries the subjunctive -*e*, but interestingly the prohibitive is not simply a negative subjunctive, for the two constructions have different tonal patterns.

(32) Kela (Forges 1977: 115)
  a. kádáŋ-á-Ø
     roast-IMP-SG
     'roast!'
  b. *po*-kádáŋ-é
     NEG-roast-IRR.2SG
     'Don't roast!'

The Mexican language Comaltepec Chinanteco illustrates a construction that allows both a realis progressive and a prohibitive reading.

(33) Comaltepec Chinanteceo (Anderson 1989: 91)
  Ha^L-hiú: ^{M}-ʔ   lú^{LM}!
  NEG-blow.PROGR-2 instrument
  'You are not playing an instrument.'
  'Don't play an instrument!'

One may assume the progressive sense to be the older one and if the two meanings ever demand a different formal realization, starting from the structure illustrated in (33), we will have a grammaticalization of a prohibitive out of a realis construction.

Yet another type of prohibitive grammaticalization relates to warnings, generally called "admonitive" and "preventive" constructions (e.g. Birjulin and Xrakovskij 2001: 37–40). Consider the Siberian language Yakut (also known as Sakha).

(34) Yakut (Pakendorf and Schalley 2007: 519)
  Sarsïn   kiehe   muosta-nï su: y-a: ya-yït!
  tomorrow evening floor-ACC wash-VPOT-2PL
  'Tomorrow evening don't wash the floor!'

The affix *-yït* is part of what is called a positive 'Voluntative-Potential' paradigm but it is only the second person that has a negative, more particularly prohibitive meaning; in the other persons and originally also in the second persons, the forms of the paradigm express possibility and hope. The semantic change that Pakendorf and Schalley (2007) describe takes a marker from the expression of possibility to a warning and then to a true prohibitive ('you might go' > 'see to it that you don't go' > 'don't go').

One type that does not seem to be widespread is that of a prohibitive marker that results from the univerbation of a standard negator and some sort of imperative marker. The Caucasian language Hunzib (Van den Berg 1997: 87) employs a suffix -*áq'(o)* for prohibition. Given that there is a transitive imperative suffix -*o* and an indicative present negative suffix -*at'* (Van den Berg 1997: 84, 87), one might hypothesize that -*áq'(o)* derives from a univerbation of both elements. In the remaining two types of non-standard negation univerbation is much more prominent.

3.2. Non-verbal and existential negation

In this section I briefly discuss the non-standard negation found with non-verbal and existential predicates.[18] Consider the following examples from Belorussian and Serbian. In both languages the standard negator is *ne*. In Belorussian we also find it in the non-verbal negation illustrated in (35b), but in Serbian the corresponding negative sentence (36b) has a specialized *nije* form. In the d-sentences we see an existential negator: here both languages use a specialized verbal form, *njama* in Belorussian and *nema* in Serbian.

(35) Belorussian (Veselinova 2008)

    a. Maryja ščastlivaja.
       Mary happy
       'Mary is happy.'

    b. Maryja *ne* ščastlivaja.
       Mary NEG happy
       'Mary is not happy.'

    c. Dzik-ija kat-y (ests').
       wild-PL.NOM cat-PL.NOM (there are)
       'There are wild cats.'

    d. Dzikix katoj *njama*.
       wild-PL.GEN cat-PL.GEN not.have.PRES
       'There are no wild cats.'

---

[18] Not much has been done on the typology. One can refer to Croft (1991), Eriksen (2005) and the ongoing work of Veselinova (2007, 2008).

(36) Serbian (Veselinova 2008)
- a. Tom je srećan.
  Tom be.PRES.3SG. happy
  'Tom is happy.'
- b. Tom *nije* srećan.
  Tom not.be.PRES.3SG. happy
  'Tom is not happy.'
- c. Ima divl-jih mač-aka.
  have.PRES.3SG. wild-PL.GEN cat-PL.GEN
  'There are some wild cats.'
- d. *Nema* divl-jih mač-aka.
  not.have.PRES.3SG. wild-PL.GEN cat-PL.GEN
  'There are no wild cats (e.g. in the garden).'

The diachrony of the special forms is still transparent: Serbian *nije* results from the univerbation of a negator and a copula, and Belorussian *njama* and Serbian *nema* result from the univerbation of a negator and a 'have' verb. Univerbation processes yielding special negative copula and existence possession constructions seem to be common (cp. also Croft 1991). Veselinova (2008) also points out that the source may be a whole phrase (as in Russian *net*, deriving from *ne ě tu* 'not is here' – with possibly a *-t'< ti* 'DAT.2SG added to *tu*, thus meaning 'not is here to you'). On the semantics, Veselinova argues that possession first yields location and only then existence and McGregor (2009) shows that there is some variation in the source construction: next to a '[possessor] has [possessum]' pattern, there is also '[possessum] belongs to [possessor]' pattern', as McGregor argued for the Australian language Nyulnyul. As already mentioned in 2.1.1, for the existential construction Croft (1991) argued that it can enter an emphasis-based Jespersen cycle, ultimately leading to the negative existential negator bleaching into a standard negator – see the Wintu example (8). However, not every standard negator that has a negative existential source may be assumed to have gone through a Jespersen cycle. Veselinova (2008) shows this with Bulgarian. In this language, the negative existence *njama* negator, the counterpart to Belorussian *njama* and Serbian *nema*, is also used to negate an ordinary declarative future. So whereas we do see a non-standard negator expanding its terrain to standard negation, the reason is very specific: in earlier Bulgarian, the positive future exploited a 'have' strategy, the negation of which also used 'have', and it is this construction that survived in present-day Bulgarian.

(37) Bulgarian (Veselinova 2008)
   a. Todor shte xodi na kino.
      Todor will go.3.SG to movie
      'Todor will go to the movies.'
   b. Todor *njama* da xodi na kino
      Todor not.have.PRES.3SG COMPL go.3SG to movie
      'Todor won't go to the movies.'

A final point: existential negators may not only turn into standard negators, but also into prohibitive ones. This has been documented for Middle Egyptian and Coptic dialects by Grossman (2008).

3.3. Negative pronouns and adverbs

How do languages form inherently negative adverbs like *nothing* and inherently negative adverbs like *nowhere* (38a–b)?

(38) a. I saw *nothing*.
    b. I saw him *nowhere*.
    c. I didn't see anything.
    d. I didn't see him anywhere.

A further question, related to the preceding one, is how languages develop strategies that combine the standard negator with non-negative constituents like English *anything* and *anywhere* (see (38c–d)). I will embed the answer to the first question in the answer to the second question and the discussion will be limited to pronouns (and thus exclude adverbs).

One way of expressing *nothing* or *not anything* is by using an ordinary clausal negator and a normal positive pronoun (or pronominal phrase) meaning 'something'. This is illustrated with the East Papuan language Nasioi.

(39) Nasioi (Kahrel 1996: 39, based on Rausch 1912: 134)
   a. Nanin nánu-i.
      someone go-RCTPST
      'Someone went.'
   b. Nanin ninu-*aru*-i
      someone go-NEG-RCTPST
      'No one went.'

Literally (39b) says 'Someone didn't go', which is grammatical in English, too, of course, but only in the sense, irrelevant here, that gives the indefinite scope over the negation ('There is someone that didn't go'). Cross-linguistically, this pattern may well be the most frequent one (Kahrel 1996: 36–39; van der Auwera et al. 2006: 314–318). I suspect that constructions such as (39b) usually do not pose any particular diachronic problem (one merely combines a standard negator with an ordinary positive indefinite), but see the point about (46e) below.

English (38c) with *anything* rather than *something* is a second construction type. The *anything* element is a negative polarity item, which means that it does not only occur in strictly negative contexts, but also in contexts like conditionals and questions.

(40) a. If you see *anything*, let me know.
     b. Have you seen *anything*?

As an original minimizer based on the numeral 'one' (Haspelmath 1997: 228), *anything* was probably emphatic, paraphrasable as 'even one thing', just like *pas* 'even a step', discussed in 2.1.1, and then it underwent bleaching, resulting in the present-day 'anything' sense. It may then become restricted to strictly negative environments. This is happening to French *personne*, for instance. Its preferred context is the strictly negative one, as in (41a), but negative polarity uses can still be found, as in (41b).

(41) French
     a. Je *n*'ai vu *personne*.   'I saw nobody'
     b. Elle chante mieux que *personne*. 'She sings better than anyone'
        (www.college-de-france.fr/media/lit_fra/UPL19782_zinkres0405.pdf, accessed on Dec. 11, 2008)

What is happening to French *personne*, again just as with clausal *pas*, is that *personne* is becoming the only exponent of negation; in an elliptic answer *personne* must occur without *ne:*

(41) c. J'ai vu *personne*.   'I saw nobody'
     d. Qui as-tu vu?   'Who did you see ?'
        **Ne personne/Personne*. 'Nobody'

Here we have in effect a Jespersen cycle yielding a true negative pronoun (see also Kiparsky and Condoravdi 2006).

According to Haspelmath (1997: 203–210), a Jespersen cycle is one of the two processes that can generate true negative pronouns. The other is called 'negative absorption'. The 'negative absorption' hypothesis says when both the indefinite pronoun and the clausal negator occur in front of the verb, the indefinite and the clausal negator may undergo univerbation or, in his terminology, the negator may be 'absorbed' into the indefinite. This can be illustrated with Baghdad Arabic.

(42) Baghdad Arabic (Haspelmath 1997: 206, based on Ali 1972: 48, 53)
  a. Saalim *ma* šaf ʔæ-waḥid hnak.
     Saalim NEG saw INDEF-one there
     'Salim didn't see anyone there.'
  b. *Ma*-ḥad kisər il šibbač
     NEG-one broke the window
     'No-one broke the window.'

In (42a) the clausal negator is in front of the verb, but in (42b), a sentence in which the verb is preceded by an indefinite, the negator is even in front of the indefinite and combines with it. According to Haspelmath (1997: 206) it is Jespersen's 'Neg-First' principle[19], the idea that a negator tends to be positioned in the beginning of the sentence, that is ultimately responsible for the early occurrence of the negator in (42b). Whether the latter explanation is completely satisfying or not[20], (42b) does show a pronoun that is the sole exponent of negation.

Note that Haspelmath (1997) uses a very restricted notion of 'absorption'. He does not use it to explain a construction such as *nobody*, though in a straightforward sense, the word *nobody* is no less of a univerbation of a negator and a positive or perhaps neutral *body* element than *ma-ḥad* is. Haspelmath's reason is double. First, the negator of his absorption has to be the clausal negator, and at least for a form like *nothing* it isn't. Haspelmath's

---

[19] The term is due to Horn (1989: 293) but the idea, once again, is Jespersen's (1971: 5).

[20] Note that the Neg-First principle does not by itself explain why the negator is not in the absolutely first position in (42a). More generally speaking, however, the relevance of a Neg-First principle is undisputed. We also need it to explain why in a language like Italian postverbal *nessuno* 'nobody' needs the clausal negator but preverbal *nessuno* dispenses with the clausal negator (Horn 1989: 450; Haspelmath 1997: 211). The contrast illustrated in (43) may also relate to the Neg-First principle.

second reason is that "in earlier English the old negative particle *ne* co-occurred with the *no*-indefinites, contrary to what we would expect if it had been absorbed by them" (Haspelmath 1997: 209). Haspelmath here refers to the possibility of so-called 'negative concord', and this is indeed a property of Old English. However, there are two problems. The first is that though in Old English negative concord was the preferred construction, it was not obligatory (Traugott 1992: 170; Mazzon 2004: 36–38). Compare the sentences in (43) from *Beowulf*.

(43) Old English (Beowulf 242–243, 692–693)

　a. þe　on　land　Dena　　 laðra　　 *nænig* mid　scip-herge
　　　That　on　land　of.Danes　of.enemies　none　with　ship.army
　　　sceðþan　*ne*　meahte
　　　ravage　 not　 might
　　　'that none of our enemies might come into Denmark
　　　do us harm with an army, their fleet of ships'

　b. *Nænig* heora　 þohte　 þæt he þanon scolde eft　　eardlufan
　　　none　 of. them thought that he thence should again earth.loved
　　　æfre　 gesecean
　　　ever　 see
　　　'None of them thought he would ever return to his native land'

The second problem is that at least in a word like *nænig* the negative element was in fact the clausal negator, which would be evidence for an absorption analysis. Whatever the correct analysis of the infrequent single exponence construction (43b) and of the frequent negative concord construction in (43a) – and of course similar constructions in the older stages of other Germanic languages (see e.g. Jäger 2008) – it is clear that when English negative concord constructions dropped the clausal negator *ne*, the negation found its expression only in the negative pronoun, and in this respect, we are again dealing with a Jespersen cycle.

According to Haspelmath (1997: 203), constructions such as Old English (43b) and Modern English (44a) constitute a mismatch between meaning and form. The form suggests that the negation has constituent scope rather than clausal scope and, that, in other words, (44a) would actually be a positive sentence, just like (45a). The continuations and variations of these clauses in (44b-f) and (45b-f) (all of which use tests first formulated by Klima 1964) show that this is not the case.

(44) a. I have seen *nobody*.
 b. *I have seen *nobody* and so has Mary.
 c. I have seen *nobody* and *neither* has Mary.
 d. I have*n't* seen anybody.
 e. *I have*n't* seen anybody and so has Mary.
 f. I have*n't* seen anybody and *neither* has Mary.

(45) a. I am *un*happy.
 b. I am *un*happy and so is Mary.
 c. *I am *un*happy and *neither* is Mary.
 d. I am *not* happy.
 e. ?*I am *not* happy and so is Mary.
 f. I am *not* happy and *neither* is Mary.

That (45a) is a positive clause is shown by the fact that it can be continued by *and so* VERB; (45d) is a negative clause and hence it can be continued by *and neither* VERB. (44a) then only allows *and neither* VERB, and so it is negative, just like (44d) and (45d). According to Haspelmath (1997: 205) the mismatch may be repaired, and in languages that underwent a Jespersen cycle, we may see new negators appearing in constructions of type (44a). The Belgian Dutch dialects are interesting in this respect. The standard language is like standard English, but the Belgian dialects either retain the concord pattern with the old *en* negator or they have installed a new concord pattern, this time with the new negator *niet*. In the area where the two zones meet, there is even a concord pattern with both *en* and *niet* (van der Auwera and Neuckermans 2004), and in a small subarea of the *niemand niet* zone, the negativity of the pronoun has been undone and we get the positive pronoun *iemand* 'someone' (van der Auwera et al. 2006).

(46) Belgian Dutch
 a. Ik   heb   niemand   gezien.
 b. Ik en heb   niemand   gezien.
 c. Ik   heb   niemand   niet   gezien.
 d. Ik en heb   niemand   niet   gezien.
    I   NEG have nobody   NEG   seen
 e. Ik   heb   iemand   niet   gezien
    I   have somebody   NEG   seen
    'I haven't seen anybody.'

Thus the variation manifested in Belgian Dutch gives a fairly good picture of the variation worldwide. Of course, merely listing the alternatives does not say anything about the conditions under which each strategy is used.

Also, the listing in (46) is not quite complete. At least one basic strategy is missing. A language may lack or at least avoid indefinite pronouns, both positive and negative, and instead use an existential construction (Kahrel 1996: 35–67; Haspelmath 1997: 54–57). This is illustrated with a negative existential in Tagalog (47).

(47) Tagalog (Haspelmath 1997: 54, based on Schachter & Otanes 1972: 521).
Wala-ng   dumating kahapon.
NEG.exist-LK   come-AG   yesterday
'No one came yesterday.' (lit. "There is not having come yesterday")

A final remark: in connection with the univerbation patterns of negative pronouns and adverbs, one can study not only what has been attested, but also what seems unattestable, because one has reason to think that no language would ever develop it. Horn (already Horn 1972, but most prominently Horn 1989: 252–267)[21] has dealt with this problem in great detail. Consider the lists in (48).

(48) a. *all, some, no, *nall*
     b. *always, sometimes, never, *nalways*
     c. *and, or, nor, *nand*
     d. *both, either, neither, *noth*

The argumentation of Horn's and of many others following him relate the univerbation pattern to the (post-)Aristotelian logical square or versions thereof (Horn 1990, van der Auwera 1996) and hypotheses about how lexicalization is either prompted or obstructed by pragmatics, Gricean or other

---

[21] As Gast (2008) points out, for dual quantifiers the puzzling absence of the fourth member was already noticed for Old English by Einenkel (1904: 72). However, as is also discussed by Gast (2008), the development from Old to Modern English has other puzzles. For one, Middle English only had two members: the ancestor to *neither* and the ancestor to Modern *either*, but the latter also covered the terrain of Modern *both*. For another, when *either* is used as a negative disjunctive clause-final particle (as in *I don't like it either*), it is the successor to *neither*: so we see a change from *not ... neither* to *not ... either*, with a *n*-drop possibly similar to the change from Brabantic Belgian Dutch *niemand niet* to *iemand niet* illustrated in (46).

(e.g. 'Relevance theoretical' in Moeschler 2009). Interestingly also, the modal notions of necessity, possibility, and their negations, are as amenable to approaches starting from the logical square as the quantifiers and conjunctions shown in (48) and it should therefore be as difficult to lexicalize the modal counterpart to *nall etc., i.e. the expression of no necessity. But this is not the case, for many languages do lexicalize the fourth member (van der Auwera 2001, van der Auwera & Bultinck 2001), English being a case in point with its dedicated *needn't* construction.[22]

(48) c. must, may, mustn't/can't, *needn't*

4. Conclusion

In this study I have provided illustrations, hypotheses and references to the literature about the processes through which negators come into being. For standard negation I analyzed what has come to be known as the 'Jespersen cycle'. I argued that the process is complex enough to warrant pluralizing the noun and speaking about 'Jespersen cycles'. I also linked up these issues to Miestamo's work on the asymmetry of negation. For prohibitive negation I focused on dedicated prohibitive markers and I showed how they correspond to a multiplicity of origins. On the subject of non-verbal and existential negators and on the subject of the origin of negative pronouns and negative adverbs, the focus was on processes of univerbation.

Abbreviations

ACC 'accusative', ACT 'active', AG 'agent', COMPL 'complementizer', DAT 'dative', EX 'existential', FUT 'future', GEN 'genitive', INDEF 'indefinite', IRR irrealis', L 'low tone', LK 'linker', M 'medium tone', NEG 'negation', NPST 'non-past', POT 'potential', PRES 'present', PROGR 'progressive', PROH 'prohibitive', RCTPST 'recent past', SG 'singular', VPOT 'voluntative-potential', x>y 'x is subject and y object', 2 'second person', 3 'third person'.

---

[22] That modality is more complicated is also suggested by the fact that the third member of the modal quadruple has two different dedicated expressions.

References

Ali, Latif H.
  1972	Observations on the preverbal negative particle *ma* in Baghdad Arabic. *Studia Linguistica* 26: 67–83.
Anderson, Judi Lynn
  1989	*Comaltepec Chinantec syntax*. Arlington: Summer Institute of Linguistics.
Barbiers, Sjef, Johan van der Auwera, Hans Bennis, Eefje Boef, Gunther De Vogelaer and Margreet van der Ham
  2009	*Syntactische atlas van de Nederlanse dialecten, deel II / Syntactic atlas of the Dutch dialects, volume II*. Amsterdam: Amsterdam University Press.
Beowulf. A dual-language edition
  1977	Translated with an introduction and commentary by Howel D. Hickering Jr. New York: Doubleday.
Bernini, Giuliano and Paolo Ramat
  1992	*La frase negativa nelle lingue d'Europa*. Bologna: Il Mulino.
Bernini, Giuliano and Paolo Ramat
  1996	*Negative sentences in the languages of Europe*. Berlin/New York: Mouton de Gruyter.
Biberauer, Theresa
  2008	Doubling vs omission: Insights from Afrikaans negation. In *Microvariations in syntactic doubling*, eds. Sjef Barbiers, Margreet Van der Ham, Olaf Koeneman, and Maria Lekakou. London: Emerald.
Birjulin, Leonid A. and Victor S. Xrakovskij
  2001	Imperative sentences: theoretical problems. In *Typology of imperative constructions*, ed. Victor S. Xrakovskij, 3–50. München: Lincom EUROPA.
Blancquaert, E.
  1923	Over de dubbele ontkenning en nog wat. *Handelingen van het Vlaamse filologencongres* 6: 60–69.
Bréal, Michel
  1897	*Essai de sémantique (Science des significations)*. Paris: Hachette.
Breitbarth, Anne
  2009	A hybrid approach to Jespersen's cycle in West Germanic. *Journal of Comparative Germanic Linguistics*.
Broschart, Jürgen
  1999	Negation in Tongan. In *Negation in Oceanic languages. Typological studies*, eds. Even Hovdhaugen and Ulrike Mosel, 96–114. München: LINCOM Europa.
Buridant, Claude
  2000	*Grammaire nouvelle de l'ancien français*. Paris: Sedes.

Capell, A. and H. E. Hinch
    1970   *Maung grammar. Texts and vocabulary*: The Hague: Mouton.
Corne, Chris
    1977   *Seychelles Creole grammar: Elements for Indian Ocean proto-Creole.* Tübingen: Narr.
Cornyn, William
    1944   *Outline of Burmese grammar.* Baltimore: Linguistic Society of America.
Croft, William
    1991   The evolution of negation. *Journal of Linguistics* 27: 1–27.
Dahl, Östen
    1979   Typology of sentence negation. *Linguistics* 17: 79–106.
De Clerck, Bernard
    2006   *The imperative in English: a corpus-based, pragmatic analysis.* Doctoral dissertation, University of Ghent.
De Cuypere, Ludovic, Johan van der Auwera and Klaas Willems
    2007   Double negation and iconicity. In *Insistent images*, Elżbieta Tabakowska, Christina Ljungberg and Olga Fischer (eds.), 301–320. Amsterdam: Benjamins.
De Cuypere, Ludovic
    2008   *Limiting the iconic. From the metatheoretical foundations to the creative possibilities of iconicity in language.* Amsterdam: Benjamins.
Dryer, Matthew S.
    1988   Universals of negative position. In *Studies in syntactic typology*, Michael Hammond, Edith Moravcsik and Jessica Wirth (eds.), 93–124. Amsterdam: Benjamins.
Early, Robert
    1994a   Lewo. In *Typological studies in negation*, Peter Kahrel and René van den Berg (eds.), 65–92. Amsterdam: Benjamins.
Early, Robert
    1994b   *A grammar of Lewo, Vanuatu.* Doctoral dissertation, Australian National University.
Eckardt, Regine
    2003   Eine Runde im Jespersen-Zyklus. Negation, emphatische Negation und negativ-polare Elemente im Altfranzösischen. Konstanzer Online-Publikations-System.
Einenkel, Eugen
    1904   *nahwæder* > *neither*. *Anglia* 27: 72–80.
Eriksen, Pål Kristian
    2005   *On the typology and semantics of non-verbal predication.* Doctoral dissertation, University of Oslo.
Forges, Germaine
    1977   *Le kela, langue bantoue du Zaïre (Zone C). Esquisse phonologique et morphologique.* Paris: Selaf.

Floricic, Franck and Lucia Molinu
   2008   L'Italie et ses dialectes. In *Lalies 28. Actes des sessions de littérature et linguistique (La Beaume-les-Aix, 28–31 août 2007)*, 5–107. Paris: Presses de l'École Normale Supérieure.
François, Alexandre
   2003   *La sémantique de prédicat en mwotlap (Vanuatu)*. Leuven: Peeters.
Gardiner, Alan H.
   1905   The word ⸱⸱⸱⸱. *Zeitschrift für Ägyptische Sprache und Altertumskunde* 41: 130–135.
Gast, Volker
   2008   *(N)either (n)or – quantifiers, coordinators and focus particles in historical perspective*. Handout, Free University of Berlin.
Givón, Talmy
   1984   *Syntax: a functional-typological introducton. Volume 1*. Amsterdam: Benjamins.
Greenberg, Robert D.
   1996   *The Balkan Slavic appellative*. München: Lincom Europa.
Grossman, Eitan
   2008   Typology, dialectology, and diachrony: the Coptic prohibitive system in typological perspective. Handout, Conference on *Language Typology and Egyptian-Coptic Linguistics*. Leipzig.
Hansen, Björn
   2004   The grammaticalization of the analytical imperatives in Russian, Polish and Serbian/Croatian. *Die Welt der Slaven* 49: 257–274.
Hansen, Maj-Britt Mosegaard
   2009   The grammaticalization of negative reinforcers in Old and Middle French: a discourse-functional approach. In *Current trends in diachronic semantics and pragmatics*, Maj-Britt Mosegaard Hansen and Jacqueline Visconti (eds.), 227–251. Bingley: Emerald.
Haspelmath, Martin
   1997   *Indefinite pronouns*. Oxford: Oxford University Press.
Heine, Bernd and Tania Kuteva
   2002   *World lexicon of grammaticalization*. Cambridge: Cambridge University Press.
Hoeksema, Jack
   2009   Jespersen recycled. In *Cyclical change*. Elly van Gelderen (ed.).
Hoeksema, Jack
   this vol. Negative and positive polarity items: an investigation of the interplay of lexical meaning and global conditions of expression.
Honda, Isao
   1996   *Negation: a cross-linguistic study*. Doctoral dissertation, University of New York at Buffalo.

Horn, Laurence R.
1972 *On the semantic properties of logical operators in English*. Doctoral dissertation, University of California at Berkeley.

Horn, Laurence R.
1989 *A natural history of negation*. Chicago: The University of Chicago Press.

Horn, Laurence R.
1990 Hamburgers and truth: why Gricean explanation is Gricean. In *Proceedings of the sixteenth annual meeting of the Berkeley linguistics society*. Kira Hall, Jean-Pierre Koenig, Michael Meacham, Sondra Reinman and Laurel A. Sutton (eds.), 454–471. Berkeley: Berkeley Linguistics Society.

Horn, Laurence R.
2001 Flaubert triggers, squatitive negation, and other quirks of grammar. In *Perspectives on negation and polarity items*, Jack Hoeksema, Hotze Rullmann, Victor Sanchez-Valencia and Ton Van der Wouden (eds.), 173–200. Amsterdam: Benjamins.

Jäger, Agnes
2008 *History of German negation*. Amsterdam: Benjamins.

Jespersen, Otto
1917 *Negation in English and other languages*. København: A. F. Høst & Søn.

Kahrel, Pieter Johannus
1996 *Aspects of Negation*. Doctoral dissertation, University of Amsterdam.

Kawaguchi, Yuji
in print Particules négatives du français: *ne, pas, point* et *mie* – Un aperçu historique. In *Mélanges Yves-Charles Morin*: Presses de l'Université Laval.

Kiparsky, Paul and Cleo Condoravdi
2006 Tracking Jespersen's cycle. In *Proceedings of the 2nd international conference of Modern Greek dialects and linguistic theory*, Mark Janse, Brian D. Joseph and Angela Ralli (eds.), 172–197. Patras: University of Patras.

Klima, Edward S.
1964 Negation in English. In *The structure of language. Readings in the philosophy of language*, Jerry A. Fodor and Jerrold J. Katz (eds.), 246–323. Englewood Cliffs, NJ: Prentice-Hall.

Kroskrity, Paul V.
1984 Negation and subordination in Arizona Tewa: discourse pragmatics influencing syntax. *International Journal of American Linguistics* 50: 94–104.

Li, Renzhi
2004 *Modality in English and Chinese: a typological perspective*. Boca Raton: Dissertation.com.

Lucas, Christopher
   2007   Jespersen's cycle in Arabic and Berber. *Transactions of the Philological Society* 105: 398–431.
Lucas, Christopher
   2008   Contact as catalyst: the case for Coptic influence in the development of Arabic negation. Manuscript Cambridge University.
Martineau, France and Raymond Mougeon
   2003   A sociolinguistic study of the origins of *ne* deletion in European and Quebec French. *Language* 79: 118–152.
Mazzon, Gabriella
   2004   *A history of English negation.* Harlow: Longman.
McGill, Stuart, Samuel Fembeti, and Mike Toupin
   1999   *A grammar of Sisaale-Pasaale.* Tamale: University of Ghana.
McGregor, William B.
   2008   Two verbless negative constructions in Nyulnyul (Nyulnyulan, Kimberley, Western Australia). Manuscript.
Meillet, Antoine
   1912   L'évolution des formes grammaticales. *Scientia* 12: 384–400. [Reprinted in Meillet, Antoine. 1926. *Linguistique historique et linguistique générale.* 130–148. Paris: H. Champion.]
Miestamo, Matti
   2005   *Standard negation. The negation of declarative verbal main clauses in a typological perspective.* Berlin: Mouton de Gruyter.
Miestamo, Matti and Johan van der Auwera
   2007   Negative declaratives and negative imperatives: similarities and differences. In *Linguistics festival. May 2006, Bremen*, Andreas Ammann (ed.), 59–77. Bochum: Brockmeyer.
Moeschler, Jacques
   2009   Pourquoi n'y a-t-il pas de particuliers négatifs? La conjecture de Horn revisitée. *Actes du colloque sur la quantification.* Electronic publication, University of Strasbourg.
Muller, Claude
   1991   *La négation en français. Syntaxe, sémantique et éléments de comparaison avec les autres langues romanes.* Genève: Droz.
Mukash Kalel, Timothée
   1982   *Le Kanyok, langue bantoue du Zaire. Phonologie, morphologie, syntagmatique.* Doctoral dissertataion, University of Paris III.
Neuckermans, Annemie
   2008   *Negatie in de Vlaamse dialecten volgens de gegevens van de Syntactische Atlas van de Nederlandse dialecten (SAND).* Doctoral dissertation, University of Ghent.
Norman, Jerry
   1988   *Chinese.* Cambridge: Cambridge University Press.

Overdiep, G. S.
1933 Dialectstudie en syntaxis. *Onze Taaltuin* 2: 18–23.
Overdiep, G. S.
1937 *Stilistische grammatica van het moderne Nederlandsch*. Zwolle: Tjeenk Willink.
Pakendorf, Brigitte and Ewa Schalley
2007 From possibility to prohibition: a rare grammaticalization pathway. *Linguistic Typology* 11: 515–540.
Parry, M. Mair
1997 Preverbal negation and clitic ordering, with particular reference to a group of North-West Italian dialects. *Zeitschrift für romanische Philologie* 113: 243–270.
Patz, Elisabeth
2002 *A grammar of the Kuku Yalanji language of North Queensland*. Canberra: Australian National University.
Pauwels, J. L.
1958 *Het dialect van Aarschot en omstreken*. Brussel: Belgisch interuniversitair centrum voor Neerlandistiek.
Pauwels, J. L.
1974 Expletief *nie* en andere herhalingswoorden als zinsafsluiters. In *Taalkunde – 'n lewe. Studies opgedra aan prof. W. Kempen by geleentheid van sy 65st verjaardag*, F. F. Oldendal (ed.), 73–76. Kaapstad: Tafelberg.
Payne, John R.
1985 Negation. In *Language typology and syntactic description. Vol. 1. Clause structure*, Timothy Shopen (ed.), 197–242. Cambridge: Cambridge University Press.
Pitkin, Harvey
1984 *Wintu grammar*. Berkeley: University of California Press.
Pott, A. F.
1833 *Etymologische Forschungen auf dem Gebiete der Indo-Germanischen Sprachen*. Vol. 1. Lemgo: Meyer.
Postal, Paul M.
2004 *Skeptical linguistic essays*. Oxford: Oxford University Press.
Ramat, Paolo
2006 Italian negatives from a typological/areal point of view. In *Scritti in onore di Emanuele Banfi in occasione del suo 60° compleanno*, Nicola Grandi and Gabriele Iannàccaro (eds.), 355–370. Cesena: Caissa Italia.
Rausch, P. J.
1912 Die Sprache von Südost-Bougainville, Deutsche Salomonsinseln. *Anthropos* 7: 105–134, 585–616, 964–994.

Roberge, Paul T.
  2000 Etymological opacity, hybridization, and the Afrikaans brace negation. *American Journal of Germanic Linguistics & Literatures* 12: 101–176.
Schachter, Paul and Fe T. Otanes
  1972 *Tagalog reference grammar.* Berkeley: University of California Press.
Schwegler, Armin
  1990 *Analycity and syntheticity. A diachronic perspective with special reference to Romance languages.* Berlin/New York: Mouton de Gruyter.
Schwegler, Armin
  1991 Predicate negation in comtemporary Brazilian Portuguese. *Orbis* 34: 187–214.
Schwegler, Armin
  1996 La doble negación dominicana y la génesis del español caribeño. *Hispanic Linguistics* 8: 247–315.
Schwenter, Scott A.
  2006 Fine-tuning Jespersen's cycle. In *Drawing the boundaries of meaning: Neo-Gricean studies in pragmatics and semantics in honour of Laurence R. Horn*, Betty Birner and Gregory Ward (eds.), 327–344. Amsterdam: Benjamins.
Traugott, Elizabeth Closs
  1992 Syntax. In *The Cambridge history of the English language*, Richard M. Hogg (ed.), 168–289. Cambridge: Cambridge University Press.
van den Berg, Helma
  1997 *A grammar of Hunzib. With texts and lexicon.* München: LINCOM EUROPA.
van der Auwera, Johan
  1996 Modality: the three-layered modal square. *Journal of Semantics* 13: 181–195.
van der Auwera, Johan
  2001 On the typology of negative modals. *Perspectives on negation and polarity items.* Jack Hoeksema, Hotze Rullmann, Viktor Sánchez-Valencia and Ton van der Wouden (eds.), 23–48. Amsterdam: Benjamins.
van der Auwera, Johan
  2009 The Jespersen cycles. In *Cyclical change.* Elly van Gelderen (ed.), 35–71. Amsterdam: Benjamins.
van der Auwera, Johan and Bert Bultinck
  2001 On the lexicalization of modals, conjunctions, and quantifiers. *Perspectives on semantics, pragmatics, and discourse.* Istvan Kenesei and R. M. Harnish (eds.), 173–186. Amsterdam: Benjamins.

van der Auwera, Johan and Annemie Neuckermans
    2004   Jespersen's cycle and the interaction of predicate and quantifier negation. In *Dialectology meets typology. Dialect grammar from a cross-linguistic perspective*, Bernd Kortmann (ed.), 453–478. Berlin/ New York: Mouton de Gruyter.

van der Auwera, Johan, Ludo Lejeune and Valentin Goussev
    2005   The prohibitive. In *The world atlas of language structures*, Martin Haspelmath, Matthew S. Dryer, David Gil and Bernard Comrie (eds.), 290–293. Oxford: Oxford University Press.

van der Auwera, Johan
    2006   Why languages prefer prohibitives. *Wai guo ju [Journal of Foreign Languages]* 161: 2–25.

van der Auwera, Johan, Ludovic De Cuypere and Annemie Neuckermans
    2006   Negative indefinites: A typological and diachronic perspective on a Brabantic construction. In *Types of variation. Dialectal, diachronic and typological interfaces*, eds. Terttu Nevalainen, Juhani Klemola. and Mikko Laitinen, 305–319. Amsterdam: Benjamins.

Van Gelderen, Elly
    2008   Negative cycles. *Linguistic Typology* 12: 195–243.

Van Olmen, Daniel and Johan van der Auwera
    2008   *Ne chante pas* ou, tout simplement, *arrête*? Sur la fréquence des constructions prohibitives rétrospectives, In *Linguista sum: Mélanges offerts à Marc Dominicy à l'occasion de son soixantième anniversaire*, Emmanuelle Danblon, Mikhail Khissine, Fabienne Martin, Christine Michaux and Svetlana Vogeleer (eds.), 225–233. Paris: L'Harmattan.

Veselinova, Ljuba
    2007   Towards a typology of negation in non-verbal and existential sentences. Paper presented at the 7th International Conference of the Association for Linguistic Typology, Paris.

Veselinova, Ljuba
    2008   Standard and special negators in the Slavonic languages: synchrony and diachrony. Paper presented at the conference on Diachronic syntax in Slavonic languages, Regensburg.

Wallage, Phillip
    2008   Jespersen's cycle in Middle English: parametric variation and grammatical competition. *Lingua* 118: 643–674.

Multiple negation in English and other languages

Laurence R. Horn

> But grammar's force with sweet success confirm
> For grammar says (O this dear Stella weigh,)
> For grammar says (to grammar who says nay)
> That in one speech two negatives affirm.
>
> Sir Philip Sidney, "To his Mistress who has Said 'No, No'", from *Astrophel and Stella*, c. 1580

The most influential monograph on negation ever written, Jespersen's *Negation in English and Other Languages* (1917), will soon observe its centenary. Its influence extends from countless citations (see, inter multa alia, Horn 1989, McCawley 1995) to later titles deferentially evoking it, from Karl Zimmer's *Affixal Negation in English and Other Languages* (1964) to the current study.

One central concern for Jespersen as for us is the reach of the law of *Duplex negatio affirmat,* adapted to English grammar by Bishop Lowth (1762: 126) under the edict "Two negatives in English destroy one another, or are equivalent to an affirmative." As has been recognized for centuries, this law is regularly abrogated under two different sets of circumstances. In HYPERNEGATION, at least one of the duplex (or multiplex) negations in question is apparently superfluous and uninterpreted; two (or more) morphosyntactically negative elements thus combine to co-express a single semantic negation. This description encompasses two distinct but related phenomena: negative concord (NC) within the clause and a range of other cases involving involve relations across clause boundaries that I will collect under the label of PN (for pleonastic negation). In other cases, the two negations do affirm, but pace Lowth and his heirs it is not necessarily obvious just WHAT they affirm.

Most discussions of multiple negation (including Horn 1991, 2002c) begin with the premise that the major distinction to be drawn is that between true *Duplex negatio affirmat* (DNA) phenomena and the typologically more robust categories reflecting *Duplex negatio negat* (DNN), including negative concord itself and the related phenomena of pleonastic (a.k.a. expletive, paratactic, sympathetic, or abusive) negation. Essentially, the map would look like this:

112 Laurence R. Horn

(1)   Multiple negation
                      /   \
        Logical double neg.   Hypernegation
             (DNA)   (DNN)
                                    /   \
                                   NC   PN

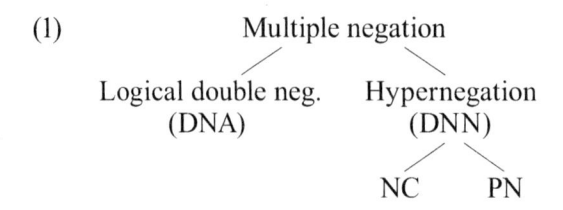

But another approach would begin by asking WHY: Why, in the DNA cases, would a speaker expend the effort on two negations to affirm when a simple less costly affirmative is available? Why, in the DNN cases, expend the effort on marking negation twice when the semantics only requires one? What motivates the twin negations in the former contexts, and what motivates the extra negation(s) in the latter?

Both logical double negation and NC are recurrent targets of traditional grammarians, who have often attributed the latter tendency to logical immaturity (doubtless a troubling condition for speakers of such primitive languages as Spanish, Italian, Russian, etc.):

> It is natural in early forms of language to strengthen the negative...but as the users of a language grow more logical they come to feel that doubling a negative in a sentence is negating a negation and there that it is not correct except in the rare cases where it is desired to express such a round-about affirmative. (*No* one who is *not* familiar with it...)   (Earle et al. 1911: 54)

But other grammarians have been more tolerant of this foible:[1]

> When a boy in the street declares that he "hain't seen no dog", it is not true that 'two negatives make one affirmative', for he intends simply an emphatic negation...In other words, two negatives may make an affirmative in logic, but they seldom do in English speech...The somewhat artificial 'not

---

[1] Martin (1748: 93) provides a cogent defense of DNN on arithmetical grounds:

> The two negatives as used by the Saxons and French must be understood by way of apposition...which way of speaking is still in use among us; and in this case the two negatives answer to the addition of two negative quantities in Algebra, the sum of which is negative. But our ordinary use of two negatives (in which the force of the first is much more than merely destroyed by the latter) corresponds to the multiplication of two negative quantities..., the product of which is always affirmative.

For more on double negation in the eighteenth century grammatical tradition, cf. Tieken-Boon von Ostade (1982) and Austin (1984).

unnecessary', 'not impossible' and the like (imitated from the Latin) are almost the only exceptions in English. (Greenough & Kittredge 1901: 220)

In older, popular English...2 or 3 negatives were felt as stronger than a single negative, on the same principle that we drive in two or three nails instead of one. (*I can't see no wit in her,* letter from Lamb to Coleridge, 1797)...Under Latin influence, we have come to feel that two negatives make an affirmative statement. (Curme 1931: 139)

1. DNA: *Duplex negatio affirmat* – but what?

Despite their differences over the status of DNN (where two negs are stronger than one, or where many negs simply express sentential negation without necessarily strengthening it), grammarians on both sides of the aisle converge to condemn the use of logical double negation as a marginal, superfluous, and suspiciously Latinate phenomenon.

Despite early dissenters[2], this view is widely shared by linguists and essayists from Lowth (1762) to Tesnière (1959: 233), who excoriates *nec non dixit* ('nor did s/he not say', i.e. 'and s/he said') as *"une des fausses élégances du latin",* and Orwell (1946: 357, 365), whose prey was the *not unjustifiable assumption* and its ilk:

> Banal statements are given an appearance of profundity by means of the *not un-* formation...It should be possible to laugh the *not un* formation out of existence...One can cure oneself of the *not un* formation by memorizing this sentence: *A not unblack dog was chasing a not unsmall rabbit across a not ungreen field.*

---

[2] One early dissenter was the best-selling American grammatical authority Lindley Murray, who began by echoing Lowth's party line on condemning all double negation –

> Two negatives, in English, destroy one another, or are equivalent to an affirmative: as, "Nor did they not perceive him"; that is, "they did perceive him." "His language, though inelegant, is not ungrammatical", that is, "it is grammatical". It is better to express an affirmation, by a regular affirmative, than by two separate negatives, as in the former sentence. (Murray 1803: 136–137)

– but then suspended the sentence on *not un*-type DNAs in his second, "improved" edition:

> ...but when one of the negatives is joined to another word, as in the latter sentence, the two negatives form a pleasing and delicate variety of expression.
> (Murray 1814: I. 187)

Along the same lines, Marchand (1960: 151–152) comments that 'Natural linguistic instinct would not make the sophisticated detour of negativing a negative to obtain a positive.' *Not uncommon, not unhappy,* and (somewhat less convincingly) *not bad,* while possible collocations, would thus fall outside what is permitted by 'natural linguistic instinct.'

Orwell's prescribed laughing-cure exploits the strict distributional constraints on doubly negative prenominal adjectives (cf. Langendoen & Bever 1973, Horn 1991) as well as on specimens of *not unA* where the negated adjective itself *(unblack, unsmall, ungreen)* is itself excluded. More plausible instances of the *not un* construction are:

(2) a. He's not an unhappy boy.
     b. It's not an impossible job.

where the outside negator has sentential scope and the sentence constitutes a predicate denial, nexal negation, or sentence negation in the terminology of Aristotle, Jespersen, and Klima respectively. In such cases, full redundancy would demand the equivalence of *not unA* and *A*, but no such equivalence can be semantically sustained for examples like those in (2a), involving the contradictory of a contrary, which does not reduce to the simple *He's a happy boy.* The standard diagnosis is provided by Dr. Jespersen (1924: 332):

> Language has a logic of its own, and in this case its logic has something to recommend it. Whenever two negatives really refer to the same idea or word the result is invariably positive; this is true of all languages...The two negatives, however, do not exactly cancel one another so that the result is identical with the simple *common, frequent;* the longer expression is always weaker: "this is not unknown to me" or "I am not ignorant of this" means 'I am to some extent aware of it', etc. The psychological reason for this is that the *détour* through the two mutually destructive negatives weakens the mental energy of the listener and implies...a hesitation which is absent from the blunt, outspoken *common* or *known.*

Even if Orwell's *not unblack* dogs and *not unsmall* rabbits are beyond the pale, in other contexts the DN detour is recommended by the earlier (and subtler) stylist Erasmus as "graceful", while Sharma (1970: 60) lauds its use as "often extremely useful and by no means superfluous", as when *not impolite* is used "to convey the fact that the person in question was not polite either." But what factors could render a given construction simultaneously laughable, faux-elegant, graceful, pleasing and delicate, and extremely useful, depending on the context and the evaluator?

Crucially, the playing field for *not unA* tokens is not level. On Aristotle's theory of opposition, CONTRARY opposites are mutually inconsistent but not mutually exhaustive while CONTRADICTORY opposites mutually exhaust their domain (see Horn 1989 for elaboration). When a neg-prefixed adjective is a contrary of its stem, a contradictory negation will not simply destroy it: what is *not unlikely* may well be likely, but it may also fall within what Sapir (1944) calls the ZONE OF INDIFFERENCE, that which is neither likely nor unlikely. I may be *not unhappy* because I'm happy, or because I'm feeling blaah. But if something is *not inconceivable* or *not impossible,* what else CAN it be but *conceivable* or *possible?* Where is the zone of indifference, the unexcluded middle, in these cases? Why don't these doubly negated forms, amounting to the contradictory of a CONTRADICTORY, result in complete redundancy?[3] What motivates a sophisticated detour, when the through road is (not im)passable? Note that the violation of least effort seems especially perverse here, since double negation violates not only Grice's Brevity maxim, the R-based principle of syntagmatic economy in Horn 1984, but also the Q-based informativeness criterion: a double negative is typically both longer and weaker than its simpler affirmative counterpart.

If the *not unA* collocation is, as Zimmer (1964) notes, 'logically quite justified' when *unA* is the contrary of *A,* it must be RHETORICALLY justified elsewhere – or at least not unjustified. Indeed, it appears that a doubly negated adjective is often perceptibly weaker or more hesitant (à la Jespersen) than the corresponding simple positive, whether the weakening is identifiable in the semantics *(not unhappy, not unintelligent, not impolite)* or is only pragmatic or rhetorical. For Seright (1966: 124), the use of double negation 'results from a basic desire to leave one's self a loophole: certainly it is much easier to get out of a situation, to equivocate, if one has said "it is not unlikely" instead of "it is not likely" or "it is likely."' With the conscious or tacit goal of loophole-procurement, the speaker describes something as *not unA* in a context in which it would be unfair, unwise, or impolitic to describe

---

[3] Such examples are hardly rare: the majority of the *not un-* forms cited in the OED (under **not**, 10c), including *not unuseful, not inconsiderable,* and *not unclever,* are negations of logical contradictories. Someone with 'a certain air of dignity, not unmingled with insolence' (OED cite, 1900) is not situated in a zone of indifference where dignity cavorts neither mingled nor unmingled with insolence. Nor is it clear how Jespersen's description (1917: 70) of Kant's table of categories as 'not unobjectionable' fails to reduce to declaring it objectionable. See van der Wouden (1996) for valuable discussion.

that entity as *A*. We see this in the attestations in (3), contributed respectively by an essayist, a cartoonist, and a poet:

(3) a. I do not pretend to be a "pure" bachelor. I was married for five years, and it was, to use a cowardly double negative, not an unhappy experience.　　(Phillip Lopate, introduction to *Bachelorhood*, 1981)

　　b. [Scene: Couple standing before doormat inscribed ***NOT UNWELCOME***]
　　Wife to Husband: "See what I mean? You're never sure just where you stand with them."　　(Cartoon caption, *New Yorker*, 2/8/71)

　　c. Nothing shows why
　　At this unique distance from isolation,
　　It becomes still more difficult to find
　　Words at once true and kind,
　　Or not untrue and not unkind.
　　　　　　　　　　(Philip Larkin, from "Talking in Bed")

The implication in each case is clear: a not unhappy marriage is not precisely a happy one, Mr. and Mrs. Dinner-Guest are left feeling not exactly welcome, and so on.

A related motivation is endorsed by Fowler, who first consigns the doubly negated predicate to the Orwellian ash-heap of "faded and jaded elegance" but then pulls a Lindley Murray to pardon it on nationalistic grounds:

> The very popularity of the idiom in English is proof enough that there is something in it congenial to the English temperament, & it is pleasant to believe that it owes its success with us to a stubborn national dislike of putting things too strongly. It is clear that there are contexts to which, for example, *not inconsiderable* is more suitable than *considerable;* by using it we seem to anticipate & put aside, instead of not foreseeing or ignoring, the possible suggestion that so-&-so is inconsiderable.　　(Fowler 1926: 383)

A more recent usagist diagnoses DNA *("They are not unskilled")* and ironic understatement in general as "marks of class – symptoms of 'talking posh'" (Nash 1986: 91).

Derived from the countervailing Q and R Principles (essentially "Say enough" and "Don't say too much" respectively), the Division of Pragmatic Labor (Horn 1984, 1989), stipulates that given the use of a longer, marked expression in lieu of a briefer, more lexicalized equipollent alternative involving less effort tends to signal that speaker was not in a position to have employed the briefer expression. From this perspective, we would expect to find a range of contexts in which the use of the double negative might be motivated by a desire to avoid the simple positive description. When a prefixal negative is itself negated so as to yield a positive, any one of a number of motivations may be at work, not all of which are subsumable under a single metaphor, be it Jespersen's weakened mental energy, Fowler's British reserve, Nash's posh style, Marchand's sophisticated detour, or Seright's loophole. When a simple positive description gives way to the prolixity and potential obscurity of a double negation, there is always (given the Division of Pragmatic Labor) a sufficient reason, but it is not the same reason in each case. In Horn (1991), I outline a taxonomy of motives for double negation which can be briefly itemized here:

(4) a. **Quality:** S is not sure A holds, or is sure it doesn't (where *unA* is contrary of *A*).
   b. **Politeness or diffidence:** S knows (or strongly believes) A holds, but is too polite, modest, or wary to mention it directly.
   c. **Weight or impressiveness of style**: S violates brevity precisely to avoid brevity.
   d. **Absence of corresponding positive:** *not unA* is motivated by the non-existence of *A*, or by the impossibility of using *A* appropriately in the context.
   e. **Parallelism of structure:** *not unA* is in juxtaposition with earlier *unB*, as in the construction B_{neg} *{if/but}* *B'*, where *B'* is more naturally realized as a DN.
   f. **Minimization of processing, in contexts of direct rebuttal or contradiction:** S's assertion *x is not unA* is triggered by an earlier assertion (or suggestion) to the effect that x is unA.

Leaving aside those cases in which specific discourse considerations motivate a DN, the prototypic exemplars of *not un-* are invoked to assert more weakly, tentatively, or circumspectly the content of the simple positive. As shown by the scalar diagnostics in (5), even double contradictories pattern as weaker than their corresponding simple affirmatives, whether by infection from the parallel negation-of-contrary cases or through the iconic connection between circumlocution and attenuation:

(5) a. She's happy, or at least not unhappy. (*not unhappy, or at least happy)
   b. It's possible he can do it,   (*not impossible, or at least possible)
      or at least it's not impossible.
   c. Not only is it not UNtrue, it's true! (*not only true, but not untrue)
   d. It's not even not untrue, let alone true.
                                      (*not even true, let alone not untrue)

The considerations we have observed with the double contradictory *not un-A* construction apply as well to double particle negation. Here too the double negatives may fail to completely cancel out, instead amounting to a weaker positive than their target would have provided.[4] A doubly negated predicate may be primed by the previous ascription of the corresponding simple predicate by the speaker or an interlocutor (**emphasis** added below):

(6) Bart: Dad, are you licking toads?
    Homer: I'm **not ᵛNOT licking toads**. [ᵛ marks fall-rise or L*+H L H⁰₀]
                         (Exchange from 2000 episode of "The Simpsons")

    Lucy: Are you friends with Mary?
    Robbie: I'm **not ᵛNOT friends with her**.
                         (Dialogue from ABC TV family comedy, "Seventh Heaven", 2001)

"So what do you think, is Tori connected to that other girl?"
Milo's lie was smooth. "I can't say that, Mr. Giacomo."
"But **you're not *not* saying it**."   (Jonathan Kellerman 2006, *Gone*, p. 145)

"Bandini croaks, Patty's got a DB to deal with, she drags him out to the street, waits awhile, calls it in…guess it fits."
Milo said, "**It sure doesn't *not* fit.**"
She smiled faintly. "You and your grammar, Mr. English Major."
                                      (Jonathan Kellerman 2007, *Obsession*, p. 277)

I have a dog. I got him because I am allergic to cats and I wanted my children to be happy…I love my children. I love my husband. They love the dog. **I don't not-love him, exactly**.
                         (Judith Warner, "Dogged By Guilt", NYT 28 August 2008)

---

[4] The mutual destruction of twin contradictories is captured by the ancient Law of Double Negation, established millennia ago in both Stoic and Buddhist propositional logic (although challenged by the Intuitionists). Describing the standard equivalence $\neg(\neg p) \equiv \neg\neg p \equiv p$, Frege (1919: 130) observes that whatever the packaging, "Wrapping up a thought in double negation does not alter its truth value." But going about unwrapped is not always politic, even for a naked thought – especially an unappealing one that a double negation can discreetly conceal while leaving the inner form intact.

Evidently, being "not not friends" with someone is weaker than being friends with her, while "not not licking" the putatively hallucinogenic red toads of Micronasia [sic] is more innocent than actually licking them. The DN coerces a reading of "not friends" and "not licking" as virtual contraries rather than pure contradictories, so that the normally excluded middle is explicitly unexcluded.

Similarly, when Sen. Trent Lott (R-Miss.) commented on a deficit-reduction package "We're not endorsing it or not endorsing it" (as quoted in the New York Times, 17 April 1985), he meant "We're neither endorsing it nor opposing it"; evidently, to [not endorse] a bill goes beyond merely failing to endorse it. (The Washington Post evidently found his proclamation so obscure it resorted to glossing Lott as declaring "We're not [either] endorsing it or not endorsing it.") Similarly, to not not support a movement is not necessarily to support it:

> "Do you support the [anti-vivisection] movement?" Hathor asked.
> "No," I said. "But I don't *not* support it, either. I want to understand its dynamics, get a fix on who's attracted to the cause and why."
> (Blaire French, *The Ticking Tenure Clock*, 1998: 104)

The cases above illustrate the tendency to strengthen apparent contradictory negation, whence the "virtual contrariety" effect noted here (cf. also Horn 1989, Chapter 5; Horn 2008b). As Bosanquet (1885: 281) put it well over a century ago, "The essence of formal negation is to invest the contrary with the character of the contradictory," so that "From 'he is not good' we may be able to infer something more than that 'it is not true that he is good'" (1885: 293). Across a wide range of languages we find a tendency, surveyed in Horn (1989: Chapter 5), for the speaker to weaken the force of her intended negative judgments, counting on the hearer to fill in the intended stronger negative. In English, the resultant contrary negatives in contradictory clothing include not only affixal negation (cf. (5)) and litotes (cf. (6)) but also the so-called "neg-raising" phenomenon, wherein a higher negation is conventionally associated with a lower clause across certain classes of predicates (cf. Horn 1989: §5.2). This too was recognized by Bosanquet (1885: 319), who recognized

> ...the habitual use of phrases such as *I do not believe it,* which refer grammatically to a fact of my intellectual state but actually serve as negations of something ascribed to reality...Compare our common phrase 'I don't think that' – which is really equivalent to 'I think that __ not'.

Given that "not think that p" is standardly used to convey "think that not-p", it follows that "I don't not think that p" does not reduce to "I think that p", and we obtain non-fully-cancelling double sentential negation:

> "For one thing, what makes you think Charley Sleet sexually abused Jillian Wints?" — "I don't think that or not think that. I just think it's possible that Leonard Wints was not alone in this."
> (Stephen Greenleaf, *Past Tense*, 1997: 104)

Note also the application of De Morgan's Law in action here: ¬φ & ¬(¬φ) → ¬(φ ∨ ¬φ).

As we have seen, the effect of virtual contrariety prevents double negations in apparent DNA contexts from completely "destroying" each other; what the *duplex negatio* affirms is not simply the content of the corresponding affirmative. When we turn to DNN contexts, we will see that the motivation for the "extra" negative marker is often even more elusive.

2. DNN: from negative concord to non-canonical hypernegation

The most familiar variety of hypernegation is negative concord, when the expression of sentence negation spreads to indefinites within the clause, a grammatical process in many Romance and Slavic languages as well as in most varieties of vernacular English (and also the bane of the "double negative"-intolerant prescriptivist). Familiar examples include (7a, b) (here and below, **highlighting** is added).

(7) a. It **ain't no** cat **can't** get into **no** coop. (Labov 1972: 773)
       (= standard Eng. 'No cat can get into any coop')
    b. I **can't** get **no** satisfaction. (Jagger & Richards 1965)

As the http://en.wikipedia.org/wiki/(I_Can't_Get_No)_Satisfaction wiki-entry on the Rolling Stones' rock anthem observes, "The title line is an example of a double negative resolving to a negative, a common usage in colloquial English."

The cross-linguistic distribution of negative concord (NC) constructions has received a good deal of attention in recent work, much of it devoted to the relation of NC to negative polarity, the interpretive status of N-WORDS (indefinites with negative force, e.g. *nessuno* 'nobody' and *niente* 'nothing' in Italian) and the diachronic reanalyses encapsulated by Jespersen's cycle. These studies include Zanuttini 1997, Martins 2000, Herburger 2001, Corblin

& Tovena 2001, de Swart & Sag 2002, Alonso-Ovalle & Guerzoni 2002, Larrivée 2004, Schwenter 2005, Eckardt 2006, Falaus 2007, Espinal 1992, 2007, and Floricic & Mignon 2007 on Romance; Giannakidou 2006 on Greek; Postal 2004 on English; Zeijlstra 2007 and Hoeksema 2009 on Dutch; Bayer 2006 on German; Abels 2005 and Inkova 2006 on Russian; and the general accounts in van der Wouden 1997, Zeijlstra 2004, Watanabe 2004, and Kiparsky & Condoravdi 2006. Typically (although not invariably), the effect of DNN in these cases is not emphasis, any more than in the agreement of verb with subject or adjective with head noun. Nor does the question of motivation arise for *I don't want none* any more than it does for *The cat **sits** on the mat* vs. *The cats **sit** on the mat*.

2.1. Pleonastic/sympathetic/paratactic/abusive negation: the classic cases

Our focus here will be on the subtype (or constellation of subtypes) of DNN lying beyond the purview of NC. Within this genus, Jespersen (1917: 75) begins by singling out a species he terms PARATACTIC NEGATION, in which

> a negative is placed in a clause dependent on a verb of negative import like 'deny, forbid, hinder, doubt.' The clause is treated as an independent sentence, and the negative is expressed as if there had been no main sentence of that particular type.

Such cases, variously termed pleonastic, expletive, or – in Smyth (1920)'s evocative term – sympathetic negation, are often attributed to the mental fusion or blend of two propositions, a positive clause in the scope of higher negation (hypotaxis) and a clause whose negative import is directly signaled (parataxis). This construction was frequent in Old English, in which such verbs as *tweo-* 'doubt', *forbeod-, forber-, geswic-* 'stop', and *wiðcweð-* 'refuse' all govern paratactic or pleonastic negation (PN); Joly (1972: fn. 3) provides a list of 20 such predicates, most with negative *for-* and *wið-* prefixes. PN also persists, although not as robustly, into Middle English and Early Modern English:

(8) Nature **defendeth** and **forbedeth** that **no** man make himself riche.
 (Chaucer)
 First he **denied** you had in him **no** [= any] right.
 (Shakespeare, *Com. Errors*)
 You may **deny** that you were **not** [= you were] the mean
 of my Lord Hastings late imprisonment.   (Shakespeare, *Richard III*)

Romance languages typically allow or require PN after verbs of fearing or forbidding, certain inherently negative adverbs (= 'unless', 'before', 'since', 'without'), and comparatives, as in (9), from French.

(9) a. avant qu'elle **ne** parle   'before she speaks'
     b. à moins qu'il **ne** vienne 'unless he comes'
     c. plus que je **ne** pensais   'more than I thought'
     d. depuis que je **ne** t'ai vu 'since I've seen you'

In modern French, embedded clauses with (optional) pleonastic *ne* are grammatically distinct from those with full sentential negation expressed by *ne...pas*.

(10) a. Je crains qu'il **ne** vienne.   'I'm afraid he's coming'
     b. Je crains qu'il **ne** vienne **pas**. 'I'm afraid he's not coming'

Compare the situation in Latin (10'), where the same distinction is expressed by the negative subjunctive complementizer *ne* ('lest') in the presence or absence of the canonical negative *non;* alternatively, the two negatives of (10'b) may cancel out to yield the positive subjunctive complementizer *ut*.

(10') a. Timeo **ne** veniat.   'I fear lest he come'   [= (10a)]
      b. Timeo {**ne non**/ut} veniat. 'I fear lest he not come' [= (10b)]

In his classic dissection of what he diagnoses as *"négation abusive,"* Vendryès (1950: 1) rejects the more standard label of *"négation explétive"* for such cases on the grounds that these negations are not grammatically on all fours with, say, the expletive dative pronoun in *Regardez-moi ça* (roughly 'Look at that for me'), which he sees as extrinsic to the primary assertion; such a sentence telescopes the two distinct clauses 'Look at that' + 'you will give me pleasure'.[5] The negatives in (9) and (10), on the other hand, are inherent parts of their clauses. While his label will not be adopted here (given the rather prescriptive import of *abusive* in English), Vendryès offers a useful catalogue of "abusive" pleonastic negation in a range of languages (Sanskrit, Ancient Greek, Latin, German, Baltic, Slavic, Amharic). His exposition of the logic of the relevant constructions recapitulates Jespersen's and Benveniste's: *X is Aer than Y* = 'X is A to an extent that Y is not', *unless*

---

[5] See Horn (2008b) for discussion of and references on non-argument datives in French, English, and other languages.

= 'if not', *before* = 'when still not', and so on. Typically, a main clause governing PN will imply the non-accomplishment, falsity, or undesirability of the embedded clause.[6]

In addition to the contexts surveyed above, we find expletive negatives in other environments, including concessive (un)conditionals in Yiddish and Russian (provided in Linguist List postings by Ellen Prince and Martin Haspelmath respectively):

(11) Es iz mir gut vu ikh zol **nit** zayn.
 it is to.me good where I SBJV NEG *be*
 'I'm fine wherever I am'

 Mne xorosho gde by ja **ni** byl.
 to.me good where SBJV *I* NEG *be*.PAST
 'I'm fine wherever I am'

Similarly, PN appears in main clause exclamatives and interrogatives in a variety of languages, as in the Paduan examples of (12) from Portner & Zanuttini (2000) and their now somewhat quaint-sounding English counterparts like (13a), cited by Jespersen. PN can still be found in embedded indirect question and modal contexts, as in (13b,c).

(12) a. **No** ga-lo magnà tuto!
  NEG *has*-S.CL *eaten everything*
  'He ate everything!'

 b. Cosse **no** ghe dise-lo!
  What NEG *him say*-S.CL
  'What things he's telling him!'

(13) a. **How often have** I **not** watched him!

 b. Keep in mind that I also have only fished this river from the Bushkill River mouth (in the Pocono Mountains) to the wing dam at

---

[6] This formulation suggests the robust non-veridicality (Giannakidou 1998) of the complement, as when verbs of doubting and forbidding are negated or questioned. (A monadic propositional operator **Op** is non-veridical if from **Op**(p) the truth of p cannot be validly inferred; see also Hoeksema's chapter in this volume.) Non-veridicality is also implicated in the use of subjunctive mood (often required in PN contexts) and the distribution of negative polarity items; indeed, much recent research on negation in Romance and other languages is devoted to the exploration of the interaction of mood with PN and NPIs.

Lambertville/New Hope!! **Who knows what I haven't** seen north and south of that stretch! (Web posting)

c. The Church of England...was so fragmented that there was **no knowing what** some sects **might not** have come to believe, but he doubted whether the christening of animals was encouraged.
(P. D. James, *The Children of Men,* on the christening of kittens)

2.2. Pleonastic negation and *parole* violations

In (9), (10), and (13), the use of "sympathetic" negation tends to mark regional usage or higher register, whence the formal and/or archaizing feel often associated with such turns. On the other hand, sympathetic negation is alive and well in colloquial speech, to the consternation of authorities like Fowler (1926: 383–384), who considers a negative "evoked in a subordinate clause as a mere unmeaning echo of an actual or virtual negative in the main clause" to be "wrong and often destructive of the sense". Alas, he concedes, "We all know people who habitually say *I shouldn't wonder if it didn't turn to snow* when they mean *if it turned.*" Nor would current usage have provided Fowler much comfort:

(14) a. **Don't** be **surprised** if it **doesn't** rain.
(standard-issue weather warning)
b. I **won't** be **shocked** if every single game is **not** a sellout.
(radio sports talk host Craig Carton, predicting that fans **would** fill Yankee Stadium during its last year of operation in 2008)
c. I would **not** be **surprised** if his doctoral dissertation committee is **not** composed of members from several departments within a university. (Letter of recommendation for applicant to graduate school)
d. **Don't** be **surprised** if the Suns **don't** come back and push the series to five games. (NYT story on NBA playoffs, 2 May 1991)

The role of the complementizer is crucial here; we do not find pleonastic negation in *I wasn't surprised that it didn't rain* (= 'that it rained'), with a veridical complement.[7]

---

[7] It's also worth noting, as Pierre Larrivée (p.c.) observes, that only sentential negation can be involved here; *Don't be surprised if nobody shows up* lacks any pleonastic interpretation.

Equally anathema to those of a Fowlerite persuasion are the pleonastic negations embedded under *keep (NP) from* and *miss;* (15a–c) are googled examples and (15d) was uttered by Lori Laughlin to Bob Sager on a Full House reunion get-together several years after the show's run was over:

(15) a. Well, really, how can I **keep from not** worrying?
　　 b. I can't **keep from not** thinking about the impending doom of it all.
　　 c. doing yoga every day...**kept me from not** thinking about dying
　　 d. We sure **miss not** seeing you every day, Bob.

That far from exotic *miss not* turn of (15d) comes in for particular opprobrium:

> Let's look at a number of familiar English words and phrases that turn out to mean the opposite or something very different from what we think they mean: **I really miss not seeing you**. Whenever people say this to me, I feel like responding, "All right, I'll leave!" Here speakers throw in a gratuitous negative, *not*, even though *I really miss seeing you* is what they want to say.
> 　　　　　　　　　　　　　　　　　　　　　　　　　　　(Lederer 2008)

A Jespersen-style fusion or parataxis analysis is eminently plausible here: (15d) = 'I miss seeing you' + 'I regret not seeing you'. Despite its "gratuitous" nature, the truth-conditionally pleonastic negative is not without its grammatical effect in that the weakly negative *miss* cannot license NPIs without its help, so that (16) cannot occur without its embedded negation for speakers outside the positive *anymore* dialect area.

(16) 　I miss *(not) seeing you around anymore.

While PN is arguably grammaticalized (at least in some dialects) with *surprised, keep (from),* and *miss,* it occurs productively elsewhere in spontaneous speech:

(17) a. I'm going to try to **avoid not** getting bogged down.
　　　　　　　　　　　　　　　　　　　　　　　(J. S. Horn, 1/5/09)
　　 b. I don't know if I can **hold** myself **back** from **not** watching it.
　　　　(Boomer Esiason on torture of watching anticipated Mets debacle,
　　　　　　　　　　　　　　　　　　　　　　　　　　　　　9/17/08)
　　 c. It's been ages **since** I haven't posted **anything** here.
　　　　　　　　　　　　　　　　　　　　　　　　(googled example)

Note the parallel between the contexts in which such PNs occur in English *parole* and those in which pleonastic *ne* is formally required or preferred in French, e.g. after verbs like *éviter* 'avoid' or adverbs like *depuis* 'since'. And while PN in modern French takes the form of solo *ne*, Vendryès (1950) observes that its counterpart for the classical 17th century dramatists Racine, Corneille, and Molière did indeed employ full sentence negation with *ne...pas* or *ne...point*, yielding the full ambiguity such constructions allow in modern English. To be sure, context in general helps distinguish compositional and PN readings of such embedded negations; [8] even in modern French we get triples like *Prends garde de tomber, Prends garde de ne tomber*, and *Prends garde de ne pas tomber*, all of which can only be taken as warning the hearer to take care not to fall.[9]

The hackles of prescriptivists, already raised by the pleonastic negatives of (14)–(17), are lifted to the rafters by the tendency of speakers and writers to lose track of the number of negations in a sentence. If *Duplex negatio affirmat*, we would predict that *Triplex negatio negat*, and indeed we do find instances in which three semantically autonomous negatives yield the force of a single negation:

> The Mets did **not not** re-sign Mike Hampton because they **didn't** want to pay him the money. (Suzyn Waldman, WFAN sports radio host, 10 May 2001)
> (= It wasn't because they didn't want to pay him that they didn't re-sign him)

> Even Susan Sontag, a former PEN president, who supports the leadership against Ms. Komisar, ...hesitated when asked about Mr. Ovitz's role. "I'm **not** saying I'm **not un**happy, " she said, but added that quibbles might be "frivolous." (NYT, 12 March 1997)

But given the conceptual markedness of even simple negation and its concomitant difficulty for the language processor, as verified in extensive empirical studies by Clark, Wason, and others (see Horn 1989: Chapter 3 for an overview), the interaction of the three negations is more accurately char-

---

[8] Thus, the opening line of the Hootie and the Blowfish song "I'll Come Running", "It's been such a long time since I haven't seen your face", must be interpreted compositionally, given the verses that follow, but only a pleonastic interpretation is possible when virtually the same line occurs in a song by Ras Midas; lyrics posted at http://www.rasmidas.com/suchalongtime.html.

[9] Larrivée (1996) discusses other cases in which expletive negation, while less robust than with simple *ne*, is possible in modern (European and Québecois) French with full negators like *pas* and *point*.

acterized as *Triplex negatio confundit*. The tendency for a triple negation to convey a positive is especially prevalent when at least one of the negatives is incorporated into an adverb like *too* or *beyond* or as expressed in an inherently negative predicate like *surprised, avoid, deny,* or *doubt,* as seen in the examples below (cf. also the links at Liberman 2004 and Whitman 2007). In these cases, the reinforcing ("illogical") double negation cancels out an ordinary negation, effectively yielding a positive; alternately, (exactly!) one negation can be viewed as pleonastic.

"I would **not** ever want to say that there are **not** people on our campus that at first are **not** hard to understand, at least until students get used to them."
(Bloomington (IL) Pantagraph, cited in *The New Yorker*)

I **can't** remember when you weren't there,
When I **didn't** care/For **anyone but** you...
(Opening lines of 1981 Kenny Rogers pop song "Through the Years")

"There **isn't** a man there who **doesn't** think they **can't** do it.
(Radio commentator Suzyn Waldman on Yankees' confidence, 2 Oct. 2000)

No detail was **too** small to **overlook**.
(*New Yorker* 12/14/81, Words of One Syllable Department)

People knew **too** little about him **not** to vote **against** him.
(Bill Moyers on why voters in 1984 primaries voted FOR Gary Hart)

Nothing is **too** small or too mean to be **disregarded** by our scientific economy.
(R. H. Patterson, *Economy of Capital* (1865), cited in Hodgson 1885: 219)

It is **not impossible** that some aspect of sound-making efficiency might **not** have played into the mechanism of natural selection during the history of the species.   (Eric Lenneberg, *Biological Foundations of Language*, 1967)

It's a deed that should **not** go **unforgotten.**
(Lewiston (ID) Morning Tribune, cited in *The New Yorker*)

Citing an **unwillingness** to **not** go **quietly** into retirement as the crowned princesses of the East, No. 1-seeded Connecticut battered No. 5 Notre Dame, 73–53.   ("UConn's Big Message: Rout Is Just First Step", NYT 3/3/98)

"Let's go out and see what the boys have in store for us tonight. **They've never ceased to let us down.**"
(Buck Showalter, Texas Rangers manager, quoted in NYT 10 Aug. 2004)

It is not for nothing that Hodgson (1885: 218) observes that "Piled-up negatives prove easy stumbling-blocks",[10] while Fowler (1926: 375) less genially concedes that "Blunders with negatives are extremely frequent." Nor is English unique in hosting such "blunders"; cf. Larrivée 2004: 139-140 on the related French instances of *Triplex negatio confundit*, including the classic *Vous n'êtes pas sans ignorer* (lit., 'You do not lack ignorance of'), for *Vous n'êtes pas sans savoir* (= 'You know'). But while, as Venrdyes (1950: 15) notes of pleonastic negation, "il est naturel que les puristes la proscrivent au nom de la logique", it is equally natural for PN to occur within the received grammar of some languages and for "ungrammatical" PN weeds to erupt regularly in linguistic gardens elsewhere.

2.3. Resumptive negation

A related but distinct variety of apparently redundant negation is the construction that Jespersen (1917: 72-73; 1924: 334) dubs RESUMPTIVE negation:

> [A]fter a negative sentence has been completed, something is added in a negative form with the obvious result that the negative effect is heightened...[T]he supplementary negative is added outside the frame of the first sentence, generally as an afterthought, as in "I shall never do it, not under any circumstances, no on any condition, neither at home nor abroad", etc.

Such "supplementary negatives" do not always follow the negative matrix as an afterthought, but may precede it as a forethought or qualifier; if (18a,b) involve resumptive negation, (18c,d) instantiate what we might call PRE-SUMPTIVE negation.

---

[10] The resulting stumbles don't stop at the courthouse door. In an Alabama case from 1912 recounted by Bryant (1930: 264), Aletha Allen, a 80-year-old deaf woman, was killed by a train after having been warned not to go onto the track. When her estate sued the railway company involved, the jury acquitted the defendant, but a new trial was granted on the grounds that the jurors had been instructed to find for the defendant **unless** the evidence showed that the engineer **did not** discover the peril of the woman in time to **avoid** injury. The higher court invoked *Triplex negatio confundit*, ruling that both the original charge to the jury and its interpretation by the jurors was based on what was meant, not what was said. (Cf. Horn 1991 for additional details.)

(18) a. He **cannot** sleep, **not** even after taking an opiate. (Jespersen)
 b. **Not** a creature was stirring, **not** even a mouse. (C. C. Moore)
 c. **Not** with my wife, you **don't**. (cf. Lawler 1974)
 d. **Not** that I know of, it **isn't**.

Insightful discussions of this phenomenon include those of Lawler 1974, van der Wouden 1997, and especially Dowty 2006, who argues persuasively that the original clause and "afterthought" constitute separate assertions:

> Resumptive Negation is an elliptical form of assertion revision: that is, it indicates a new assertion which is intended to replace the assertion made in the core clause; it may be either a strengthening or a weakening of the original assertion...Neither negation is in the scope of the other, nor is one of the negations merely pleonastic. (Dowty 2006: 5–6)

The possible role of (p)resumptive negation in the development of embracing negation in languages like French, Brazilian Portuguese, and Brabantic Dutch is discussed in van der Auwera's chapter in this volume.

A related construction is that of negative parentheticals, in which negation appears pleonastically within a parenthetical based on one of a range of neg-raising propositional attitude predicates *(believe, suppose, imagine)* following the expression of main clause negation (cf. Horn 1978: 190ff.):

(19) a. The Republicans **aren't**, I (**don't**) think, going to support the President's bill.
 b. The Republicans, I (***don't**) think, aren't going to support the President's bill.

The usual crusaders curiously seem inclined to take both the resumptive and parenthetical negatives of (18) and (19a) in stride. Indeed, Jespersen (1924: 333–334) invokes the tacit tolerance of prescriptivists toward resumptive negation in his defense of negative concord against the standard charge of faulty logic:

> No one objects from a logical point of view to "I shall never consent, not under any circumstances, neither at home nor abroad"; it is true that here pauses, which in writing are marked by commas, separate the negatives, as if they belonged to so many different sentences, while in "he never said nothing" ... the negatives belong to the same sentence. But it is perfectly impossible to draw a line between what constitutes one, and what constitutes two sentences; does a sentence like "I cannot goe no further" (Shakespeare) become more logical by the mere addition of a comma: "I cannot goe, no further"?

Jespersen's citation in this passage refers to a scene in the very same Forest of Arden that serves as a locus classicus for exhibiting the relatively free variation often attested in the distribution of hyper- and standard negation:

(20) a. CELIA. I pray you, bear with me. **I cannot go no further.**
[*As You Like It,* from opening of **II. iv** in Arden Forest]
b. ADAM. Dear master, **I can go no further.**
[*As You Like It,* from opening of **II. vi**, in "another part of the forest"]

Relatedly, code-switching between variants is not uncommon, as when the now defunct electronics chain "The Wiz" presumably promulgated its commercial jingle "Nobody beats the Wiz/Ain't nobody gonna beat the Wiz" to capture the full range of potential consumers. Another illustration is Dylan's celebrated apothegm with its hide-and-seek negative concord. Pinker (2009) has recently lambasted the legal opinions of Chief Justice John Roberts for having "altered quotations to conform to his notions of grammaticality, as when he excised the 'ain't' from Bob Dylan's line 'When you ain't got nothing, you got nothing to lose'." But in the version of "Like a Rolling Stone" published in *The Definitive Dylan Songbook* (Dylan 2001), the original version of the line appears as "When you got nothing, you got nothing to lose." Perhaps it's not a matter of Justice Roberts striking out NC to meet his strictures of decorum as much as Mr. Zimmerman inserting NC to conform to metrical considerations and/or the perceived sociolinguistic expectations of his listeners.[11]

2.4. Pleonastic negation: the whys and wherefores

As we have seen, DNN constructions – both in general and in the NC form in particular have long aroused the purist's wrath for the arrant violation of *Duplex negatio affirmat*. For the linguist, negative concord – as it name suggests – is essentially no more illogical than subject-verb agreement or vowel harmony, a spreading of the negative force of a statement to indefinites within the same clause. But what of other varieties of pleonastic negation, which do not instantiate concord and yet seem to be semantically

---

[11] In the other direction, Emmylou Harris's 1992 cover of Jimmy Davis's 1926 bluegrass classic "Nobody's Darling But Mine" alters the hypernegative line of the original, "Promise me that you will **never**/Be **nobody's** darlin' but mine," to "Promise me that you will **always**/Be **nobody's** darlin' but mine" in two of her three renditions of the refrain.

empty? Such expletive markers, often resulting in two diametrically opposed readings for an embedded clause, would appear to violate both the "Avoid ambiguity" and "Be brief" submaxims of manner (or the Q and R principles that respectively subsume them; cf. Horn 1989), so why do these constructions not just subsist but flourish?

Psychological explanations have been offered, from Damourette and Pichon (1928) on the motivation for PN in French to Wason and Reich (1979) on the pragmatic factors facilitating the plausible (if wrong) interpretation of statements like *No head injury is too trivial to be ignored*. A particularly subtle investigation of so-called expletive negation in Old English is offered by Joly (1972), who also criticizes Jespersen's "paratactic" analysis, noting that a truly paratactic treatment would relate (10a) to (21a) when its meaning is much closer to that of (21b).

(10) a. Je crains qu'il ne vienne. 'I'm afraid that he's coming'

(21) a. Il ne viendra, je le crains. 'He isn't coming, I fear'
    b. Il viendra, je le crains. 'He is coming, I fear'

An alternative approach, suggested by Tesnière (1959), would switch verbs and relate (10a) to a distinct proposition representing the content of my fear, e.g. *J'espère qu'il ne viendra pas* or *Je souhaite qu'il ne vienne pas* ('I hope/wish he doesn't come'). But any such derivation would appear to be ad hoc, while (as Joly points out) also failing to offer an account of the optionality of PN in (10a) and the analogous cases in (9).[12]

Another unwelcome prediction of the paratactic account is that the clause hosting a PN should behave more like an independent sentence than an ordinary embedded clause; in fact, the opposite is true. Haegeman (2005: 161) provides the minimal pair in (22),

(22) a. Comment crains-tu qu'il   ne se   comporte pas?
       how   *fear-you that he* NE *self behaves not*
      'How much do you fear that he won't behave?'
    b. Comment crains-tu qu'il   ne se   comporte?
       how   *fear-you that he* NE *self behaves*
      'How do you fear he will behave?'

---

[12] After pointing out the problems with previous accounts, Joly (1972: 43) concludes that an embedded PN is a "retrospective" negation that reinforces a negative virtually present in the main clause.

along with parallel data from West Flemish (which retains the expletive *en* dropped centuries ago in standard Dutch). Crucially, the manner adverb *comment* cannot be construed with the negative embedded clause in (22a), while it can in (22b), "suggesting that the intervention of the expletive *ne* does not give rise to a blocking effect." (See Espinal 1997: 92 for a similar argument based on Spanish exclamatives.) This would be unexpected were the embedded clause truly paratactic in the latter case.

We have seen that the use of PN may affect NPI licensing (recall (16) above) and may also raise or lower the register, as in modern French or colloquial English. But in addition, PN – although, by definition, irrelevant to truth conditions – is not always semantically and pragmatically inert.[13] This is not a new observation; Joly (1972: 33) mentions that in describing PN as semantically empty or "inutile au sens", the standard account fails to recognize that there may be different levels of sense. One such level is emphasis; Espinal (1997) and Zanuttini & Portner (2000) show that in Basque, Greek, French, Dutch, English, Spanish, and Paduan, an optional PN within an exclamative like (12b) conveys affective or evaluative force, typically intensification. Relatedly, Zeijlstra (2008) notes that while standard Dutch is a double negation rather than NC language, it contains certain NP-internal constructions – e.g. *niemand niet* (lit. 'nobody not'), *nooit geen* ('nothing no') – that are understood as single but emphatic negations, rather than simply canceling out. He analyzes these as lexical items, not as true instances of NC.

In addition to such emphatic functions, PN may sometimes introduce presuppositions, although the data are subtle and interpretations differ. In their influential paper on Italian comparatives, Napoli and Nespor (1976) explore the constraints on the use of what they term *non*, in comparatives and indirect questions as illustrated in (23):

---

[13] Indeed, it is this property that renders PN a happier fit than EN for our category, given the OED's glosses for *pleonasm* ('redundancy of expression either as a fault of style or as a rhetorical figure **used for emphasis or clarity**' [emphasis added]) and *expletive* ('a word or phrase...used for filling up a sentence, eking out a metrical line etc. without adding anything to the sense'). Another problem with EN is that while frequently used for expletive negation (Espinal 1997, Yoon 2009), Zeijlstra (2009) uses it to abbreviate embracing negation, as in French *ne...pas* or similar constructions in Catalan and Brazilian Portuguese (cf. Schwenter 2005, 2006).

(23) a. Maria è piu intelligente di quanto {è/**non** sia} Carlo.
 'Maria is more intelligent than Carlo is'
 b. Maria è piu intelligente di quanto tu {credi/**non** creda}.
 'Maria is more intelligent than you believe'

In these pairs, the utterer of the latter variant – with PN and subjunctive mood – presupposes that her comparative contravenes a previously held belief, while in (23'),

(23') Chissà che {ti sposi/**non** ti sposi}.
 'Who knows if {he'll/he might not} marry you'

the PN version "is used when the speaker expects the negated proposition to surprise someone, or to be contrary to prior expectations" (Napoli and Nespor 1976: 836; cf. also (13c) above). Zanuttini (1997) offers a similar take on *non V mica* in Italian. Alternatively, Schwenter (2005, 2006) argues that the relevant property for licensing secondary, non-canonical double sentential negation in Brazilian Portuguese is the discourse-old status of the proposition being denied. Visconti generalizes Schwenter's finding to the distribution of *mica*, arguing that the belief in the truth of the denied proposition is neither necessary nor sufficient for the the occurrence of *mica* in Italian (or, equivalently, for the *no V pas* construction in Catalan), which instead requires that the negated proposition be discourse-old (Prince 1992) or activated (Dryer 1996).

Other non-truth-conditional differences between unmarked and PN-bearing embedded clauses have been posited in various languages. Thus Espinal (1992: 336, 1997: 75) cites minimal pairs in Catalan like (24) in which the speaker of (24a) regards the arrival as likely, while the speaker of (24b) – with PN and subjunctive mood – is more neutral or doubtful about the occurrence of the event, as the glosses suggest.

(24) a. Tinc por que arribaran tard.
 have fear that arrive+FUT *late*
 'I'm afraid they will arrive late'
 b. Tinc por que **no** arribin tard.'
 arrive+SUBJ
 'I'm afraid they might arrive late'

More controversially, Yoon (2008) has argued that in Korean and Japanese, PN under verbs of hoping as well as verbs of fearing is associated with the speaker's uncertainty and/or her assessment of the undesirability of the event denoted by the complement; this would refute the generalization that "nonveridicality is a necessary condition for licensing [PN]" (Espinal 2007: 58). But Yoon's data, including the veridical *kitayha/kitaisi* examples in (25), get a mixed reception from native speakers.

(25) a. John-un   Mary-ka   oci-**an**-ul-ci   {**kekcengha/kitayha**}-kooista.
      John-TOP Mary-NOM come-PN-FUT-COMP *{fear/hope}*-ASP
      'John fears/hopes that Mary might come'   (Kor.)

   b. John-wa   Mary-ga   ko-**nai**-ka(-to)   {**sinpaisi/kitaisi**}-te iru.
      John-TOP Mary-NOM come-PN-NF(-COMP) *{fear/hope}*-ASP
      'John fears/hopes that Mary might come'   (Jap.)

To be sure, as pointed out some time ago by Vendryès (1950: 5), the specific conditions on the use of PN are subtle and hard to pin down. Given that a natural account of such conventional but non-truth-conditional components of speaker meaning would invoke Gricean conventional implicature, this result should not be surprising; cf. Potts 2007 and Horn 2008b on the "descriptive ineffability" of conventional implicatures across a wide range of construction types.

2.5. Hypernegation and regional variation

Besides conveying emphasis and introducing presuppositions, another function – or in any case, another result – of DNN-type hypernegation is to reinforce regional identity within a speech community. A shibboleth of New England speech is *so don't I*, or more generally *so AUXn't NP*. The most complete inventory for this "Massachusetts negative positive" (see 1999 cite below) is the draft entry for *so* that will appear in the forthcoming 5th volume of the *Dictionary of American Regional English*.[14]

---

[14] I am grateful to *DARE* editor Joan Houston Hall for allowing me access to the data included here; see also the archived ads-l threads [http://listserv.linguistlist.org/cgi-bin/wa?S1=ads-l] from 1998 to 2001 (in particular the postings by Frank Abate, Jason Eisner, Beverly Flanigan, Bryan Gick, Dan Johnson, and Mark Liberman) for remarks on the origin, development, and distribution of *so don't I* and its cousins.

So adv **chiefly N[ew]Eng[land]**. In neg constr following positive constr: used to express agreement with the positive const – often in phr *so don't I* 'so do I'.

1962 *NYT Book Rev.* 28 Jan. 16/1, This expression [= 'don't be surprised if he doesn't visit you one of these days] is akin to the old jocular negative in the following piece of dialogue: "I wish I had an orange." "So don't I." Here again, the speaker means a strong "So do I."

1980 *Daily Hampshire Gaz.* 9 Sept 16/2, And just as the mood of the once-solemn convocation has changed over the past few years, so hasn't the opening address by President Jill Ker Conway.

1998 *NADS Letters* nwPA (as of c. 1980), The standard response indicating agreement was "so don't I" (as in A: "I like ice cream." B: "Mmmm. So don't I!" Also "so didn't I", "so doesn't she", etc.)

1999 *DARE* File – Internet [Boston Online *The Wicked Good Guide to Boston English*], *So don't I* – An example of the Massachusetts negative positive. Used like this: "I just love the food at Kelly's." "Oh, so don't I!"

While this construction is frequently described as ironic or jocular, both cites and speaker intuitions falsify this claim, as noted by John Lawler (p.c.), who also points out its extension to speech communities as far from its New England base as Illinois. In recent unpublished work, Jim Wood uses elicited data to argue that *so don't/can't NP* cancels the exhaustiveness implicature of the corresponding affirmative.

Another question regarding this construction is whence it arises, a question to which I can only provide a non-answer. Recent work by Freeman (2004: 36–37) and Pappas (2004: 57) tracks *so do not I* and the like back four centuries to Shakespeare (*Twelfth Night* III. iv; *Love's Labour's Lost* V. ii; *Richard III* I. iv, II. ii) and Beaumont and Fletcher, with – so it is claimed – pleonastic (non-)force sometimes intended. But these arguments are unconvincing. Even a superficially plausible candidate, the cite from *Twelfth Night* —

> VIOLA: Methinks his words do from such passion fly,
> That he believes himself: so do not I.

— turns out to involve a negation with its ordinary force. As the full context of Viola's remark indicates, her meaning here is contrastive: 'I, unlike Antonio, do not (believe him).' No fronting or inversion would be possible in this context in Present-Day English, where the only possible syntax would be *I do not do so*.

Midland's negative positive answer to *so AUXn't NP* is the regionally restricted reading of *don't care to* reflected in this entry from Montgomery's (2004) compendium of Smoky Mountain English:

> **care** *verb* To be willing or agreeable to (usu. in phr. *I don't care to*, a response to a suggestion or invitation). The verb may range in sense from the understatement "not to mind if one does" or "to be pleased if one does."
>
> **1929** Chapman *Mt Man* 510 "I don't care for work" means "I like to work – I don't mind working." And "I'd not care to drive a car" means "I am not afraid to – I'd like to drive a car." Yet outlanders who have lived years in the mountains are still taking these comments in the modern sense, and advertising that the mountain man is lazy and that he is shy of modern invention.
>
> **1939** Hall *Recording Speech* 7 Examples of *not to care to* for *not to mind*, as in a sentence spoken by an Emerts Cove man, "She don't care to talk," meaning "She doesn't mind talking," are found in both the 16th and 17th centuries.
>
> **1998** Brewer *Words of Past* Another East Tennesseism is the practice, when asking somebody to do something, of adding "if you don't care to" when the meaning is exactly opposite of the plain English. An example would be, "Would you carry me to work, if you don't care to?"

As the cites warn, this construction is as likely as *so AUXn't NP* to be misinterpreted by outlanders. Yet in retrospect, the (non-)development reflected here should not be surprising; essentially, *don't mind* and *don't care*, compositional twins, went their different ways for standard dialect speakers, just as *horrific* and *terrific* (or *awful* and *awesome*) have become antonyms. Midland and related varieties simply preserve the status quo ante. Here is the historical record as displayed in the OED entry:

> **care**, 4b: Not to mind (something proposed); to have no disinclination or objection, be disposed to. Now only [sic – but see above] with *if, though*.
>
> **1526** *Pilgr. Perf.* (W. de W. 1531) 18 Some for a fewe tythes, with Cayn, careth not to lese the eternall rychesse of heuen.
>
> **1597** SHAKES. *2 Hen. IV*, I. ii. 142, I care not if I be your Physitian.
>
> **1611** FLORIO, *Scrócca il fuso*. . a light-heeled trull that cares not to horne hir husband.
>
> **1748** RICHARDSON *Clarissa* (1811) Will you eat, or drink, friend? I dont care if I do.

Note in particular the 1611 cite, in which the (alleged) complaisant trollop doesn't object in the least to cuckolding her husband.[15]

2.6. DNN in the lexicon

Hypernegation extends from the syntactic cases we have surveyed here to lexical instances of the phenomenon. This catalogue can open with double negative affixes with single negative meaning, as in the adjectival cases in which prefixal and suffixal negation do not cancel out. This occurred more productively in earlier (16th and 17th century) times, as seen in redundant formations like *unmatchless* [= 'unmatched', 'matchless'], *unguiltless, unhelpless, unmerciless* and so on (OED **un**[1], 5a; cf. also Horn 1988: 224), but survives in present-day *irregardless* or German *unzweifellos* 'doubtless', lit. 'undoubtless'.

While hypernegated adjectives are now marginal, their verbal counterparts, involving an *un-/de-/dis-* prefixed verb that reinforces rather than reverses the meaning of the stem, are robust. Standard examples include such (ir)reversatives as *unthaw (= thaw, unfreeze), unloose(n) (= loose(n), untighten), unravel, dissever,* and *disannul* and denominals like *unpeel, unshell, unpit, deworm,* and *debone*. Once again, we seem to be confronted with a case in which redundancy is tolerated, and prescriptivists have not hesitated to excoriate *unloosen, unthaw* and their ilk as illogical and contrary to good sense. I have argued (Horn 1988, 2002b) that these forms are motivated by the fact that *un*-verbs unambiguously signal a source-oriented reading in which the object or theme is returned to a state of nature (helping entropy along, as it were). A speaker may not know whether boning a chicken involves inserting or removing the bones or whether raveling the threads of a fabric entangles or disentangles them, but with *deboning* and *unraveling* only one meaning is possible. Ironically, this disambiguating function of redundant affixation has been compromised by technological advances: with the emergence of means to reverse deletions and erasures, previously entropic verbs like *unerase, undelete,* or *unsort* can now be interpreted compositionally, although the redundant readings are still attested.

---

[15] These "negative positives" should not be confused with the "positive negative" described by Frazier (1997), which is (like *don't care* but unlike *so AUXn't NP*) a Midland phenomenon: the use of a jack-in-the-box negative as when "you go into a convenience store…and you ask the salesperson if they have any cat food, he or she will reply, cheerful as can be, 'We sure don't!'"

3. Hypernegation, hyponegation, and the inverse reading of proximatives

The other side of the hypernegative coin is HYPONEGATION, in which there are more (rather than fewer) negatives available for interpretation than are actually expressed. Examples surveyed in Horn 2009 include lexical items like *unpacked* (= 'not yet unpacked'), sarcastic indirect negatives like *That'll teach you* or *fat chance* and the locus classicus *could care less*. The negative status of such expressions is attested by the occurrence of negative polarity items within their scope. Indeed, the phenomenon of implicit negation as a licenser of NPIs extends broadly across both ironic and "straight" contexts; see the discussion of "Flaubert polarity" in Horn 2001, 2009b.

A rich crosslinguistic vein for mining both hypernegation (which doesn't count when it "should") and hyponegation (which counts when it "shouldn't") is that of the inverse readings of proximative adverbs like *almost* and *barely*. While *almost* doing something entails coming close to doing it (the "proximative" component) and not doing it (the "polar" component), I have argued (Horn 2002a, to appear; cf. Schwenter 2002, Ziegeler 2006, Amaral 2007) that the former entailment is not asserted but rather ASSERTORICALLY INERT. One argument for this asymmetry is the existence of INVERSE READINGS, in which a given adverb retains a proximative meaning but switches its polar meaning, so that it can yield either 'almost' or 'barely' (= 'almost not') interpretation depending on the context. The result is that an overt negation may be understood pleonastically or a non-existent negation may be pragmatically supplied. Examples include the following:

[Mandarin Chinese; cf. Li 1976, Biq 1989]
(26) Wo chadianr   mei chi.   a. 'I almost didn't eat', 'I barely ate'
     I   miss-a-little not eat b. 'I almost ate' [= Wo chadianr chi le]

[Spanish; cf. Schwenter 2002, Pons Bordería & Schwenter 2005]
(27) a. Por poco sale.   'She almost left'
     b. Por poco no sale.   'She almost didn't leave', 'She barely left'
(27') a. Por poco se mata.   'She was almost killed'
      b. Por poco no se mata.   ≠ (i) 'She almost wasn't killed'
                                = (ii) 'She was almost killed'

[Valencia Spanish; cf. Schwenter 2002]
(28) a. ¡Casi salgo!"   'I almost didn't get out'   [lit. 'I almost get out']
     b. ¡Casi llegas!   'You barely made it!'   [lit., 'You almost arrive!']

Thus, while the effect in (27'b) – a construction whose pleonastic understanding was recognized by Farrar (1867: 185) – is hypernegation, that in (28) is hyponegation, as it is in Swiss German where *fasch* normally corresponds to standard German *fast* 'almost' but also allows an 'almost not, barely' sense. In English too, we can use *near miss* in a similar Janus-faced way both for disasters just avoided and goals just attained, to the consternation of the late William Safire and his fellow language mavens.

In such cases, the polar semantics – did she or didn't she? – is determined by the utterance pragmatics rather than the overt syntax. Conversely, a given negative effect may be obtained in the presence or absence of overt negation, as with the SQUATITIVES – *squat* and its kin, including *doodly-squat, jack shit, zilch, beans,* et al.), a set of minimizers with peculiar properties explored in Horn 2001 (where the sources for (29) and (30) are provided, along with related examples), Postal 2004, and Hoeksema 2009; see also Postma 2001 for a related class of drecative NPIs in Dutch. A licensed squatitive is essentially an NPI like *anything* or *a damn thing:*

(29)   He then looked into a career as a newspaper reporter but discovered writing didn't pay **squat**.

   The designated hitter or DH: A player who is designated to bat for the pitcher, who, with rare exceptions, can't hit for **squat**.

   We're all professionals, we understand the season's over. We happened to be 15-3, that doesn't mean **squat** now.

Unlicensed *squat* is an n-word like *nothing, nada,* or *niente:*

(30)   All the talk of a resurrected Yeomen football program the past two seasons will mean **squat** if the team fumbles its opportunity to make the playoffs.

   When the more sophisticated students complain that they are learning **squat**, I would direct the professor to remind them that tutoring builds the self-esteem of both tutor and tutee.

   My dad got 'em [football tickets] for free. He works at the university. They pay him **squat** so they give him perks.

A unified analysis of squatitives would take the sentences in (29) to be hypernegations or those in (30) to be hyponegations – but not both.

Essentially (as semi-seriously proposed in Horn 2001), we are dealing here with a mid-stage cross-section of Jespersen's cycle (on which see Horn 1989, Eckardt 2006, Schwenter 2006, and Dahl's and van der Auwera's papers in this volume). Twenty years *avant la lettre*, Bréal (1897) had depicted the process of CONTAGION, whereby etymologically positive indefinite elements like Fr. *pas, rien, jamais,* and *personne* 'serve to reinforce the only genuine negative' until they eventually 'by their association with the word *ne* became themselves negatives' and can now 'dispense with their companion' (1897: 200–202). While we cannot yet foretell whether the future development of *squat* will track the evolution of the French indefinites (or, more recently, that of Spanish *en absoluto* 'not at all' and *en mi vida* 'never'), it would be fitting if reinforcers in the form of *jack-shit, doodly-squat,* Du. *sodemieter,* or the ur-drecative Ger. *Dreck* were to provide the next turn of the wheel on the cycle of negative "contagion."[16]

4. Last words

What morals can we draw from the wide range of data we have explored here? Returning to the question with which we began, we have noted that the presence of apparently extraneous negative particles, when not attributable to grammatical concord as such, seems at first to violate general considerations of avoidance of redundancy, whether in *Duplex negatio affirmat* or *Duplex negatio negat* contexts. But, as with other cases in which the Division of Pragmatic Labor is activated, there is always a reason for a speaker (or a construction) to utilize marked forms – but it is not always the same reason. DNA tends to be used when the speaker is not in a position to express a simple positive directly, and typically (although not invariably)

---

[16] The open-ended productivity of the class is demonstrated by attested examples like (i) and (ii); in the latter, the squatitive object is negative enough to attract its own NPI satellite:

(i) Chris, my best friend at the time, sat at the head of the table and seemed more disinterested in the conversation than Sandy and I, so I figured that he **gave two shits** about what was going on under the table. [Web posting; **gave two shits** = 'didn't care']

(ii) His parents, he said, were at the opera, the ballet, the symphony, several nights a week, "while understanding ***dogshit*** about any of it." [Sue Miller, *While I Was Gone* (1999 novel: 49)]

indicates a qualified or weakened assertion. DNN – and the optional presence of pleonastic negation in particular – tends to reinforce or highlight the speaker's attribution of undesirability, uncertainty, or emotive affect toward the proposition in question. As non-truth-conditional features of meaning, such addenda are plausibly analyzed as conventional implicatures.

More generally, as shown by examples drawn from both *langue* and *parole*, the appearance of an overt negative marker or the nonappearance of a functional one may be a poor guide to the syntactic and semantic polarity of the statement in which it occurs – or fails to occur. After all, as Jespersen (1917) well knew, negation is the un-wizzywig of grammatical categories, where all too often what you see is what you don't get – and vice versa.[17]

References

Abels, Klaus
    2005   "Expletive negation" in Russian: A conspiracy theory. *Journal of Slavic Linguistics* 13: 5–74.
Alonso-Ovalle, Luis and Elena Guerzoni
    2002   Double negatives, negative concord and metalinguistic negation. *CLS* 38(1): *The Main Session*, 15–31.
Amaral, Patricia
    2007   The Meaning of Approximative Adverbs: Evidence from European Portuguese. Ohio State University dissertation.
Austin, Frances
    1984   Double negatives and the Eighteenth Century. In N. F. Blake and C. Jones (eds.), *English Historical Linguistics: Studies in Development*, 138–148. Sheffield: CECTAL.
Bayer, Josef
    2006   *Nothing/nichts* as negative polarity survivors? In *Between 40 and 60 Puzzles for Krifka: a web festschrift for Manfred Krifka*, H.-M. Gärtner et al. (eds.). Berlin: ZAS.
Barbiers, Sjef et al. (eds.)
    2008   *Microvariations in Syntactic Doubling*. Bingley, UK: Emerald Group.

---

[17] I am grateful to Patricia Amaral, David Dowty, Kai von Fintel, Joan Houston Hall, Jack Hoeksema, Beverly Flanigan, Jack Hoeksema, Yasuhiko Kato, Pierre Larrivée, John Lawler, Mark Liberman, Michael Montgomery, Yoshiki Mori, Paul Postal, Chris Potts, Seungja Choi, Jim Wood, Ton van der Wouden, Suwon Yoon, Ben Zimmer, Arnold Zwicky, and the audiences at the American Dialect Society (Horn 2008a), the Brussels LNAT (Logic Now And Then) workshop (Horn 2008c), and the Berkeley Linguistics Society (Horn 2009a) for their examples, inspiration, and attention. The usual disclaimers apply.

Biq, Yung-O.
    1989   Metalinguistic negation in Mandarin. *Journal of Chinese Linguistics* 17: 75–94.

Bosanquet, Bernard
    1885   *Essentials of Logic*. London: Macmillan & Co.

Bréal, Michel
    1897   *Essai de sémantique*. Translated as *Semantics*, Mrs. H. Cust, trans. London: W. Heinemann, 1900.

Bryant, Margaret
    1930   *English in the Law Courts*. New York: Columbia University Press.

Corblin, Francis and Lucia Tovena
    2001   On the multiple expression of negation in Romance. In *Romance Languages and Linguistic Theory 1999*, Y. D'Hulst et al. (eds.), 87–115. Amsterdam: John Benjamins.

Curme, George O.
    1931   *A Grammar of the English Language, Vol. III: Syntax*. Boston: Heath.

Damourette, Jacques and Édouard Pichon
    1928   Sur la signification psychologique de la négation en français. *Journal de psychologie normale et pathologique* 25: 228–254.

Dowty, David
    2006   Resumptive negation as assertion revision. Unpublished ms., The Ohio State U.

Dryer, Matthew
    1996   Focus, pragmatic presupposition, and activated propositions. *Journal of Pragmatics* 26: 475–523.

Dylan, Bob
    2001   *The Definitive Dylan Songbook*. New York: Amsco Publications.

Earle, Samuel, Howard Savage and Frank Seavey
    1911   *Sentences and Their Elements*. New York: Macmillan.

Eckardt, Regine
    2006   From step to negation: The development of French complex negation patterns. *Meaning Change in Grammaticalization: An Inquiry into Semantic Reanalysis*, Ch. 5. Oxford: Oxford U. Press.

Espinal, M. Teresa
    1992   Expletive negation and logical absorption. *The Linguistic Review* 9: 333–358.

Espinal, M. Teresa
    1997   Non-negative negation and *wh*-exclamatives. In *Negation and Polarity: Syntax and Semantics*, D. Forget et al. (eds.), 75–93. Amsterdam: John Benjamins.

Espinal, M. Teresa
    2007   Licensing expletive negation and negative concord in Catalan and Spanish. In Floricic (ed.), 49–74.

Falaus, Anamaria
  2007   Double negation and negative concord: the Romanian puzzle. In *Selected Papers from the 36th Linguistics Symposium on Romance Languages*, J. Camacho and V. Déprez (eds.), 135–48. Amsterdam: John Benjamins.
Farrar, Frederic William
  1867   *A brief Greek syntax, and hints on Greek accidence; with some reference to comparative philology, and with illustrations from various modern languages*. London: Longmans, Green.
Floricic, Franck (ed.)
  2007   *La négation dans les langues romanes*. Amsterdam: John Benjamins.
Floricic, Franck and Françoise Mignon
  2007   Négation et réduplication intensive en français et en italien. In Floricic (ed.), 117–136.
Fowler, H. W.
  1926   *Modern English Usage*. Oxford: Clarendon Press.
Frazier, Ian
  1997   The positive negative: Saying no with a smile. *Atlantic Monthly*, June 1997, 24–26.
Freeman, Jason
  2004   Syntactical analysis of the "so don't I" construction. *Cranberry Linguistics* 2 (University of Connecticut Working Papers in Linguistics 12): 25–38.
Frege, Gottlob
  1919   Negation. In *The Frege Reader*, M. Beaney (ed.), 346–361. Oxford: Blackwell [Reprint 1997].
Giannakidou, Anastasia
  1998   *Polarity Sensitivity as (Non)Veridical Dependency*. Amsterdam: John Benjamins.
Giannakidou, Anastasia
  2006   N-words and negative concord. In *The Syntax Companion, vol. 3*, M. Everaert et al. (eds.), 327–391. London: Blackwell.
Greenough, J. B. and G. L. Kittredge
  1901   *Words and their Ways in English Speech*. New York: Macmillan.
Haegeman, Liliane
  2005   *The Syntax of Negation*. Amsterdam: John Benjamins.
Herburger, Elena
  2001   The negative concord puzzle revisited. *Natural Language Semantics* 9: 289–333.
Hodgson, W. B.
  1885   *Errors in the Use of English*. New York: Appleton.
Hoeksema, Jack
  2009   Jespersen recycled. In *Cyclical Change*, E. van Gelderen (ed.), 15–34. Amsterdam: John Benjamins.

Hoeksema, Jack, Hotze Rullmann, Victor Sanchez-Valencia, and Ton van der Wouden (eds.)
    2001   *Perspectives on Negation and Polarity Items*. Amsterdam: John Benjamins.

Horn, Laurence R.
    1978   Some aspects of negation. In *Universals of Human Language, Vol. 4: Syntax*, J. Greenberg et al. (eds.), 127–210. Stanford: Stanford University Press.

Horn, Laurence R.
    1984   Toward a new taxonomy for pragmatic inference: Q-based and R-based implicature. In *GURT '84: Meaning, Form, and Use in Context*, D. Schiffrin (ed.), 11–42. Washington: Georgetown University Press.

Horn, Laurence R.
    1988   Morphology, pragmatics, and the *un*-verb. *ESCOL '88*: 210–233.

Horn, Laurence R.
    1989   *A Natural History of Negation*. Chicago: University of Chicago Press. Reissued with new preface, Stanford: CSLI, 2001.

Horn, Laurence R.
    1991   *Duplex negatio affirmat...:* The economy of double negation. *The Parasession on Negation: CLS 27, Part Two*, 78–106.

Horn, Laurence R.
    2001   Flaubert triggers, squatitive negation and other quirks of grammar. In Hoeksema et al. (eds.), 173–202.

Horn, Laurence R.
    2002a   Assertoric inertia and NPI licensing. *CLS 38, Part 2*, 55–82.

Horn, Laurence R.
    2002b   Uncovering the un-word: A study in lexical pragmatics. *Sophia Linguistica* 49: 1–64.

Horn, Laurence R.
    2002c   The logic of logical double negation. In *Proceedings of the Sophia Symposium on Negation*, Y. Kato (ed.), 79–112. Tokyo: Sophia University.

Horn, Laurence R.
    2008a   Reneging: Hypernegation and hyponegation in vernacular and regional English. Paper presented to the American Dialect Society.

Horn, Laurence R.
    2008b   *"I love me some him":* The landscape of non-argument datives. In *Empirical Issues in Formal Syntax and Semantics* 7, O. Bonami & P. Cabredo Hofherr (eds.), 169–192. At http://www.cssp.cnrs.fr/eiss7.

Horn, Laurence R.
    2008c   On the contrary. Paper presented at LNAT – Logic Now and Then workshop, Brussels.

Horn, Laurence R.
2009a *Almost* et al.: Scalar adverbs revisited. In *Current Issues in Unity and Diversity of Languages* (Papers from CIL 18, Seoul, Korea). Seoul: Linguistic Society of Korea.

Horn, Laurence R.
2009b Hypernegation, hyponegation, and *parole* violations. Paper presented at BLS 35. To appear in proceedings of the conference.

Inkova, Olga
2006. La negation explétive: un regard d'ailleurs. *Cahiers Ferdinand de Saussure* 59: 107–129.

Jagger, Mick and Keith Richards
1965 *(I Can't Get No) Satisfaction.*

Jespersen, Otto
1917 *Negation in English and Other Languages.* Copenhagen: Høst.

Jespersen, Otto
1924 *The Philosophy of Grammar.* London: Allen & Unwin.

Joly, André
1972 La negation dite "expletive" en vieil anglais et dans d'autres langues indo-européennes. *Foundations of Language* 25: 30–44.

Kiparsky, Paul and Cleo Condoravdi
2006 Tracking Jespersen's cycle. In *Proceedings of the 2nd International Conference of Modern Greek Dialects and Linguistic Theory*, M. Janse et al. (eds.), 172–197. Patras: University of Patras Press.

Labov, William
1972 Negative attraction and negative concord in English Grammar. *Language* 48: 773–818.

Langendoen, D. T. and Thomas G. Bever
1973 Can a not unhappy person be called a not sad one? In *A Festschrift for Morris Halle*, S. Anderson and P. Kiparsky (eds.), 392–409. New York: Holt, Rinehart & Winston.

Larrivée, Pierre
1996 *Pas* explétive. *Revue romane* 31: 19–28.

Larrivée, Pierre
2004 *L'association negative.* Geneva: Droz.

Lawler, John
1974 Ample negatives. *CLS 10,* 357–377.

Lederer, Richard
2008 Our crazy English expressions. *Vocabula Review* June 2008. Cited on http://www.vocabula.com/.

Li, Charles
1976 A functional explanation for an unexpected case of ambiguity (S or ~S). In *Linguistic Studies Offered to Joseph Greenberg*, Vol. 3, A.M. Devine and L. Stephens (eds.), 527–535. Saratoga: Anma Libri.

Liberman, Mark
   2004   Caring less with stress. Language Log post, July 8, 2004. http://158.130.17.5/~myl/languagelog/archives/001182.html.

Lowth, (Bishop) Robert
   1762   *A Short Introduction to English Grammar.* London: J. Hughs.

Marchand, Hans
   1960   *The Categories and Types of Present-Day English Word Formation.* Wiesbaden: O. Harrassowitz.

Martin, Benjamin
   1748   *Institutions of Language.* Facsimile reprint: Menston, GB: The Scolar Press, 1970.

Martins, Ana Maria
   2000   Polarity items in Romance: Underspecification and lexical change. In *Diachronic Syntax: Models and Mechanisms*, S. Pintzuk (ed.), 191–219. Oxford: Oxford University Press.

McCawley, James D.
   1995   Jespersen's 1917 monograph on negation. *Word* 46: 29–39.

Murray, Lindley
   1803   *An English Grammar.* Philadelphia: B. Johnson.

Murray, Lindley
   1814   *An English Grammar, 2d edition, improved.* New York: Collins & Co.

Montgomery, Michael
   2004   *Dictionary of Smoky Mountain English.* Knoxville: University of Tennessee Press.

Napoli, Donna Jo and Marina Nespor
   1976   Negatives in comparatives. *Language* 52: 811–838.

Nash, Walter
   1986   *English Usage.* London: Routledge & Kegan Paul.

Orwell, George
   1946   Politics and the English language. Reprinted in *Collected Essays*, 353–367. London: Secker & Warburg, 1961.

Pappas, Dino Angelo
   2004   A sociolinguistic and historical investigation of the "so don't I" construction. *Cranberry Linguistics* 2 (University of Connecticut Working Papers in Linguistics 12), 53–62.

Pinker, Steven
   2009   Oaf of office. New York Times Op-Ed Column, 1/22/09.

Pons Bordería, Salvador and Scott Schwenter
   2005   Polar meaning and 'expletive' negation in approximative adverbs: Spanish *por poco (no). Journal of Historical Pragmatics* 6: 262–282.

Portner, Paul and Raffaela Zanuttini
   2000   The force of negation in wh exclamatives and interrogatives. In *Negation and Polarity*, L. Horn & Y. Kato, 193–231. Oxford: Oxford University Press.

Postma, Gertjan
　2001　　Negative polarity and the syntax of taboo. In Hoeksema et al. (eds.), 283–330.
Postal, Paul
　2004　　The structure of one type of American English vulgar minimizer. In *Skeptical Linguistic Essays*, 159–172. New York: Oxford University Press.
Potts, Christopher
　2007　　The expressive dimension. *Theoretical Linguistics* 33: 165–198.
Prince, Ellen
　1992　　The ZPG letter: Subjects, definiteness and information status. In *Discourse Description: Diverse Analyses of a Fund Raising Text*, S. Thompson and W. Mann (eds.), 295–325. Amsterdam: John Benjamins.
Sapir, Edward
　1944　　Grading: A study in semantics. Reprinted in D. Mandelbaum (ed.), *Selected Writings of Edward Sapir*, 122–49. Berkeley: University of California Press, 1951.
Schwenter, Scott
　2002　　Discourse context and polysemy: Spanish *casi*. In *Romance Philology and Variation: Selected Papers from the 30th Linguistic Symposium on Romance Languages*, C. Wiltshire & J. Camps (eds.), 161–175. Amsterdam: John Benjamins.
Schwenter, Scott
　2005　　The pragmatics of negation in Brazilian Portuguese. *Lingua* 115: 1427–1156.
Schwenter, Scott
　2006　　Fine-tuning Jespersen's Cycle. In *Drawing the Boundaries of Meaning: Neo-Gricean Studies in Pragmatics and Semantics in Honor of Laurence R. Horn*, B. Birner & G. Ward (eds.), 327–344. Amsterdam: John Benjamins.
Seright, Oren D.
　1966　　Double negatives in standard modern English. *American Speech* 41: 123–126.
Sharma, Dhirendra
　1970　　*The Negative Dialectics of India*. Leiden. [No publisher given.]
de Swart, Henriëtte, and Ivan Sag
　2002　　Negation and negative concord in Romance. *Linguistics and Philosophy* 25: 373–417.
Smyth, H. W.
　1920　　*A Greek Grammar*. Cambridge: Harvard University Press.
Tesnière, Lucien
　1959　　*Éléments de syntaxe structurale*. Paris: C. Klincksieck.
Tieken-Boon van Ostade, Ingrid
　1982　　Double negation and eighteenth-century English grammars. *Neophilologus* 66: 278–285.

Vendryès, Joseph
    1950   Sur la négation abusive. *Bulletin de la société de linguistique de Paris* 46: 1–18.
Visconti, Jacqueline
    2009   From "textual" to "interpersonal": On the diachrony of the Italian particle *mica*. *Journal of Pragmatics* 41: 937–50.
Wason, Peter and Shuli Reich
    1979   A verbal illusion. *Quarterly Journal of Experimental Psychology* 31: 591–597.
Watanabe, Akira
    2004   The genesis of negative concord: Syntax and morphology of negative doubling. *Linguistic Inquiry* 35: 559–612.
Whitman, Neal
    2007   Literal-Minded Archive for the over-negation category. http://literal-minded.wordpress.com/category/semantics/overnegation/.
van der Wouden, Ton
    1996   Litotes and downward monotonicity. In *Negation: A Notion in Focus*, H. Wansing (ed.), 145–67. Berlin/New York: Mouton de Gruyter.
van der Wouden, Ton
    1997   *Negative Contexts: Collocations, Polarity and Multiple Negation.* London: Routledge.
Yoon, Suwon
    2008   Expletive negation in Japanese and Korean. In *Japanese/Korean Linguistics*, M. den Dikken & W. McClure (eds.), *Vol. 18*. Stanford: CSLI.
Zanuttini, Raffaella
    1997   *Negation and Clausal Structure: A Comparative Study of Romance Languages.* New York: Oxford University Press.
Zeijlstra, Hedde
    2004   *Sentential Negation and Negative Concord.* Utrecht: LOT.
Zeijlstra, Hedde
    2007   Negation in natural language: On the form and meaning of negative elements. *Language and Linguistics Compass* 1: 498–518.
Zeilstra, Hedde
    2008   Emphatic multiple negative expressions in Dutch – a product of loss of negative concord. In *Microvariations in Syntactic Doubling*, S. Barbiers et al. (ed.). Bingley, UK: Emerald Group.
Zeilstra, Hedde
    2009   On French negation. Paper presented at BLS 35. To appear in proceedings of the conference.
Ziegeler, Debra
    2006   *Interfaces with English Aspect.* Amsterdam: John Benjamins.
Zimmer, Karl
    1964   *Affixal Negation in English and Other Languages.* Supplement to *Word*, Monograph No. 5.

Quantifier-negation interaction in English: A corpus linguistic study of *all...not* constructions[1]

Gunnel Tottie and Anja Neukom-Hermann

1. Introduction

1.1. Negation and corpus linguistics

Much of the work on negation in various disciplines such as philosophy, psychology and linguistics has been based on intuition and introspection, and empirical studies have probably not constituted the majority. In psychology as well as in linguistics, empirical studies have often been experimental ones, based on subjects' manipulation or acceptance of made-up sentences. The study of the actual use of negation in context has long been mostly restricted to historical linguistics; historical treatments of negation obviously cannot rely on introspection but must of necessity use bodies of text as data. For English this has also been done to a great extent, as is clear from Mazzon (2004), the bibliography supplied in that work, and other historical works on negation like Ingham (2006a, 2006b, 2007) and Kallel (2007). Historical scholars have for a long time made use of their own selections of text editions, available in printed publications or, in recent times, in electronic form. These selections have – of necessity – been idiosyncratic and have not always yielded comparable or replicable data. Publicly available, representative historical corpora have only existed since the 1980s, when the pioneering Helsinki Corpus was first made available by Matti Rissanen and his collaborators at Helsinki University. This corpus enabled such work as Fischer's on negative raising, Rissanen's on word order with *not*, and Tottie's on affixal negation, all in Tieken-Boon van Ostade et al. (1999). Other corpus-based historical work includes Laing (2002) on variation in early ME negation between *ne, ne...not* and *not*, Fitzmaurice (2002)

---

[1] We thank Sebastian Hoffmann for helping us with retrieval and for helpful comments on an earlier version of our work. We are also indebted to members of the Stockholm University English linguistics seminar and participants at the ISLE Conference at Freiburg for valuable suggestions. The responsibility for the final version is our own.

on negative infinitive clauses, and Warner (2005) on *do*-support in negative sentences.

For Present-Day English, computerized corpora have been available since the 1960s, providing valuable sources of data. Tottie (1983) used the pioneer one-million-word Brown Corpus for a study of the variation between synthetic and analytic forms of negation ("*no*-negation" and "*not*-negation") in American English. Tottie (1991) dealt with that topic as well as the pragmatics of negation and the variation between affixal and non-affixal negation in British English, based on two other early corpora, the London-Lund Corpus (500,000 words) and the Lancaster-Oslo/Bergen Corpus (LOB, 1 million words). *Not*-contraction has been treated by several authors using large corpora, e.g. Westergren-Axelsson (1998), Gasparrini (2001), Kjellmer (1998), Tagliamonte and Smith (2002), Walker (2005), and Yaeger-Dror et al. (2002). Palacios Martínez (1999) deals with negative polarity items; ambiguity of negative sentences is treated by Kjellmer (2001). Negative concord (a. k. a multiple negation) is the topic of Palacios Martínez (2003) and Anderwald (2005); other non-standard phenomena are discussed by Pust (1998) and Anderwald (2003, 2004). Negation of lexical *have* is the topic of Nelson (2004) and Takizawa (2005). Biber et al. (1999) also provide data on many aspects of negation based on the Longman Spoken and Written Corpus, comprising c. 40 million words of British and American English.

1.2. Quantifier-negation interaction

One important topic that has so far not received much attention from corpus linguists is the interaction between quantifiers and negation. This has long been a puzzle both for logic and linguistics – cf. Horn and the literature reviewed by him (1989: 226–231 and 490–503) and later works like e.g. Büring (1997). Things are especially complicated in sentences where a universal quantifier like *all* precedes the negator *not,* as demonstrated by (1) and (2), which can have different readings. As frequently pointed out in the literature, the negative may have wide scope, over the quantifier, as shown in (1a) and (2a), or narrow scope, over the verb, as shown in (1b) and (2b):

(1) *All* the arrows did*n't* hit the target
 a. *Not all* the arrows hit the target ~A NEG-Q
 b. All the arrows *'not-hit'* (*no* arrows hit the target) A~ NEG-V

(2) *All* the boys did*n't* leave
   a. *Not all* the boys left   ~**A**   NEG-Q
   b. All the boys *'not-left'* (*no* boys left, all stayed)   **A**~   NEG-V

Wide scope of negation over the quantifier can be symbolized as ~**A**, and narrow scope over the verb as **A**~. In what follows, we will use the terms NEG-Q and NEG-V, respectively.[2] NEG-Q can normally be paraphrased by *not all* with *not*-negation and NEG-V by *no*-negation, i.e. words like *no, nothing, none, no one, nobody*. (But see the discussion in section 6.1.)

Logicians like e.g. Lepore (2000: 312) have argued that the only (correct) readings of sentences like (1) and (2) are (1b) and (2b), NEG-V. Lepore does admit that ambiguity can arise in certain cases but dismisses them as unusual exceptions, which are disambiguated by "placing focal stress on the quantifier words", thereby forcing their interpretations as NEG-Q, as in (1a) and (2a). Prescriptive grammarians have castigated the use of *all...not* constructions with NEG-Q readings, whereas they have been defended by other scholars (cf. Russell 1934, 1935). Linguists (e.g. Jespersen 1917: 86ff.) have been inclined to recognize and accept NEG-Q readings. Horn (1989: 228) even states, presumably based on intuition, that "it is sometimes – in fact *usually* – the case that '**A**...~' = '~**A**'" (our italics) and that "the NEG-V reading is a priori less accessible" (Horn 1989: 229).

However, there has been little empirical research concerning how speakers and writers actually use and interpret *all...not* constructions. Within the generative framework Carden (1970, 1973a, 1973b, 1976) carried out some elicitation experiments to ascertain the existence of NEG-Q, NEG-V and "AMB" dialects (i.e. dialects of subjects recognizing both readings of *all...not* constructions), finding that most speakers preferred NEG-Q readings. Carden's methodology was dubious, and his work was criticized by Heringer (1970), Stokes (1974), Labov (unpublished, mentioned by Horn 1989: 229) and Baltin (1974). Horn (1989: 229) finds support for Labov's and Baltin's views when they state that the dialect split posited by Carden "in fact represents a differential ability to contextualize the two possible readings." The question what speakers and writers actually do has thus not been answered.

Our purpose here is to try to throw light on the problem by presenting the results of a large-scale corpus study of the interaction of a quantifier

---

[2] The terms NEG-Q and NEG-V are adopted from Carden (1970) and also used by Horn.

152 *Gunnel Tottie and Anja Neukom-Hermann*

and negation in sentences where *all* precedes *not,* to be referred to as *all...not* constructions in what follows. We will describe their use and meaning in the British National Corpus, BNC, consisting of some 100 million words of British English from around 1990. We begin by restating the problem and presenting three – not two – different readings of *all...not* constructions in section 2. We then describe our methodology and our results concerning the frequency and distribution of the three readings in BNC in section 3. In section 4, we briefly present their distribution in speech and writing. We then treat factors forcing different readings of *all...not* constructions in section 5. In section 6, we discuss reasons for choosing the ambiguous *all...not* constructions rather than unambiguous *not all* or *no/none* versions. We first treat structural and stylistic factors, and continue with discourse motivations and historical factors. Finally, in section 7, we summarize our findings and discuss their implications for the description of English and how they may be accommodated within linguistic theories.

2. Three readings of *all...not* constructions

The inspiration for our work was a pioneering unpublished study from 1985 by the late Josef Taglicht. Taglicht based his study on corpus evidence from the small Brown and Lancaster-Oslo-Bergen (LOB) corpora of written English (one million words each of American and British English, respectively) and the London-Lund Corpus (about half a million words of spoken British English). In his study Taglicht first points to some interesting facts. He recalls the work of Stebbing (1948) and Jespersen (1917: 89–90), reminding us of something that many philosophers and linguists have overlooked or ignored, viz. that *all...not* constructions can in fact have not just two, but three different interpretations, as shown by (3).[3]

(3)   *All* the bi**lls** do*n't* amount to $50
    a. *Not all* the bills amount to $50   NEG-Q
    b. All the bills '*not-amount*' to $50 (*no* bills do)   NEG-V
    c. *All* the bills taken *together* do *not* amount to $50   COLL[ective]

---

[3] The COLL reading seems to have been first observed by Tobler (1902). Jespersen mentions the reading 'the sum of' for *all,* as in *On me, whose all not equals Edwards moytie* (Shakespeare, *Richard III*, I, 2, 250).

The third reading in (3c) is collective, meaning that the sum of the bills, the bills taken together, does not amount to fifty dollars, whereas the readings in (3a) and (3b) are distributive, i.e. they refer to the individual bills. ((3) is a very clear example of a sentence with three different readings, but it can sometimes be difficult to determine whether a NEG-V or a COLL reading should be assigned, as will be demonstrated in what follows.)

The collective reading is familiar from the nursery rhyme and riddle about Humpty Dumpty in (4), quoted by Lewis Carroll in *Alice in Wonderland* (all italics in quoted examples are ours):

(4) *All* the King's horses and *all* the King's men
Could *not* put Humpty Dumpty together again[4]

Another COLL example is (5) from Shakespeare:

(5) *All* the perfumes of Arabia
Will *not* sweeten this little hand (*Macbeth*, V, 1, 53–54)

Taglicht found a total of 23 instances of the *all...not* construction in the written Brown and LOB corpora (11.5 per million words, henceforth pmw) but none in the spoken London-Lund corpus. The distribution in his material was 52% NEG-Q, 35% COLL, and only 13% NEG-V.[5] But as Taglicht was aware, his material was too small to yield conclusive results; in particular, the fact that he found no instances of *all...not* constructions in his spoken sample at all shows the need for a larger study.

3. *All...not* constructions in the British National Corpus

3.1. Aims and limitations

We therefore decided to use the British National Corpus, containing approximately 90 million words of written text and 10 million words of transcribed speech, seeking to answer the following questions:

---

[4] Carroll's version has *put Humpty Dumpty in his place again*; as Alice remarks, that makes the line far too long.

[5] Taglicht's original figures were: NEG-Q: 13/21, NEG-V: 0/21 and COLL: 8/21; we include two NEG-V instances that were excluded by Taglicht because they formed part of coordinated structures (cf. for example sentences (33) and (34), section 5.4), and we changed one of Taglicht's NEG-Q examples to NEG-V as we found that it had been analyzed incorrectly.

a) How frequent are *all...not* constructions in natural language?
b) What are the proportions of the three readings, NEG-Q, NEG-V or COLL?
c) Are there differences between speech and writing?
d) Can any factors influencing the readings of *all...not* constructions be uncovered?
e) Why are *all...not* constructions used rather than unambiguous ones with *not all* or *no*-negation?

Our study has the following limitations: We will only be reporting on cases where the lexical item *all* precedes the negator *not* (except for examples where *all* follows a NP, as in *the men all didn't leave*). We thus do not include constructions with the universal quantifier *every* or combinations with this element like *everyone, everybody, everything*, or instances where other negators such as *never* are used.[6] We also do not report on the use of constructions where *not* precedes the universal quantifier, as in *Not all the boys had money,* or examples like *All the boys had no money,* with *no*-negation, as these are not ambiguous.[7] However, in a couple of cases, preliminary results concerning the use of *not all* constructions will be cited for comparison.

Moreover, we cannot take prosody into account. As has often been pointed out (e.g. Liberman and Sag 1974: 417, Horn 1989: 229, Ladd 1980: 145–162, Büring 1997), intonation and stress can be crucial for the disambiguation of *all...not* constructions. Thus a combination of fall-rise and stress on *all* produces a NEG-Q reading in (6a), and the falling intonation and normal stress on the noun *men* produces a NEG-V reading in (6b).

---

[6] According to Biber et al. (1999: 278, Table 4.15), *all* is the most frequent quantifier in all registers.

[7] A strange exception from Tolkien's *Lord of the Rings* occurs in (i) (*not all* with a COLL reading):

(i) Only Sam was doubtful: he at any rate still thought boats as bad as wild horses, or worse, and *not all* the dangers that he had survived made him think better of them.

Jespersen (1917: 90) quotes some historical examples of this type, e.g. (ii):

(ii) *Not all* the water in the rough rude sea
Can wash the balme from an anoynted king.
(Shakespeare, *Richard II*, III, 2, 54–55)

(6) a. \/ *All* the men did*n't* go   NEG-Q
    b. \ *All* the men did*n't* go   NEG-V

However, the spoken part of BNC does not contain information about prosody, and the written part obviously cannot do so. This does not seriously invalidate our study, as the spoken material accounts for only a fraction of our sample (7%; see below), and as far as interpreting the written material is concerned, we are obviously no worse off than the original readers of the texts under investigation. Our interpretations are thus based on contextual information, which is accessible in BNC, and world knowledge. Furthermore, as Horn (1989: 231) points out, "any general account of the *all-that-glitters* phenomenon [i.e. NEG-Q *all...not* constructions] must extend well beyond the ups and downs of the fall-rise contour." It is to such a general account that we are aspiring in the present work, and we intend to show that many factors other than intonation are of crucial importance.

3.2. Methodology

We searched for constructions of the form *all (NP) V n't/not* in main and subordinate clauses, by means of a Perl script prepared for us by Sebastian Hoffmann. This is not a trivial task; finding the right restrictions is difficult, and we therefore chose to optimize recall rather than precision, i.e. we made sure that we did not exclude relevant examples by making search criteria too rigorous. The first search therefore yielded a large number of examples: 2,416. These sentences were then checked manually, yielding 495 relevant cases or just under 5 instances per million words. *All...not* constructions are thus not a frequent phenomenon in English.[8] The relevant sentences were then classified as NEG-Q, NEG-V or COLL.

We show one clear example of each from BNC in (7)–(9):

(7) We live in a fallen world. *All* is *not* true, so not everything should be
    believed ...   [C8V: 578; NEG-Q]
    'not all is true'

---

[8] A caveat is necessary here: The frequency of syntactic constructions cannot be compared with the frequency of individual lexical items, and there seems to be little information available on the frequency of constructions. But at least Biber et al. (1999: 1013) do not list any four-word "lexical bundles" containing *all* plus negation under the heading *Quantifiers* in their survey of frequent collocations.

(8)   ...that small degree of compassion...which *all* men in whatever circumstance or however degraded should *not* be denied. [ADA: 1641; NEG-V]
'compassion which no men should be denied'

(9)   If *all* that money we gave to Band Aid did*n't* do the trick, it must be because there are just too many of them. [HH3: 200; COLL]
'the total sum of money'

The analysis was time-consuming and frequently difficult. Sentences with adverbial modifications often caused problems, like (10). Here we settled for NEG-Q:

(10) ...the hunting fraternity is not blameless and indeed the, *all* their arguments are *not quite* correct *in every sense* ...
'not all their arguments are correct'   [JNB: 544; NEG-Q, SPOKEN]

Sentences containing more than one negative item like (11) were also problematic; in this case we decided on NEG-V, based on the extended preceding context provided in BNC.

(11) ...there was *no* positive proof that *all* this had *not* in fact happened, and that it did not belong to a mental lapse ...   [B0U: 1762; NEG-V]

Very complex or abstract contexts, where features in other clauses influenced the interpretation of the *all...not* construction, also caused problems. (12) with its confusing array of negatives is a good example, where we finally settled for NEG-Q:

(12) ...there are two circumstances. Not all of either circumstance was required for the effect, given that the other whole circumstance existed. Each, however, was alternatively required. This is not to say that if the situation had been different then it would still be true that if *all* of either had *not* existed, then if the other had not, the effect would not have occurred.   [EVX: 0183; NEG-Q]

We also had to assign some of our examples to a residual category of unclear (UNCL) examples that could not be assigned a particular reading with any certainty because they were either ambiguous, underspecified or presented other problems, like (13) and (14):

(13) It seemed *all* of her father's old acquaintances had *not* been too impressed by his choice of wife. [AD9: 1347; UNCL (NEG-Q or NEG-V?)]

(14) ...*all* that we've done over the last 40 years is *not* as important as what we do next week. [CHA: 1470; UNCL (NEG-V or COLL?)]

In (13), was everyone unimpressed by the father's choice of wife or were just some of them impressed? And what about the scope of negation and the adverb *too?* In (14), was it the sum of everything or every single thing done in the last 40 years that was less important than upcoming actions? (14) is also a typical example of the difficulty of deciding between NEG-V or COLL interpretations – there is often a fuzzy area here. Moreover, (13) and (14) demonstrate that in natural language, exactness is not always of prime importance – very often, underspecification does not seem to bother either speakers/writers or listeners/readers. Note that the adjectival participle in (13) and the adjective in (14) are both gradable and that imprecision might be deliberate in such cases.

We had to dismiss 43 out of the total sum of 495 examples as UNCL, or 9% of the total, ending up with a sample of 452 instances. Because of the large number of anacolutha, unfinished sentences, and transcription problems in speech, a higher proportion of spoken examples than written had to be rejected – 13/44 or 30%. The remainder, 31 analyzable spoken sentences out of the final 452, thus accounts for only 7% of the total of our examples. We thus still have very little quantitative evidence of what happens in spoken communication, and the bulk of our findings therefore concerns written language. When spoken examples are quoted in what follows, this will be indicated in the BNC references.

3.3. Overall results

Based on the 452 examples that we could analyze we found the distribution of the three different readings shown in Table 1.

Table 1. The distribution of readings of *all...not* constructions in the BNC

	NEG-Q	NEG-V	COLL	TOTAL
n	243	83	126	452
%	54%	18%	28%	100%

We see that the NEG-Q reading is the most frequent of all, with 54%, and that the COLL reading comes second, with 28%. NEG-V is the least frequent type, with only 18%. The proportions are thus similar to Taglicht's results, but it will be remembered that he found no instances at all in spoken language. Our findings also support Horn's claims concerning the preference for NEG-Q readings and Carden's finding that most speakers preferred NEG-Q readings in his elicitation experiments.

4. *All...not* constructions in speech and writing

We also compared the use of *all...not* constructions in speech and writing and found the distributions shown in Table 2. Based on the entire sample (including UNCL examples), we see that *all...not* constructions are slightly more frequent in writing than in speech, averaging 5.2 pmw in writing, compared with 4.3 pmw in spoken language.

Table 2. The use of the *all...not* construction in speech and writing in BNC

	SPEECH	WRITING	TOTAL
n	44	451	495
pmw	4.3	5.2	5

Table 3 shows the distribution of the different readings in speech and writing in the BNC after UNCL instances were removed.

Table 3. Frequencies of the different readings in speech and writing in BNC (clear instances only)

	SPEECH		WRITING	
	n	%	n	%
NEG-Q	12	39	231	55
NEG-V	12	39	71	17
COLL	7	23	119	28
TOTAL	31	100	421	100

The enormous difference in size between the speech and writing samples necessitates caution when comparing proportions. Bearing this caveat in mind, we can observe that there is a higher proportion of NEG-Q in writing than in speech – 55% vs. 39% – and more NEG-V in speech than in writing – 39% vs. 17%. This is perhaps somewhat surprising; we might have expected writers to have more time to produce more *all...not* constructions with the "logical" NEG-V reading, and to avoid "illogical" NEG-Q sentences. The difference between speech and writing is significant at the 0.01 level, but a much larger spoken sample is needed for a serious comparison and analysis, and we will leave the matter here, returning only briefly to it in 5.3.[9]

5. Factors influencing the reading of *all...not* constructions

When analyzing our data, we found that some factors that have not been previously noted in the literature strongly correlate with the use and meaning of *all...not* constructions, viz. the function of *all* as a pronoun (NP head) or a predeterminer, the length of NPs premodified by *all,* and the use of *all...not* in formulaic expressions.

5.1. The function of all – head or predeterminer?

The most influential of these factors is probably the function of *all* as NP head or predeterminer. Notice that we classified *all* in constructions like *All of the boys* as a predeterminer, as shown below: [10]

All is *not* lost	NP Head, pronominal use, "bare *all*"
All the boys, *All of* the boys	Predeterminer

Table 4 shows that when *all* is used as a NP head, only 2/162, or just over 1% of all cases, have a NEG-V reading. The overall tendency is thus for these "bare *all*" sentences to be NEG-Q. The situation is totally different in sentences where *all* functions as a predeterminer. Here we have real three-ways ambiguity, and the majority reading is COLL, with 43%. NEG-Q and NEG-V account for almost equal proportions of these constructions, just under 30% each.

---

[9] For differences between written text domains, see Neukom-Hermann (2006).
[10] As pointed out by Quirk et al. (1985: 381): "Technically, *all* is a pronoun when followed by *of*." However, from a functional point of view, *all of* can be viewed as a predeterminer.

Table 4. The distribution of readings of sentences with pronominal *all* and *all* as a predeterminer. Column percentages

	all as NP head		*all* as predeterminer		TOTAL
	n	%	n	%	
NEG-Q	160	99	83	29	243
NEG-V	2	1	81	28	83
COLL	0	0	126	43	126
TOTAL	162		290		452

(15) and (16) are typical NEG-Q examples where *all* is a NP head:

(15) I recognized that *all* would *not* be plain sailing. [ABU: 1417; NEG-Q]

(16) It usually dawns on you slowly that *all* is *not* as it would appear on the surface. [B2F: 320; NEG-Q]

The two exceptional NEG-V examples are quoted with extensive contexts in (17) and (18):

(17) The rather tacky set, the lucklustre performances, the script from David Straun and Heather Williams that lurches from trite audience participation to over-the-head jokes (would any primary-school child get the one about water privatisation?), *all* did*n't* seem to matter as the company of four scampered around with their well-intentioned tale of how the white man destroyed the American Indians. [AA9: 180; NEG-V]

(17) is a borderline case: *all this* could have been used instead of bare *all*, and it resembles the many cases where *all this* must be given a COLL interpretation; however, *none of this mattered* is a better paraphrase than *all these things taken together didn't matter*, which made us decide on a NEG-V reading.

(18) When Eva had gone and I lay for the first time in the same house as Charlie and Eva and my father, I thought about the difference between the interesting people and the nice people. And how they can't always be identical. The interesting people you wanted to be with – their minds were unusual, you saw things freshly with them and *all* was *not* deadness and repetition. I longed to know what Eva made of things, what she thought of Jamila, say, and the marriage of Changez. [C8E: 905; NEG-V]

(18) is taken from the novel *The Buddha of Suburbia* by Hanif Kureishi. The context makes it clear that the main character finds everything connected to Eva and Charlie very exciting – they are obviously interesting rather than nice – and that the meaning of the sentence is that 'nothing was deadness and repetition'; it cannot mean 'not all was deadness and repetition (but some things were).' We thus settled for a NEG-V reading.

In both of these exceptional cases, it takes an extended context to make the meaning clear, but we see that NEG-V is not impossible, even if highly unlikely, when *all* functions as a pronominal NP head.

The kind of construction where ambiguity is most likely (when *all* is a predeterminer) is exemplified by (19) – (21), which show the three different readings:

(19) There are a few well-rehearsed cases in which *all* the information provided by a text is *not* used to interpret it. [B2X: 412; NEG-Q]

(20) Here *all* infringements of that space from external sources are *not* permitted. [FE6: 732; NEG-V]

(21) "It's a tragedy that *all* those millions of pounds of investment are *not* going to create many jobs." [B7F: 42; COLL]

5.2. Complexity of NPs with *all* as a predeterminer

Furthermore, we found a correlation between the complexity of the quantified expressions and their readings. We have already seen that non-complex examples where "bare" *all* functions as a NP head normally have the NEG-Q reading. We classified the cases where *all* functions as a predeterminer into three types: simple, one-way complex (with either pre- or post-modification or coordination) and two-way complex (with at least two of these features) as follows:

All boys, all the boys	Simplex
All *good* boys/all the boys *in town*	One-way complex
All *good* boys *in town*	Two-way complex

Looking first at the rows of Table 5, we see that in NEG-Q sentences, short constructions prevail: *all* tends to either be bare (66%) or occur as a predeterminer in simplex constructions without additional modifiers (23%). NEG-V and COLL are both mostly of the simplex type (48% and 51%, respectively), both have a fair share of one-way complex constructions (41%

and 29%, respectively), but COLL sentences show a higher proportion of two-way complex constructions than NEG-V sentences (20% vs. 8%). If we look at the column figures, we see that two-way complex NPs have a marked tendency to have COLL readings: 25/35, or over 70%, of the two-way complex sentences are COLL. As we shall show below, this tendency is related to the formulaic use of COLL *all...not* constructions.

Table 5. Complexity and reading of sentences with *all...not* constructions

	NP Head	Predeterminer			TOTAL
	Bare *all*	Simplex	Complex 1	Complex 2	
NEG-Q	160 (66%)	56 (23%)	24 (10%)	3 (1%)	243
NEG-V	2 (2%)	40 (48%)	34 (41%)	7 (8%)	83
COLL	0	64 (51%)	37 (29%)	25 (20%)	126
TOTAL	162	160	95	35	290

5.3. Formulaic uses of *all...not* constructions

Structural factors are not the only ones to influence the readings of *all...not* constructions. We found that a large proportion of our examples were formulaic expressions, and that these could have either a NEG-Q or a COLL reading, but not a NEG-V reading.[11] 133/452 instances, or 29% of the total, were of the type *all is not lost/well/perfect/good/gloom (and doom)*. These expressions are always NEG-Q. Examples are presented in (22) and (23):

(22) A third aide ... insisted *all* was *not* lost between Charles and Diana. [CEN: 2108; NEG-Q]

(23) Sock Shop admitted earlier this year that *all* was *not* well with its American outlets. [AAS: 376; NEG-Q]

Evidence that such instances are indeed formulaic comes from the fact that BNC does not contain a single instance of ***not all** is/was lost*, the most readily available paraphrase for ***all** is/was **not** lost*, of which BNC contains 32 instances (or as many as 49 if we include slight variations like *all **may***

---

[11] For a discussion of formulaicity in language, see e.g. Wray (2002).

not be lost or *all is not yet lost*).[12] Similarly, there are no instances of ***not all is/was well***, but 44 instances of ***all is/was not well*** (or 52 with the modified cases). In addition to this, several of the "free" instances with variable subject complements may have been influenced by the formulaic ones. Thus there are 13 instances of *all is/was not X* followed by other complements: *needed/free/derelict/dead/trouble-free/harsh antipathy/(yet) completed/(yet) over/politics and institutions/in perfect control/as portrayed on the media/going to plan/as aunt Bertha had surmised*. It seems reasonable to assume that the frequent formulaic *all is not* constructions have influenced usage here.

However, the *all is/was not X* construction itself is not entirely frozen: there are eleven cases of ***not all is/was X*** in BNC, with the following complements: *spleen/gloom/gloom and doom/hopeless/dead/seen/forgotten or forgiven / stress / running Glaxo's way / of the type / beyond reproach*. Compared to the total number of formulaic *all...not* constructions where *all* is a NP head – 130 as shown in Table 6 – this is a drop in the ocean, and the tendency for such constructions to be formulaic is clear.

There are also 35 formulaic *all...not* constructions of a type that is always COLL. These are mostly of the type *all* NP *in the world* V *not*, but nine of them are of the form *(as if) all this wasn't enough*. Typical examples are given in (24)–(26):

(24) ... *all* the marketing and computers *in the world* won't help you.
    [A6L: 1316; COLL]

(25) ... the coroner had pointed out how *all* the instruments *in the world* could *not* have detected it. [AD1: 2783; COLL]

(26) *As if all this were not enough*, schools have started managing their own financial affairs. [ABE: 1816; COLL]

Of all our instances, 37% (168/452) are formulaic, 55% (133/243) of the NEG-Q examples and 28% (35/128) of the COLL cases, whereas there are no instances at all of formulaic NEG-V *all...not* constructions. Formulaicity and the function of *all* as a pronoun or predeterminer also correlate with readings in an interesting way, as appears from Table 6.

---

[12] A google search yielded 175,000 instances of *not all is lost*, but 1,220,000 for *all is not lost*; the *not all* construction is clearly used but makes up only 12.5% of the total here.

Table 6. All as NP head or predeterminer in the BNC; free and formulaic uses

	all as NP head Free	*all* as NP head Formulaic	*all* as predeterminer Free	*all* as predeterminer Formulaic	TOTAL
NEG-Q	30	130	80	3	243
NEG-V	2	0	81	0	83
COLL	0	0	91	35	126
TOTAL	32	130	252	38	452
GRAND TOTAL		162		290	452

Table 6 shows that among the pronominal uses of *all*, the majority occur in frozen NEG-Q expressions, 130/162 or 80% of the total. In constructions where *all* is a predeterminer, only 38/290 or 13% are formulaic, mostly with COLL readings. (Somewhat surprisingly, 29/31 or 94% of the spoken examples were free, but only 255/421 or 61% of the written ones. This difference is significant at p <.0001, but again, much more material would be required for a serious study.)

The free uses where *all* is a predeterminer are fairly evenly divided between the three readings, 80/252 (32%) NEG-Q, 81/252 (32%) NEG-V, and 91/252 (36%) COLL. It is thus the free constructions where *all* functions as a predeterminer that can have any one of the three possible readings. This is also the type that most previous writers have used to demonstrate ambiguity, based on their intuitions, and where intonation can be used to manipulate readings. We shall return to the reasons for this in our discussion below.

The syntactic function of *all* as a NP head or predeterminer, the length of the quantified expression containing *all*, and use in formulaic expressions can thus all be shown to contribute to the reading of *all...not* constructions. But as we shall show in the following section, all these factors can be overruled by contextual ones.

5.4. The influence of context – linguistic and extra-linguistic

In the theoretical literature on negation and quantifiers, made-up examples of *all...not* constructions are usually kept very simple for the sake of clarity of exposition – *All the boys didn't leave* and *All of them didn't come* are typical examples – and the possible effect of context, especially extended context, has often been neglected. This has been pointed out previously by e.g. Baltin (1974: 32), and Horn also uses the notion of a forcing context

(1989: 545, fn. 16). In experimental and empirical research, scholars have also been aware of the importance of context for the interpretation of *all... not* constructions, as is shown by the work of Heringer (1970), Stokes (1974), and Labov (1972a, 1975). However, their empirical work is mostly based on constructed sentences with isolated lexical, syntactic and pragmatic constraints added, and although this has provided valuable information, it does not tell us what real-world speakers and writers actually do.

Real sentences containing *all...not* constructions tend to be much more complex than those produced for questionnaires or other types of elicitation experiments. They are produced by speakers and writers in real-world or reported situations, requiring both linguistic and extra-linguistic contexts to be understood. Clues can be found either in the same sentence or the wider linguistic context, and they can be syntactic, semantic or pragmatic; often world knowledge or specialist knowledge is required for the interpretation of *all...not* sentences, and often more than one constraint operates at the same time.

In what follows, we give some examples of different types of forcing contexts found in BNC, beginning with lexical ones in (27) and (28), where the semantics of the adjective *aggregate* and the verb *add up* force COLL readings:

(27) The director's emoluments must also be included in the bandings in the note to the accounts unless *all* the directors' *aggregate* emoluments do *not* exceed 60, 000. [CBY: 3682; COLL]

(28) But *all* this did *not* yet *add up* to a widespread expectation in London that Anglo-American relations would or should retain their wartime intimacy. [HY8: 262; COLL]

Lexical items like *these* and *other* can also force NEG-V interpretations, as illustrated by (29)–(32):

(29) ... *all these kids* did*n't* know what two pence was, did*n't* know what twenty pence was and ten pence and fifty pence ...!
 [KE2: 8349; NEG-V, SPOKEN]

(30) A: Does it print it?
 B: No, *all these characters* are *not* printed. [KP1: 583; NEG-V, SPOKEN]

In (29) and (30), the NEG-V reading is made salient through the presence of *these* after *all*. In fact, *all* seems to function mostly as an intensifying item in these sentences since the propositional content does not change if *all* is removed.[13] As we shall see, this is a frequent characteristic of NEG-V as well as COLL examples.)

In (31) and (32) the item that forces a NEG-V reading is *other* (either as a pronoun or as an adjective):

(31) So having four children, one of them will have PKU, but *all the others* wo*n't*, they will be normal. [F8L: 845; NEG-V]

(32) I think that's what makes Curve stand out from the rest of the pack, asserts Alex, because *all these other* indie bands do*n't* seem to be interested in guitar music. [C9J: 1846; NEG-V]

As already noted by Taglicht, coordinated subjects or predicates force a NEG-V reading, as shown by (33) and (34), with negation in either the first or the second coordinated element:

(33) *All* these people that tear about on the water on a Sunday do*n't* know what they're doing **and** are in desperate need of someone to manage them. [J3W: 283; NEG-V]

(34) Check that *all* cables are firmly in their sockets **and** have *not* worked loose. [HAC: 2088; NEG-V]

NEG-V can also be forced by a non-restrictive parenthetical insertion as in (35). In fact, the inserted parenthetical clause could be argued to function in a similar way as coordination in (33) and (34), and it can also be paraphrased as such (cf. 35').

(35) *All* of these benefits, *whilst clear within the Bank*, were *not* widely known by our customers. [GX9: 121; NEG-V]

(35') *All* these benefits were clear within the Bank and/but were *not* widely known by our customers.

In many cases, a feature that occurs later in the sentence clarifies which reading is intended. As shown by example (11), repeated here for convenience, anaphoric pronouns like *it* can force a NEG-V reading of the previous clause.[14]

---

[13] This is debatable from a strictly truth conditional point of view, but we consider *all* to be just a pragmatic marker of emphasis here. Cf. further examples (55) and (56) below.

[14] The preceding context also helps here. It includes "series of disjointed actions" and many dreams that did not occur simultaneously, so it cannot be COLL.

(11) ... there was no positive proof that *all this* had *not* in fact happened, and that *it* did not belong to a mental lapse from which I had recovered. [B0U: 1762; NEG-V]

Similarly, in (36), *it* refers to the whole amount of money and shows that it was not just part of it that was not needed for war:

(36) And because *all* this money was *not* needed for war, *it* was available ... to show its subjects that their monarchy lived in style and elegance. [AE4: 336; NEG-V]

In other cases, it is the wider linguistic context that indicates the intended meaning, as in (37):

(37) I found it most questionable that Forward Publishing, having won fifteen awards – including a class winner – was not mentioned once throughout the whole ceremony. Had it been the form that *all* agencies were *not* mentioned, this may have been acceptable, but to be forced to listen to a litany of other agency names – not least Barkers Trident – without a single mention of our own was deplorable. [HAK: 64; NEG-V]

It is clear from the context that (37) must be interpreted as NEG-V and be paraphrased *no agencies (at all) were mentioned*. While the word *other* plays a role, it is not the only pointer to the intended reading. The clues are distributed in the context preceding and following the *all...not* construction; the writer is clearly angry that his own company was not mentioned and feels that it has been deliberately neglected.

Finally, there are examples that can only be disambiguated with the help of knowledge of a particular situation, a specific culture, or very specialized professional knowledge. This is the case in (38). A reader who knows that Sainsbury's is a chain of supermarkets (and possibly also that *Good Housekeeping* is a magazine) will automatically access the NEG-Q reading. Similarly, world knowledge tells the reader that the writer of (39) probably does not want any bits at all of a jigsaw puzzle to fall out of the box into the laundry bag. But (40) requires specialist knowledge for disambiguation, even if NEG-V seems likely because of the length of the noun phrase. As we are not biochemists, we had to leave it among the unclear examples.

(38) Many of you may have noticed that Good Housekeeping is now on sale at the checkout in Sainsbury's ... I can't think why *all* supermarkets do*n't* put GH at the checkout. [ED3: 19; NEG-Q]

(39) The jigsaw won't fit in the suitcase. It's back in its box now, but it's still pretty big – so I have to stick it in with the dirty washing. I hope *all* the bits do*n't* fall out of the box. [A74: 1529; NEG-V]

(40) It is obvious that *all* point mutations affecting the D-stem or the size of the extra arm did *not* cause any discriminatory effects on the identity elements for the tRNA (m 5 C49) methyltransferase. [FTC: 908; UNCL]

So far, we have shown that the meaning of *all...not* constructions can be related to structural as well as contextual factors, and that some uses are formulaic expressions in English. We now proceed to discuss the role of contrast and emphasis in *all...not* constructions.

5.5. Contrast, emphasis and metalinguistic negation

There are some major differences between *all...not* constructions with NEG-Q readings on the one hand and NEG-V on the other, and COLL sentences share some characteristics of both of these two types. Thus NEG-Q and COLL sentences tend to be used as emphatic denials of previous utterances, or as self-corrections and clarifications, but this is not the case with NEG-V sentences. An indication of this difference is the occurrence of explicitly contrastive conjunctions such as *but* or *however* in NEG-Q and COLL examples. Table 7 shows the frequency of contrastive conjunctions or adverbs with the different readings of *all...not* constructions in BNC.

Table 7. The distribution of *all...not* constructions with contrastive conjunctions/adverbs

	WITH EXPLICIT CONTRAST	%	TOTAL
NEG-Q	47	19	243
NEG-V	3	4	83
COLL	19	15	126
TOTAL	69	15	452

As appears from Table 7, 19% of all NEG-Q instances are explicitly contrastive, and 15% of the COLL examples, but very few of the NEG-V ones – a mere 4%. Examples of NEG-Q are given in (41)–(43).

(41) After so long a period of remorseless manufacturing decline in the UK, to hope for a reversal might seem wishful thinking. **But** *all* is *not* lost.
[CBU: 1022; NEG-Q]

(42) A landmine destroyed one vehicle, a patrol was ambushed as they spoke to some locals, and a foreign parachute and map were discovered on a beach after a tip-off. *All* did *not* go against the Key Company, **however**; prisoners were taken at some of the incidents, some carrying vital information. [A77: 1864; NEG-Q]

(43) More than 90% of patients with duodenal ulcer are carriers of H pylori. *All* carriers of the bacterium, **however**, do *not* suffer from duodenal ulcer. [HU3: 1317; NEG-Q]

In all of these examples (41)–(43), a bleak background is painted in the sentences preceding the *all...not* constructions – manufacturing declines, a war is not going well, or 90% of patients are carrying the H pylori bacterium. Readers may well believe the worst, but the authors hasten to reassure them that things are not so bad after all.

Contrastive COLL examples are shown in (44) and (45). In both examples, a build-up of positive circumstances seems to point to the conclusion that things are pretty good – no rationing, exemptions from military service etc. in (44), and the availability of technical devices in (45). But the authors point out that all this doesn't help.

(44) Not least, the absence of any serious rationing of consumer goods, and the large number of exemptions from military service for skilled workers and farmers made it appear that the regime was well in control of developments, did not fear a war on the 1914–18 scale, and was even rather generous in its provisioning arrangements. **Nevertheless**, *all* this could *not* completely conceal the tension.   [ADD: 25; COLL]

(45) Aided by cheap drum machines and synthesizers, these canny aspirants endeavour to construct a spectacular pop by distilling the pop essence from all past efflorescences, using the hindsight sophistication of the present day. **But** *all* the ambition, striving, graft, pop-learning in the world ca*n't* earn you pop divinity. [AB3: 1789; COLL]

The number of contrastive *all...not* sentences is likely to be even higher as contrast can be expressed without an overt marker, but we believe that the actual frequency of contrastive conjunctions is an indicator of the nature of a large proportion of NEG-Q and COLL constructions. In contrast to this, it is noteworthy that very few of the NEG-V instances contain any such explicit markers of contrast – in fact, only 6, or a mere 7% of the NEG-V instances are contrastive and express some kind of denial of a previously expressed or inferred proposition. (31) repeated from above, is an example of this infrequent type:

(31) So having four children, one of them will have PKU, *but all* the others wo*n't*, they will be normal. [F8L: 845; NEG-V]

The use of *all...not* constructions as emphatic denials (either with or without contrastive conjunctions) thus seems to be typical of NEG-Q and COLL readings.[15] It could be argued that the NEG-Q instances are instances of what has been called metalinguistic negation, defined by Horn (1989: 363) as "a device for objecting to a previous utterance on any grounds whatever, including the conventional or conversational implicata it potentially induces, its morphology, its style or register, or its phonetic realization." What we are dealing with here are the implicata rather than formal characteristics of the preceding context.

Horn (1989: 496) also reminds us that "the fall-rise contours which tend to be associated with the NEG-Q readings ...[are] in fact a general characteristic of metalinguistic negation" and that "[t]his supports the view that the wide-scope (NEG-Q) reading ... occurs most naturally in metalinguistic uses." However, although Horn goes on to say that "the fact that no special intonation is required ... suggests that this construction, with its [NEG-Q] interpretation, must also be analyzable as realizing ordinary predicate denial," he acknowledges in a note that there remains a metalinguistic tinge to the NEG-Q *all...not* construction (1989: 576, note 31).[16]

6. Why are *all...not* constructions used?

Previous studies have been concerned with how *all...not* constructions are interpreted (and sometimes why a particular interpretation is chosen), but the question why these potentially ambiguous constructions are used at all

---

[15] Jespersen (1917: 87) claims that for NEG-Q senses "very often *all* is placed first for the sake of emphasis ..."

[16] If it is true that NEG-Q *all...not* constructions always involve metalinguistic negation, then it makes sense that logicians deny the existence of the NEG-Q reading, as traditional logic only recognizes ordinary predicate denials.

has received much less attention. Why choose ambiguous *all...not* constructions when the NEG-Q reading can be clearly and conveniently expressed by *not all*, and the NEG-V reading by *no/none*?[17] (Only the COLL reading has no such immediately available and convenient paraphrase.)

In what follows, we will discuss factors found in our sample that make the use of *all...not* constructions obligatory or preferable, beginning with structural and stylistic factors and continuing with discourse factors.

6.1. Structural and stylistic factors

First of all, in sentences where there is coordination with other quantifiers, *all* is required, and *no/none* is impossible, as shown by (46) with *some* and its variant (46').

(46) But if a total consciousness is an organic whole, then *some or all* of these parts could *not* exist in the same character in another different sort of whole. [CS2: 464; NEG-V]

(46') *But if a total consciousness is an organic whole, then *some or none* of these parts could exist in the same character in another different sort of whole.

Similarly, it is difficult to construct a variant of (47); (47') is doubtful:

(47) So Structuralism, Chomskyism and *all* our official Linguistics are *not* exactly wrong but they are radically incomplete. [J7U: 26; NEG-V]

---

[17] Horn (1989: 498–99) discusses a related question, viz. why the NEG-Q reading is available with universal quantifiers but not with existential ones. He elaborates on Jespersen's discussion of the problem (1924: 327) and defines it as "a rivalry among several functional tendencies": the tendency to put the negative first (Horn's "Neg-First" condition), the tendency to put the subject first in a sentence, as well as the "preference for overt negation [*not*] to surface in its unmarked ... position, as a particle or inflection on the finite verb or auxiliary" (Horn's "Nexal Not" condition; 1989: 498). Horn assumes that NEG-Q readings are available for *all...not* constructions (in contrast to *some...not* constructions) because its alternative expression for the NEG-Q reading (*not all*) is more marked than the lexicalized *nobody* (the alternative expression for NEG-Q *some...not*). This argument might also be invoked for the question of why *all...not* constructions are used at all, rather than the *not all* alternative.

(47') ?? So Structuralism and Chomskyism are not exactly wrong, and *none* of our official Linguistics is exactly wrong ...

In some cases NEG-V *all...not* constructions must have been chosen for stylistic or rhetorical reasons, as in the echoic constructions in (48) and (49), where the repetition (verbatim or with slight changes) of the first sentence produces an ironical effect:

(48) Mr. Ashworth submitted that if *all else failed*, Community law would come to his rescue. As *all else has not failed*, I propose to deal with this aspect of the matter shortly. [FCR: 494; NEG-V][18]

(49) Craig and Wedderburn showed that while *almost all industrial workers* have to clock in, *almost all managers* do *not* have to do so; ...
[FR4: 460; NEG-V]

As already pointed out above, in a very large number of cases *all* seems to function merely as an intensifier. Thus in (50)–(54) the definite article or demonstratives already point to the universality of the statements.; the function of *all* is to emphasize the fact that there are no exceptions, and that the speaker or writer is referring to a large number of entities.

(50) ... although there is a declining performance as the age of sign language learning increases, it does not mean that *all those* over the age of 30 years can*not* learn BSL, nor that they will be unable to communicate. [CLH: 351; NEG-V]

(51) A: We've got four local dogs and they
 B: Lay down!
 A: and *all the* dogs do*n't* like my cat! [KBL: 3456; NEG-V, SPOKEN]

(52) He has implied that *all those* who are involved in taking a vehicle are *not* already committing an offence, but they are. Everyone who is involved in taking a vehicle is implicitly committing an offence.
[HHX: 7007; NEG-V]

(53) *All these* setbacks did*n't* matter, old chap, because the referee only has eyes for the big boys. [CHV: 164; NEG-V]

---

[18] As pointed out by the editor, this could be formally an instance of NEG-Q, i.e. "an assertion that contrary to Mr. Ashworth it is **not** the case that all else has failed, with any stronger implication *(all else has not-failed)* derived pragmatically."

(54) *All the* galleries of New York do *not* contain its equal –.
[HJH: 3873; NEG-V]

For COLL *all...not* constructions there are no obvious variants, but some typical uses can be identified. As mentioned above, formulaic constructions make up a substantial proportion of the COLL instances with 35/126 (28%). Secondly, an even higher proportion 45/126 (36%) contain anaphoric *all this* referring to a previous context; cf. (55) and (56):

(55) I shall associate with others when they are useful to me ... nothing in sex will concern me except the pure physical pleasure; I shall be incapable of love ... I shall reject all arts and entertainments which depend on participation in the feelings of other people ... All my pleasures will be solitary ... Clearly *all this* is *not* going to be much fun, but my aim is egoism and not hedonism.   [CB1: 432; COLL]

(56) In Spain ... the Communist-dominated United Left ... has reservations about the treaty ... Meanwhile, in Italy, some of the ... politicians who used to support closer European co-operation have suddenly swung round to oppose it ... *All this* does *not* mean, of course, that France, Spain and Italy have gone anti-European.   [CRA: 2051; COLL]

In examples of this kind, *all this* always refers back to the totality of the things mentioned in the previous context, not individual factors; although sentences with *no/none* would be idiomatic, they would not have a collective meaning.[19]

Thirdly, as in the NEG-V examples cited in (50)–(54), COLL *all* as predeterminer can also be semantically redundant – again *all* seems to function as an intensifier, as illustrated by sentences (57) and (58).

(57) She expected him to keep looking over his shoulder to make sure that *all the fuss* was *not* intended for the man behind.   [A68: 2608; COLL]

(58) *All this modelling* is*n't* half as glamorous as it's made out to be.
[AJJ: 69; COLL]

Roughly 20% of the COLL instances contain such an emphatic *all*, which is always followed by a definite noun phrase and occurs mostly with uncount-

---

[19] Anaphoric *all this* often occurs with factual verbs (cf. Quirk et al. 1985: 1180ff.) such as *mean, say*, or *suggest*, as in (56). It could perhaps be argued that the 13 examples with such factual verbs that occurred in our corpus are also formulaic.

able nouns. Typically, *all the* or *all this* precedes a noun that denotes something that the writer or speaker finds unpleasant or annoying, as in (57) and (58) above.

The rest of the COLL *all...not* constructions are mostly used to convey the sense 'not even all N can...' This meaning is comparable to that expressed by the frequent formulaic sequence *all NP in the world V not*. Typical examples are (59) and (60); as in the NEG-V examples, *all* seems to imply that there were a lot of lovely words or precautions, not just a few or a couple of them.

(59) *All* the most lovely words of love and passion could *not* express one tenth of what I feel for you. [ABL: 505; COLL]

(60) Yet *all* her precautions do *not* seem to have prevented the 26-year-old woman from abduction. [CBF: 12392; COLL]

Usually the implication is that more of the NP would make the meaning of the predicate more likely, but there are a few cases with the opposite meaning, i.e. the more of the particular NP, the less likely the predicate. This use is exemplified by (61).

(61) *All* the pressures of people having a go at me all the time *don't* help, said Kylie. [ADR: 1235; COLL]

There is also a semantic factor: although NEG-V constructions can normally be paraphrased by *no*-negation this is not always quite true. It seems to be the case that the use of the universal quantifier *all* is associated with a stronger existential presupposition than the negative universal quantifier *no/none*. Thus in (36), repeated below for convenience, there actually is money in the state coffers, but in (36'), that is not necessarily the case, as is clear from the fact that the constructed sentence in (36a) does not make sense, but (36b) does. (The existential presupposition is not created by *this*; the definite article would work as well.)[20]

---

[20] Cf. the German example in (i), taken from a broadcast. It was clear from the context that there actually were some goals of war but that none were achieved, and we thus have a NEG-V reading.

(i) Also *alle* Kriegsziele wurden *nicht* erreicht. [SWR1, 14-8-2006; NEG-V]
'So *all* goals of war were *not* achieved'

(36) And because *all this* money was *not* needed for war, it was available ... to show its subjects that its monarchy lived in style and elegance.
[AE4: 336; NEG-V]

(36') And because *no* money was needed for war, it was available ... to show its subjects that its monarchy lived in style and elegance.

(36) a. *And because *all* this money was *not* needed for war, it was OK that the government's coffers were empty.
   b. And because *no* money was needed for war, it was OK that the government's coffers were empty.

So far, we have shown that there are several different types of factors that can make speakers and writers use *all...not* constructions rather than their unambiguous equivalents with *not all* or *no*-negation: structural, stylistic and semantic considerations as well as the wish to provide emphasis by stressing totality, size and number. However, we believe that there are also other reasons for the use of *all...not* constructions. We will argue that some important reasons for using them can be found in the pragmatics of negation and in the organization of discourse, and that ultimately, there may be historical factors that account for the use of a large part of them.

6.2. Discourse factors

It is generally recognized that negative sentences are used in a context where an expectation has been raised by the speaker or writer, or arises from the situational or linguistic context. In the words of Givón (1979: 139), negative sentences "are uttered in contexts where the corresponding affirmative has been discussed, or else when the speaker assumes ... the hearer's bias toward or belief in – and thus familiarity with – the corresponding affirmative." As Chafe (1994: 180) puts it in his mostly speech-based account, "the speaker's mind necessarily includes a dynamic model of what is happening in the mind of the listener." This certainly applies to writers and readers as well: when a writer realizes that s/he may have created an unwarranted expectation in the reader, s/he hastens to tell the reader that things are not what they seem to be. Although writers are less subject to

---

If the speaker had said *Keine* Kriegsziele ('*no* goals of war'), at least the first part of the sentence could be understood to mean that there never were any goals of war, whereas *alle Kriegsziele* presupposes the existence of some goals.

production constraints than speakers, the conditions for making negative statements are, in principle, the same whether we speak or write. (For a detailed discussion of the use of negation in written texts, see Tottie 1991: 24ff.)

Givón makes his statement concerning the production of negative sentences under the heading "Discourse Presuppositions of Negation," but as Horn points out (1989: 73), Givón's notion of presupposition here "is closer to the Praguean notion of GIVEN or OLD INFORMATION ... than to either the logical/semantic or the formal pragmatic ... approaches explicitly cited." (Cf. e.g. Firbas 1992; see also Prince 1981; Halliday 1985). It is precisely such a perspective that may serve to explain the occurrence of many *all... not* constructions. *All* at the beginning of a clause usually refers back to information given in the preceding context, and this old information constitutes the theme or topic of the new clause. What has been asserted or implied, what the reader or listener might have been led to assume, is summarized by *all*. Speakers or writers use *all* to refer to what they think the listeners or readers might reasonably believe at this point of the discourse, and these assumptions are then denied by means of the negator *not*. The denial of the predicate constitutes the new information, what we may call the rheme or comment according to our theoretical framework (see especially examples (17), (18), (37)–(45) and (55), (56), where extended contexts are provided).[21]

Our suggestion that *all* is usually the old information, the topic of the sentence, and that the negated proposition is the comment or new information tallies well with Chafe's (1994) two important concepts of a LIGHT SUBJECT CONSTRAINT and a ONE NEW IDEA CONSTRAINT PER INTONATION UNIT, both motivated by processing costs to the speaker. Chafe (1994: 109) proposes that an intonation unit can express no more than one new idea and that thought proceeds in terms of one such activation at a time. Similar ideas have been advanced by Givón, based on Bantu languages, (1975, quoted from Chafe), who proposes that "the information per a certain unit of *message-transaction* is restricted – say 'one unit per proposition'." Du Bois (1987: 826), working on Sacapultec, found evidence for the existence of a "one new argument constraint." Although writers with time at their disposal can obviously cram information more densely into their sentences than speakers, we would like to argue that Chafe's one new idea constraint is also valid for much writing. It seems likely that writers as well

---

[21] We take an eclectic approach – our purpose is not to vindicate any one theory, but to underline the fact that a communicative or functional perspective is useful for the understanding of the use of *all...not* constructions.

as speakers use *all...not* constructions because they are easier to produce than sentences beginning with *not all* and that there is thus a cognitive constraint operating here. The reason why *all...not* constructions are sometimes chosen instead of *no*-negation is possibly that the existential presupposition of *all* is stronger than that of *no, nothing,* etc., as suggested above.

6.3. Historical factors

It is likely that historical factors can also contribute to the explanation of the use of *all...not* constructions, especially the formulaic ones where *all* functions as a NP head (like *All ist not lost/well* etc.). We know that the post-verbal negator *not* developed in Late Old English as a reinforcer of preverbal *ne*, so that we get the series OE *Ic ne secge* > ME *I ne seye not* > late ME/early ModE *I say not*, as summarized in Jespersen's classical examples (1917: 9).[22] *Not* was normally assigned a spot after the verb; although (as shown by Laing 2002: 311f.) the order *not* (...) *ne* could occur in early Middle English, this was a "minor variant." (Cf. also Iyeiri 2005.) Laing quotes no examples of negation with the quantifier *all*, and there seems to be no other research on this subject, but it seems unlikely that *not* would appear before subject *all* until quite late in the history of English. The earliest instance of an *all...not* construction that we have come across in the existing literature is (62) from Chaucer's *Canterbury Tales*, quoted by Russell (1934: 317):

(62) But *al* thing which that shineth as the gold
 Nis nat gold ...

We therefore assume that structures like *all V not* were in use for a long time and became idioms before *not all V* would occur. This could then account for the existence of the many formulaic expressions of the type *All is not lost/well/perfect/good* in Present-Day English, and also for the use of the *all...not* construction with other complements, like those quoted in 5.3, *All is not trouble-free/going to plan/in perfect control* etc.

---

[22] As Mazzon (2004: 6) points out, the Old English form is "neither the exclusive [nor the] totally predominant type," as postulated in "standard accounts", but on the whole, Jespersen's cycle is a useful simplification.

7. Summary and discussion

Our investigation of the British National Corpus has unveiled a number of new and unexpected facts concerning the interaction between the universal quantifier *all* and the negator *not*. We have shown that in spite of the attention given to *all...not* constructions in the relevant philosophical and theoretical linguistic literature, they are not frequent in real language, with an occurrence of <5 per million words. Because of the limitations of our corpus, we were able to find few examples of spoken language, but it appears that they are slightly less frequent in speech than in writing.

One important fact that has been ignored by philosophers and most linguists is that there are three possible readings of *all...not* constructions, not just two. In addition to the wide scope reading, or NEG-Q, and the narrow scope reading, or NEG-V, there is a COLLECTIVE reading, meaning 'the sum of'. We found that in BNC, NEG-Q is the most common reading, followed by COLL, but the reading accepted by most philosophers and prescriptive grammarians, NEG-V, is the least frequent one. Admittedly, there is a fuzzy line of demarcation between NEG-V and COLL, especially where *all* occurs in collocations like *all this, all these,* but the distributions are nevertheless striking.

We also found several factors triggering the different interpretations of *all...not* constructions that have not been previously observed in the literature. Perhaps the most important one is that the interaction between the universal quantifier *all* and the negator *not* tends to function in different ways depending on the use of *all* as a NP head or as a predeterminer. When *all* functions as a NP head, the scope of negation is almost always wide, or NEG-Q; thus the reading of sentences like *All is not lost* is 'Not all is lost' in over 99% of the 162 sentences of this type that we examined. True variability is found almost exclusively in (non-formulaic) sentences where *all* is a predeterminer – in our material 32% were NEG-Q, 32% were NEG-V, and 36% COLL.

We also found that the longer and the more complex the NP that *all* is part of, the more likely the COLL reading. This is certainly due to the many formulaic examples of this type: *all the N in the world V not* is frequent. Formulaicity is also an important factor in other uses of *all...not* constructions, especially where *all* is a NP head as in *All is not well* or *All is not lost. All is/was not* constructions vastly outnumber those beginning with *not all is* in BNC. But more work needs to be done on the use of *not all* in constructions where *all* is a predeterminer.

Our research has also corroborated the assertions of previous writers that context, linguistic as well as extralinguistic, is crucial for the meaning of *all...not* sentences. Lexical and structural factors (especially coordination), extended context and pragmatic factors based on knowledge of the world can override other constraints such as the preference for NEG-Q when *all* is a NP head.

An interesting question is why *all...not* constructions are used for NEG-Q and NEG-V rather than their straightforward equivalents with *not all* or with *no*-negation, which can do the same work without incurring ambiguity. There seem to be several reasons for their use: both structural and stylistic factors. There is also a semantic factor: although NEG-V constructions can normally be paraphrased by *no*-negation, this is not always quite true. It seems to be the case that the use of the universal quantifier *all* is associated with a stronger existential presupposition than the negative universal quantifier *no/none*. Another reason for using *all...not* constructions appears to be that in NEG-V and COLL constructions, *all* can be used for emphasis, as an intensifier, rather than to express totality, which is already suggested by the definite article.

However, an important reason for the use of *all...not* constructions may also be found in discourse considerations. Both as a predeterminer and as a NP head, *all* is used to sum up old information from the preceding context as the topic or theme of a new sentence where the new information, the comment or rheme, consists of the negated predicate. It is thus necessary to take a functional perspective on the use of *all...not* constructions, and presumably also a cognitive one. We have few authentic examples of our construction from spoken language, but it seems highly likely that production constraints are also at work in the creation of written language. Thus the principle of one new idea per intonation unit invoked by Chafe is presumably valid, *mutatis mutandis*, for writing as well. Writers first use *all* to sum up old information and then introduce new information in a negative predication. (To what extent writers – and speakers – actually prefer *all...not* constructions over *not all* or *no*-negation must remain an open question at present.)

Clearly, the use of a large corpus can yield interesting – and sometimes unexpected – findings concerning quantifier-negation interaction in natural language: our discovery of the importance of the syntactic functions of *all* as well as the prevalence of formulaic uses were serendipitous finds. But what we have seen here is certainly only the tip of an iceberg, as our study was limited to constructions containing the lexical item *all*, and most of our examples came from written texts. Much more remains to be done in corpus work in this field. We need to explore the use of other lexical items serving as universal quantifiers in combination with negation: *every* and its compounds *everybody, everyone, everything* are likely to show distributions and collocation patterns differing from those of *all*, and there are likely to be register differences as well.[23] The influence of formulaic language is also likely to be different. Moreover, as pointed out above, constructions where the quantifier precedes negation must be compared with those where *not* precedes the quantifier or where *no*-negation is used.

We also need much more spoken material than that available in BNC to find out what speakers really do and to test the discourse constraints on quantifier-negation interaction. Our small spoken material suggested that speakers might be more inclined than writers to use NEG-V constructions (cf. Table 3) and that surprisingly, writers may be more inclined than speakers to use formulaic constructions; if these hypotheses hold water, that would tell us interesting things about the processing of negation and quantifiers under different conditions of language production.

A historical dimension based on research using the now increasingly available historical corpora would certainly contribute interesting information concerning the development of relevant constructions, especially the formulaic ones. As the postverbal negator *not* only developed in Late Old English as a reinforcer of preverbal *ne*, it is very unlikely that it could have appeared sentence-initially until much later; we have to assume that structures like *all V not* were in use for a long time before *not all V* could occur. This may have contributed to the existence of the many formulaic expressions of the type *All is not lost/well* in Present-Day English, and also to the use of similar structures with other complements.

Finally of course, the sky's the limit: What happens in quantifier-negative interaction in other languages is likely to differ substantially from what is likely or possible in English. All is definitely not yet known about the interaction between negation and quantifiers in natural language – not even in English.

---

[23] Neukom-Hermann (2006) showed that there are register differences concerning the use of *all...not* constructions in BNC.

References

Primary source:

The British National Corpus. http://www.natcorp.ox.ac.uk/

Secondary sources:

Anderwald, Lieselotte
    2003   Non-Standard English and Typological Principles: The Case of Negation. In *Determinants of Grammatical Variation in English*, Günter Rohdenburg and Britta Mondorf (eds.), 507–529. Berlin/New York: Mouton de Gruyter.

Anderwald, Lieselotte
    2004   Local Markedness as a Heuristic Tool in Dialectology: The Case of *Amn't*. In *Dialectology Meets Typology: Dialect Grammar from a Cross-Linguistic Perspective*, Bernd Kortmann (ed.), 47–67. Berlin/New York: Mouton de Gruyter.

Anderwald, Lieselotte
    2005   Negative Concord in British Dialects. In *Aspects of English Negation*, Yoko Iyeiri (ed.), 113–137. Amsterdam: Benjamins.

Baltin, Mark
    1974   Quantifier-Negative Interaction. In *Studies in Language Variation*, Ralph W. Fasold and Roger W. Shuy (eds.), 30–36. Washington D.C.: Georgetown University Press.

Biber, Douglas, Stig Johansson, Geoffrey Leech, Susan Conrad and Edward Finegan
    1999   *Longman Grammar of Spoken and Written English.* Harlow: Longman.

Büring, Daniel
    1997   The Great Scope Inversion Conspiracy. *Linguistics and Philosophy* 20: 175–194. 26 Jan 2004. http://www.linguistics.ucla.edu/people/buring.

Carden, Guy
    1970   A Note on Conflicting Idiolects. *Linguistic Inquiry* 1: 281–290.

Carden, Guy
    1973a   Dialect Variation and Abstract Syntax. In *Some New Directions in Linguistics*, R. W. Shuy (ed.), 1–34. Washington, D.C.: Georgetown University Press.

Carden, Guy
    1973b   Disambiguation, Favored Readings, and Variable Rules. In *New Ways of Analyzing Variation in English*, C. J. N. Bailey and R. W. Shuy (eds.), 171–182. Washington, D.C.: Georgetown University Press.

Carden, Guy
    1976   *English Quantifiers: Logical Structure and Linguistic Variation.* New York: Academic Press.

Chafe, Wallace
    1994   *Discourse, Consciousness, and Time*. Chicago/London: Chicago University Press.

Du Bois, John W.
    1987   The Discourse Basis of Ergativity. *Language* 63: 805–855.

Firbas, Jan
    1992   *Functional Sentence Perspective in Written and Spoken Communication*. Cambridge: Cambridge University Press.

Fischer, Olga
    1999   On Negative Raising in the History of English. In *Negation in the History of English*, Ingrid Tieken-Boon van Ostade, Gunnel Tottie and Wim van der Wurff (eds.), 55–93. Berlin/New York: Mouton de Gruyter.

Fitzmaurice, Susan M.
    2002   The Textual Resolution of Structural Ambiguity in Eighteenth-Century English: A Corpus Linguistic Study of Patterns of Negation. In *Using Corpora to Explore Linguistic Variation*, Randi Reppen, Susan M. Fitzmaurice and Douglas Biber (eds.), 227–247. Amsterdam: Benjamins.

Gasparrini, Désirée
    2001   *It isn't, it is not* or *it's not*? Regional Differences in Contraction in Spoken British English. Unpublished M. A. thesis. The University of Zurich.

Givón, Talmy
    1975   Focus and the Scope of Assertion: Some Bantu Evidence. *Studies in African Linguistics* 6: 185–205.

Givón, Talmy
    1979   *On Understanding Grammar*. New York et al.: Academic Press.

Halliday, Michael A. K.
    1985   *An Introduction to Functional Grammar*. London: Arnold.

Heringer, James T.
    1970   Research on Quantifier-Negative Idiolects. *CLS* 6: 287–96.

Horn, Laurence R.
    1989   *A Natural History of Negation*. London, Chicago: University of Chicago Press.

Ingham, Richard
    2006a   On Two Negative Concord Dialects in Early English. *Language Variation and Change* 18: 241–266.

Ingham, Richard
    2006b   Negative Concord and the Loss of the Negative Particle *ne* in Late Middle English. *Studia Anglica Posnaniensia: An International Review of English Studies* 42: 77–97.

Ingham, Richard
    2007   A Structural Constraint on Multiple Negation in Late Middle and Early Modern English. In *To Make His Englissh Sweete upon His Tonge*, Marcin Krygier and Liliana Sikorska (eds.), 55–67. Frankfurt: Peter Lang.

Iyeiri, Yoko
    2005   "I not say" once again: A Study of the Early History of the "*not* + Finite Verb" Type in English. In *Aspects of Negation in English*, Yoko Iyeiri (ed.), 59–81. Amsterdam: Benjamins.

Jespersen, Otto
    1917   *Negation in English and Other Languages*. Historisk-filologiske Meddelelser I, 5. Copenhagen. Reprinted 1962 in *Selected Writings of Otto Jespersen*, Allen & Unwin, London.

Jespersen, Otto
    1924   *The Philosophy of Grammar*. London: Allen & Unwin.

Kallel, Amel
    2007   The Loss of Negative Concord: Internal Factors. *Language Variation and Change* 19: 27–49.

Kjellmer, Göran
    1998   On Contraction in Modern English. *Studia Neophilologica* 69: 155–186.

Kjellmer, Göran
    2001   *No Work Will Spoil a Child*: On Ambiguous Negation, Corpus Work and Linguistic Argument. *International Journal of Corpus Linguistics* 5: 121–132.

Labov, William
    1972   *Sociolinguistic Patterns*. Reprint 1991. Philadelphia: University of Pennsylvania Press.

Labov, William
    1975   *What is a Linguistic Fact?* Lisse: The Peter de Ridder Press.

Ladd, D. Robert
    1980   *The Structure of Intonational Meaning*. Bloomington: Indiana University Press.

Laing, Margaret
    2002   Corpus-Provoked Questions about Negation in Early Middle English. *Language Sciences* 24: 297–321.

Lepore, Ernest
    2000   *Meaning and Argument: An Introduction to Logic through Language*. Oxford: Blackwell.

Liberman, Mark and Ivan Sag
    1974   Prosodic Form and Discourse Function. *CLS* 10: 416–427.

Mazzon, Gabriella
    2004   *A History of English Negation*. London: Pearson Longman.

Nelson, Gerald
   2004   Negation of Lexical *Have* in Conversational English. *World Englishes* 23: 299–308.

Neukom-Hermann, Anja
   2006   Quantifier-Negation Interaction in English. A Corpus Linguistic Study. Unpublished M. A. thesis. The University of Zurich.

Palacios Martínez, Ignacio
   1999   Negative Polarity Idioms in Modern English. *ICAME Journal* 23: 65–115.

Palacios Martínez, Ignacio
   2003   Multiple Negation in Modern English: A Preliminary Corpus-Based Study. *Neuphilologische Mitteilungen* 104: 477–498.

Prince, Ellen F.
   1981   Toward a Taxonomy of Given-New Information. In *Radical Pragmatics*, Peter Cole (ed.), 223–255. New York: Academic Press.

Pust, Lieselotte
   1998   *I cannae see it*: Negation in Scottish English and Dialect Data from the British National Corpus. *Arbeiten aus Anglistik und Amerikanistik* 23: 17–30.

Quirk, Randolph, Sidney Greenbaum, Geoffrey Leech and Jan Svartvik
   1985   *A Comprehensive Grammar of the English Language*. Harlow: Pearson Education.

Rissanen, Matti
   1999   On the Order of the Post-Verbal Subject and the Negative Particle in the History of English. In *Negation in the History of English*, Ingrid Tieken-Boon van Ostade, Gunnel Tottie and Wim van der Wurff (eds.), 189–205. Berlin/New York: Mouton de Gruyter.

Russell, Willis
   1934   The *All...Not* Idiom. *American Speech* 9 (2): 115–119.

Russell, Willis
   1935   Addenda on the *All...Not* Idiom. *American Speech* 10 (4): 316–318.

Stebbing, Lizzie Susan
   1948   *A Modern Introduction to Logic*. London: Methuen.

Stokes, William
   1974   All the Work on Quantifier-Negation Isn't Convincing. *CLS* 10: 692–700.

Tagliamonte, Sali and Jennifer Smith
   2002   *Either it isn't or it's not*: NEG/AUX Contraction in British Dialects. *English World-Wide* 23: 251–281.

Taglicht, Josef
   1985   Some Uses of All or: The King's Horses and Other Curiosities or: All is Not Well in Our Grammar. The Hebrew University of Jerusalem: 1–12. MS.

Takizawa, Naohiro
  2005 A Corpus-Based Study of the *Haven't* NP Pattern in American English. In *Aspects of English Negation*, Yoko Iyeiri (ed.), 159–171. Amsterdam: Benjamins.
Tieken-Boon van Ostade, Ingrid, Gunnel Tottie and Wim van der Wurff (eds.)
  1999 *Negation in the History of English*. Berlin/New York: Mouton de Gruyter.
Tobler, Adolf
  1902 Tout ce qui reluit n'est pas or. In *Vermischte Beiträge zur Französischen Grammatik,* 2nd ed., 190–196. Leipzig: S. Hirzel.
Tottie, Gunnel
  1983 *Much about Not and Nothing. A Study of Analytic and Synthetic Negation in Contemporary American English.* Publications of the Royal Society of Letters at Lund 1983–1984: 1. Lund: Kungl. Humanistiska Vetenskapssamfundet.
Tottie, Gunnel
  1991 *Negation in English Speech and Writing: A Study in Variation.* San Diego, New York, London: Academic Press.
Tottie, Gunnel
  1999 Affixal and Non-affixal Negation – a Case of Stable Variation over Time? In *Negation in the History of English*, eds. Ingrid Tieken-Boon van Ostade, Gunnel Tottie and Wim van der Wurff, 233–265. Berlin/New York: Mouton de Gruyter.
Walker, James A.
  2005 The *Ain't* Constraint: *Not*-Contraction in Early African American English. *Language Variation and Change* 17: 1–17.
Warner, Anthony
  2005 Why DO Dove: Evidence for Register Variation in Early Modern English Negatives. *Language Variation and Change* 17: 257–280.
Westergren-Axelsson, Margareta
  1998 *Contraction in British Newspapers in the Late 20th Century.* Vol. 102: Studia Anglistica Upsaliensia. Uppsala: Acta Universitatis Upsaliensis.
Wray, Alison
  2002 *Formulaic Language and the Lexicon*. Cambridge: Cambridge University Press.
Yaeger-Dror, Malcah, Lauren Hall-Lew and Sharon Deckert
  2002 It's *not* or *isn't it?* Using Large Corpora to Determine the Influences on Contraction Strategies. *Langage Variation and Change* 14: 79–118.

Negative and positive polarity items:
An investigation of the interplay of lexical meaning and global conditions on expression

Jack Hoeksema

1. Introduction[1]

Why do languages have such odd and complicated things as negative and positive polarity items? Surely, life would be much easier without them, and to be entirely frank, I have not yet encountered a single such item that I could not do without, if forced to. They appear to be part of the stylistic icing on the linguistic cake, adding color to texts and speech, making our daily conversations not only more complex than they need to be, but perhaps also a bit more fun. The idea that polarity items are primarily rhetorical devices has been put forward by a number of people, starting with Bolinger (e.g. 1972), and culminating in the work of Michael Israel, who has pursued this idea with great vigor in a long series of publications (Israel 1996, 1998a, 2001, 2004, to appear). Other work, with a somewhat different slant, such as Kadmon and Landman's (1993) study of *any*, also suggests that negative polarity items are primarily intended to add rhetorical spice to a statement ('strengthening').

Much of the work on polarity items has circled around issues of licensing, or triggering as it is often termed. Negative polarity items are licensed in certain environments, such as the scope of negation, and ungrammatical elsewhere, whereas positive polarity items are unwelcome ("anti-licensed" or "anti-triggered") in the scope of (at least) direct negation (cf. Horn 1989, van der Wouden 1994, 1997, Zwarts 1998, Hoeksema 2000, Szabolcsi 2004). Rather less attention has been paid to matters of lexical semantics (what types of expressions, with what kind of lexical semantics, tend to become negative or positive polarity items) and even less to numerous collocation effects that appear to interfere with the licensing of polarity items (but

---

[1] I would like to thank Larry Horn for comments and for advice well taken, as well as audiences in Berlin, Tübingen, Swarthmore, and Madison, where some of the material in this paper was presented. However, the sole responsibility for all errors and mistakes rests with me.

see van der Wouden 1994, 1997, Sailer and Richter 2002). My goal, in this chapter, is to argue that lexical semantics and collocation effects should not be ignored, as they often reveal crucial information about the expressions involved. In particular, I want to make a case for the following claim: The distribution of negative polarity items results from the interplay of lexical meaning with global conditions on the proper use of these items. In addition, I want to argue that some properties of positive and negative polarity items are best understood from the perspective of *expression*, that is, the mapping from intended meaning to meaningful form.

But first let me be a bit more specific about what I mean by global conditions. As a first illustration, let me briefly summarize Kadmon and Landman's (1993) hugely influential analysis of *any*, without any doubt the world's best-known, and most intensively studied polarity item. According to this analysis, the distribution of *any* is due to the interaction of its lexical semantics with a global constraint on acceptability. On the lexical semantic level, Kadmon and Landman view *any* as a domain widener. A combination such as *any potato* denotes roughly the same thing as *a potato*, but with a weaker contextual requirement as to what counts as a (relevant) potato, yielding a wider, larger set as the denotation. Hence *Any potato will do* is a stronger requirement than *A potato will do*, since it generically quantifies over potatoes of any stripe. If you are just a little bit hungry, a potato might do the trick, but there are, of course, potatoes too small to do much good: so you might, under these circumstances, say *I'm not very hungry, a potato will do*, but not necessarily *I'm not very hungry, any potato will do*. The global requirement on the proper use of *any* is strengthening: widening the domain, say the set of relevant items counting as potatoes, should lead to a stronger statement. Normally, widening leads to weakening, in the sense of yielding a less informative statement. For instance, *I saw an animal* is less informative than *I saw a cat,* since the latter statement entails the former, but not vice versa. Widening the predicate *cat* to the predicate *animal* makes a statement more informative only in so-called entailment-reversing or downward monotone environments, such as the restriction of a universal quantifier, the scope of negation, the complement of *without*, etc. (cf. Fauconnier 1978, Ladusaw 1979, Zwarts 1981).

The reason for calling the strengthening requirement a global one is clear: Only within the larger context of the item can it be determined whether or not an occurrence leads to strengthening.[2] It is not my inten-

---

[2] I should point out here that the requirement is actually semi-global, since the strengthening requirement need not hold for the entire sentence. Cases of polarity

tion to weigh the pros and cons of the Kadmon and Landman account here. Especially for weakly stressed occurrences of *any*, the account does not seem all that plausible (cf. Krifka 1995), but perhaps it can be maintained for stressed *any*. Stressed and unstressed occurrences of *any* have different distributions anyway (cf. Sahlin 1979 for a corpus study). For the purposes of this chapter, the interesting thing about the Kadmon and Landman approach is the way it combines lexical semantics (widening) with global properties (strengthening) to derive the distributional properties of *any*.

A rather similar approach could be taken for another type of polarity item, the very weak type identified in Giannakidou (1997, 1998, 1999, 2008), exemplified by Modern Greek *kanenas* (among others). This type of item is acceptable not only in negative contexts, but also in disjunctions, in the scope of modal and other intensional verbs, in subjunctive clauses, in the scope of epistemic adverbs such as *perhaps*, and so on. The generalization that Giannakidou offered was that the very weak items have to be in the scope of a *nonveridical operator*. An operator Op is nonveridical just in case Op(p) does not entail p. *Perhaps it rains* does not entail *it rains*, and so *perhaps* is nonveridical. Disjunction, likewise, is nonveridical: p ∨ q does not entail p. The Greek weak polarity items of the *kanenas*-series are all existential pronouns. It is not too far-fetched to assume that the distribution of these items results from the interplay of lexical meaning (existential quantification) and a global requirement of nonreference: the item may not appear in a context permitting existential generalization. Only nonveridical contexts provide the kind of shelter where such items are safe from the onslaught of existential commitment.

In this chapter, I will pursue the general hypothesis that the interaction of global constraints with lexical semantics lies at the root of the often odd-seeming distributional patterns of polarity items. I do not, however, want to suggest that this interaction is the only explanatory factor. Rather, one would have to stress that distribution usually is the complex result of syntagmatic factors (such as the aforementioned type of interaction) and paradigmatic factors, in particular the competition of other items. Quite often,

---

items embedded under double negation are possible, even though the two negations cancel out (cf. Hoeksema 1986), and English *any* is no exception in this regard, cf. e.g.

(i) I can't believe that you didn't bring any money.

Kadmon and Landman (1993) maintain that it suffices for the acceptability of *any*, that it strengthens an utterance in some intermediate domain, say, in the case of example (i), that of the embedded clause.

items have odd 'holes' in their distribution, due to the interference of alternative ways of expressing a message (cf. Hoeksema 1999, Pereltsvaig 2004, de Swart 2006). Some kind of blocking account, perhaps in an OT vein, would have to be assumed here. To mention just one case, German *auch nur* 'even' is a polarity item occurring in all the usual environments, except for one: direct negation (König 1981). Actually, even direct negation is OK, unless the negator *nicht* 'not' is strictly adjacent to *auch nur*, as the examples in (1) below show:

(1) a. *Keiner hat es auch nur vermutet.*
Nobody has it even suspected
'Nobody even suspected it'
b. *Sie hat es nicht mit auch nur einem Wort erwähnt*
She has it not with even one word mentioned
'She did not mention it with so much as one word'
c. **Sie hat es nicht auch nur vermuten können.*
She has it not even suspect can
'She could not even suspect it'

Instead, German has a dedicated element *(ein)mal* 'even' for use in direct negation contexts:

(2) *Sie hat es nicht einmal vermuten können.*
She has it not even suspect can
'She could not even suspect it'

In all other contexts, *einmal* is ruled out, or rather, it reverts to its original meaning 'once':

(3) *Keiner hat es einmal vermutet*
Nobody has it once suspected
'Nobody has suspected it once'

I assume that a blocking principle such as the Elsewhere Condition (Kiparsky 1973) will have to be invoked here. The item with the more limited distribution and the more specialized use will oust the item with the wider, less specialized distribution from its sphere of influence.

I also assume, furthermore, that polarity items are the product of a process of grammaticalization (Hoeksema 1994, 1998a).[3] Many of the typical

---

[3] I also assume that once a certain type of polarity item has been established, further items may be added by a process of lexical replacement or calquing.

properties associated in the literature with the phenomenon of grammaticalization are found here as well:

(4) – layering (grammaticalized forms next to nongrammaticalized forms)
     – semantic bleaching
     – pragmatic strengthening
     – unidirectionality

Layering, in particular, is extremely prevalent. We have already seen an example of this above: German *einmal* 'once < one time' has developed an additional use as a scalar particle meaning 'even' under negation. But the original interpretation is still around. For the automatic detection of polarity items from corpora, this is a major problem (cf. Hoeksema 1998b, Lichte and Sailer 2004, Lichte and Soehn, 2007 for some discussion).

Semantic bleaching and pragmatic strengthening are likewise easy to identify. For bleaching, consider e.g. the use of swear words as polarity items, as in *Fred did not do bugger all*. Here the literal meaning has bleached to something like 'anything', while pragmatically, it has become a standard marker of affective import on the part of the speaker.

No doubt the most problematic property among the ones in (4), also in grammaticalization circles, is the last one: unidirectionality (cf. Hoeksema 1998a, and for more general discussion of the unidirectionality requirement, e.g. Traugott 1990, Traugott and Heine 1991, Janda 2000, Fischer et al., 2004).

A final assumption I will be making, and partly motivating, in this chapter is that polarity items may take part in larger constructions, and that their distributional behavior should be viewed from the perspective of these larger constructions. This is an assumption in line with much of the work by Michael Israel, and one that needs to be worked out in far more detail than can be done here.

The structure of the remainder of this chapter is as follows: In Section 2, I present some case studies of negative polarity items, instantiating different types of global conditions on acceptability, in Section 3, I discuss a number of positive polarity items from the same general perspective, and in Section 4, I present my conclusions.

E.g. early-modern Dutch *wat duivel* 'what devil' is based on French *que/qui diable*. Taboo expletives seem especially likely to undergo lexical replacement (Hoeksema and Napoli 2008).

2. Global constraints on Negative Polarity Items

2.1. The case of *ever* and *any*

The analysis of *any* by Kadmon and Landman (1993) combines global strengthening with local widening. As we have seen above, this yields the result that the item must appear in downward entailing contexts. As noted by Jackson (1995), such an account does not explain the so-called intervention effects on the licensing of *any*, discussed in Linebarger (1981, 1987) and Szabolcsi and Zwarts (1993), among others. When certain operators, such as universal quantifiers, scopally intervene between a polarity item and its trigger, the result may be degraded (cf. example (5a) below). Intervening indefinites, such as *a teacher* in example (5b) below, do not have this effect, however.

(5)   a. ?No student gave every teacher any apples.
      b. No student gave a teacher any apples.

Both occurrences of *any* in the examples in (5) appear in monotone decreasing contexts. To see this, simply note that (5a) and (5b) entail the sentences in (6a) and (6b), respectively: [4]

(6)   a. ?No student gave every teacher any red apples.
      b. No student gave a teacher any red apples.

Hence accounts such as those of Ladusaw (1979) or Kadmon and Landman (1993) have problems in explaining the intervention effects. Linebarger (1981) takes the data to be evidence for an adjacency requirement at Logical Form: polarity items may not be separated from their negative triggers by intervening quantifiers. However, Linebarger's account does not say much about polarity items that do not suffer from intervention effects, such as the modal auxiliary *need* and the adverb of degree *all that*:

(7)   a. None of us need worry about the police.
      b. Not every student need participate in the presentation.
      c. *Some of us need worry about the police.

(8)   a. None of us were all that pleased with the result.
      b. Not everybody was all that pleased with the result.
      c. *Some of us were all all that pleased with the result.

---

[4] Note that (5a), while degraded, seems to be interpretable, hence the possibility to draw entailments from it.

Also, that account does not explain why the constraint should exist. Adjacency requirements at Logical Form are not exactly a common type of phenomena in the literature on semantic restrictions.

In Jackson's view, the intervention effects follow from the fact that items such as *any* are indefinites with a global requirement, which is that they must appear in *general statements*. General or universal statements are statements with the property that they are easy to falsify. A single counterexample suffices to falsify the statement *Jones does not have any apples*. In combination with the assumption that *any* is essentially an indefinite, interpreted as an existential quantifier, this means that *any* may only be used in negative environments. Otherwise, an existentially-quantified statement would arise, and existentially-quantified statements are easy to verify, but hard to falsify. The sentence *Jones has an apple* requires us to consider all pairs consisting of Jones as its first member and *some apple* as the second member as possible members of the denotation of *have*. The sentence is not falsified until we have considered, and rejected, every such pair. Sentences involving free choice *any*, being universal in nature, are likewise easy to falsify.[5]

More precisely, we require that some subformula of a sentence in which an indefinite polarity item such as *any* or *ever* is embedded, is easily falsi-

---

[5] The status of free choice *any* as a universal is not undisputed. In the words of Vendler (1967): 'The meaning of *any* is a many-splendored thing.' In imperatives, there is nonequivalence with regular universal quantifiers: *Take any apple* ≠ *Take every apple*. Yet, as Vendler (1967: 79) states, "there is some generosity left in this offer too: generosity in the sense of generality". The offer holds for each and every apple, and so in a real sense, has a universal flavor. Geach (1962) offers the example *Tom can lawfully marry any sister of Bill's* as clearly differing in truth conditions from *Tom can lawfully marry every sister of Bill's*. Remarks to the same effect are to be found in Jennings (1994), Horn (2000a, b), Giannakidou (2001), as well as van Rooij (2008). Many authors observe that imperatives such as 'Pick any card' are not true commands, but permission-granting statements, equivalent to 'Take a card, no matter which one.' It appears, though, that pragmatic factors may sometimes yield true universal commands: consider an imperative like *Correct any spelling-mistakes before handing in your assignments*. Clearly, this is not a case of free-choice permission, where students may pick their favorite spelling mistake in order to correct it. Most likely, the pragmatic oddness of such a request blocks a free-choice permission reading. Dayal (1998) and van Rooij (2008) explicitly argue that true commands and *must* statements do not contain occurrences of FC *any*. The above example shows that they are wrong about this. Compare also: *You must correct any spelling-mistakes before handing in the assignment*.

fied (by a single counterexample, or else by a relatively small set of counterexamples – see below). This subformula may equal the entire sentence, but also a proper subpart of it, in light of the fact that double negation does not generally appear to block the licensing of polarity items (cf. fn. 1 above). Compare e.g.:

(9) a. I can't believe that nobody has ever heard of this painter.
    b. Nobody could believe that he did not want any alcohol.

The acceptability of the sentences in (9) is due to the fact that subformula's corresponding to the embedded clauses are general statements. Note, by the way, that the requirement of generality is a global one. Whether an indefinite takes part in a general statement depends on its context. In the scope of negation, an indefinite takes on universal force, but not when there is an intervening universal quantifier. The sentence *Not everybody has a cat* is verified easily: a single non-cat-owning person suffices as a counterexample, but it is hard to falsify: we have to consider every pair of a person and a cat. Only after we have satisfied ourselves that each and every such pair is in the denotation of the verb *have*, can we be sure that the statement is false.

A slight complication of the account is necessary in order to take care of cases where the sentence is not strictly universal:

(10) a. Few people ever crossed Antarctica on skis.
     b. Fewer than 5 people have ever climbed Mount Erebus.

Here, we need to weaken the requirement of generality as follows: a statement is a general statement iff it is easily falsified, and a statement is easily falsified iff it is falsified by a small number of counterexamples. What counts as small is of course context-dependent (cf. e.g. Westerståhl 1985 for discussion). If the quantifier *few people* in (10a) is understood as less than n, where n is some relatively small number n, the statement in (10a) is refuted by n counterexamples, still a relatively small number. In the case of (10b), 5 counterexamples suffice to falsify (10). As supporting evidence for this move, Jackson notes the possibility of using non-downward entailing quantifiers just in case there is an implicature of 'smallness' (the observation is originally due to Linebarger 1987):

(11) ?Exactly four people in the world have ever read that dissertation: Bill, Mary, Tom and Ed.

When, however, the number of counterexamples is not that small, even downward entailing quantifiers become less acceptable (Jackson 1995: 196), as the difference between the following two sentences suggests:

(12) a. At most one hundred Americans have any children.
b. ?At most one hundred people in this room have any children.

Like the account in Linebarger (1981), Jackson's theory makes a principled distinction between triggering by *few* and *less/fewer than 5* and triggering by *not all* or *not everybody*. The former group is acceptable for existential polarity items such as *any* and *ever*, the latter is not. This contrasts with the theory of Zwarts (1981, 1986, 1996), which treats *few* and *not all* on a par as being monotone decreasing but not anti-additive quantifiers. A noun phrase X is monotone decreasing iff for all predicates Y, Z we have:

(13) X Y or Z → X Y and X Z (X is monotone decreasing)

A noun phrase X is anti-additive iff

(14) X Y or Z ↔ X Y and X Z (X is anti-additive)

Monotone decreasing inferences are indeed generally valid for *few* and *not all*, but anti-additive inferences fail:

(15) a. Few people have climbed Mount Erebus or have climbed Mount Sidley → Few people have climbed Mount Erebus and few people have climbed Mount Sidley
b. Not everybody has climbed Mount Erebus or has climbed Mount Sidley → Not everybody has climbed Mount Erebus and not everybody has climbed Mount Sidley.

Why are the implications not valid from right to left? In the case of (15a) it is because it may be the case that the number of people having climbed Mount Erebus is small, and that the number of people having climbed Mount Sidley is likewise small, but that the two sets are disjoint and their combination just large enough to no longer count as small. In the case of (15b), just consider the case where one half of all people in the domain of discussion climbed Mount Erebus and the other half Mount Sidley. Then not everybody climbed Mount Erebus and not everybody climbed Mount Sidley, but everybody climbed either Mount Erebus or Mount Sidley.

In this connection, it is interesting to see which theory makes the better predictions. It is dangerous to rely on introspective judgments alone, since they tend to be fragile and subtle, especially regarding rare combinations of polarity items and triggers. Many years of classroom experience in teaching polarity issues have taught me how easy it is to find disagreement about the acceptability of examples from the literature. Therefore I have made an effort to supplement introspection with corpus data. In the case of English *ever*, I have collected examples of polarity-sensitive uses (excluding non-polarity uses such as *he was ever so smart* or *she was ever the lady* – cf. Israel 1998b for a discussion of the various uses of *ever*). For Dutch, I have a large collection of polarity-sensitive *ooit* (for the distinction between polarity-sensitive and nonsensitive *ooit*, cf. Zwarts 1995, Hoeksema 1998a) and for German *je/jemals* (excluding cases of nonsensitive *je* – that is, occurrences of so-called binominal *je* which functions as a distributive operator in German – cf. *die Mädchen hatten je drei Bücher* 'the girls had three books **each**'). The fact that each of these three expressions has other uses besides the polarity-sensitive use in which I am interested indicates that it won't be easy to mechanically extract data from a corpus such as the British National Corpus. It has therefore been necessary to hand-select the relevant data. I have culled data from newspapers, journals, books, TV, and electronic data sources, and classified each example in a number of ways, such as the kind of context or trigger that licenses the occurrence of the polarity item. In Table 1 below, I provide some data on occurrences of *ever, ooit, je(mals)* with weak triggers.

Table 1. EVER [N=3728], Dutch OOIT [N=17304] and German JE/JEMALS [N=838] in combinations with weak triggers

CONTEXT/TRIGGER	EVER	%	OOIT	%	JE(MALS)	%
Few/little	39	1	46	0.2	6	1
Not much/many	5	0.2	2	0.01	–	–
Seldom	1	0.03	5	0.03	–	–
Neg+universal	3	0.1	–	–	–	–

From this table, it appears that negated universal quantifiers are indeed an extremely rare group of triggers for this set of polarity items, in accordance with Jackson's account.[6] In fact, the three occurrences that were found for English were the result of googling for various strings. It is well-known that

---

[6] Jackson's account for *any* and *ever* covers a great deal of ground, but is probably only part of a larger puzzle. There are many expressions involving *any* which have idiomatic status, and a different distribution, such as *at any rate* or *in any event*. The expression *in any way* is a stricter polarity item that *any*, since it lacks a free choice reading. The free choice reading, available for (i) below, is lacking for (iv):

(i) John is interested in any sport.
(ii) John is not interested in any sport.
(iii) John is not in any way interested in sports.
(iv) *John is in any way interested in sports.

Another, more serious, type of problem for Jackson's account is what I would like to call *noncommittal any*. Unlike free choice *any* (but see fn. 5), this use of *any* does not appear to have universal force, yet it occurs in various nonveridical, but not downward entailing environments:

(v) I used to stare at this photo for minutes at a time, trying to detect within it any evidence of the trauma of the previous week (Nick Hornby, *Fever Pitch*, Riverhead Books, New York, 1998, p. 28).
(vi) I'd appreciate any comments on this paper.

Accounts that favor some kind of strengthening, such as Kadmon and Landman (1993), also fail to come to grips with such examples, which are, admittedly, relatively rare, but by no means marginal. The reason I would like to call this 'noncommittal *any*' is that the import of *any* seems to be to highlight the fact that no existential commitment is made by the speaker. Of course, if we have to admit something like a special noncommittal use, separate from free choice or negative polarity uses, the prospects for a unified theory of *any*, as advocated in various ways by Quine (1960), Partee (1986), Kadmon and Landman (1993), and Horn (2005), among others, are rather dim. (Some measure of 'splitting' will have to be admitted anyway, even by the most generous 'lumpers', given the existence of *adverbial any*, exemplified by *Is she any good? Can you be any more insulting? I don't want to discuss this any further.* Adverbial *any*, for starters, does not have a free choice use, as discussed in Horn 2000b. The fact that *any* is used as an adverb of degree is not surprising, given that more indefinite pronouns are used in that way, e.g. *somewhat, a bit*, but does not follow from its use as a determiner. Many other determiners/pronouns cannot be used as adverbs of degree, compare e.g. *She was somewhat/*something annoyed at his suggestions.*)

Google provides examples for just about anything, including ungrammatical or borderline constructions. For *any*, my data show a similar pattern:

Table 2. Weak triggers for ANY

CONTEXT/TRIGGER	ANY N=6600	%
Few/little	25	0.4
Not much/many	3	0.05
Seldom/rarely	20	0.3
Neg+universal	1	0.02

Google data make the point even stronger. The string *not all of us had any* appears once on the World Wide Web, according to Google, whereas there were 10 hits for *not many of us had any* and 220 actual hits (out of an "estimated" 17, 900) for *few of us had any*. For the string *not every one of us had any*, no hits were found. The preference for *few* over *not many* that we see here, as well as for *ever, ooit, je*, is most likely due to the fact that negative antonyms make for stronger statements, pragmatically, than negated forms of positive antonyms (cf. Horn 1989, Levinson 2000, Blutner 2002 for discussion).

At this point, the skeptical reader may object that the difference shown in Table 1 between weak existential environments such as *few/little* on the one hand and negated universals on the other hand could simply be due to the fact that the latter type of environment is less common anyway. While I do not have hard quantitative data to settle the matter once and for all, I doubt that we could explain away the difference along these lines. Negated universal quantifiers are not that rare, and we should find some in the huge sets of data collected for Dutch and German, but none were found. Among other negative polarity items, by contrast, triggering by negated universal quantifiers is by no means a rare phenomenon. In the next section, I will discuss a set of expressions that illustrate this very point.

2.2. Need, hoeven, brauchen

English, Dutch and German all have a polarity-sensitive modal verb denoting obligation (cf. van der Wouden 2001, van der Auwera and Taeymans, to appear). Although the three verbs are not etymologically related, they have a strikingly similar distribution, as we will see. In English, this modal verb is *need*, as exemplified in (16). Note that main verb *need*, which combines with infinitival *to*, is not a polarity item (cf. 16e).

(16) a. You need not worry.
 b. *You need worry.
 c. You need worry about nothing.
 d. *You need worry about your grades.
 e. You need to worry about your grades.

In Dutch, the polarity-sensitive modal verb is *hoeven* (in written Dutch also *behoeven*). There is also a main verb *behoeven* which is not polarity sensitive.

(17) a. *Je hoeft niet meedoen.*
  You need not collaborate
  'You don't have to collaborate'
 b. *Dat behoeft toelichting.*
  That needs explanation
  'That requires explanation'

In German, finally, the polarity-sensitive modal auxiliary verb is *brauchen*. Again, this verb also has a separate use, as a main verb, which is not polarity-sensitive:

(18) a. *Du brauchst nicht anzurufen.*
  You need not call
  'You don't have to call'
 b. *Wir brauchen mehr Zeit.*
  We need more time
  'We need more time'

In Table 3, I compare corpus data for the three modal verbs. Note that the difference between negated universals and expressions such as *few, little, seldom* that we found for *ever, ooit, jemals* in the preceding section, does not show up here: both types of weak triggers are represented about equally well. If we interpret Table 3, as seems reasonable, to show that there is no restriction against the modal verbs in negated universal contexts, we might venture the guess that the two types of weak triggers are roughly equally common. Hence the possibility, raised in section 2.1., that perhaps the asymmetry between negated universal quantifiers and weak existential quantifiers observed for *ever* and its counterparts might be due to a difference in frequency between the two types of triggers, should be rejected.

Table 3. Distributional data for English *Need* (N=418), Dutch *Hoeven* (N=1576), German *Brauchen* (N=380) [the fixed expression *if need be* was left out of the selection]⁷

CONTEXT/TRIGGER	NEED	HOEVEN	BRAUCHEN
Before-clauses	–	–	–
Comparatives	2%	1%	–
Conditional clauses	–	0.5%	–
Few/Little	1%	2%	1%
Hardly/scarcely	4%	1%	2%
N-word	14%	22%	25%
Not	60%	53%	51%
Negated universal⁷	2%	2%	2%
Negative predicate	–	–	–
Question	3%	2%	1%
Restrictive adverb	13%	14%	17%
Universal	2%	1%	1%
Without	–	3%	1%

What is perhaps most striking about the distributional data in this table is the relative prominence of the category *restrictive adverb*. In English, the main restrictive adverb is *only*, in Dutch the set is larger (*alleen, slechts, maar, alleen maar, enkel, pas, uitsluitend*), while German is intermediate between Dutch and English in this respect (*nur, bloss, lediglich*). Some examples of *need* etc. with restrictive adverbs are:

(19) You need bring only one thing.

(20) Je hoeft maar één ding mee te brengen.
You need but one thing along to bring
'You need bring only one thing'

---

[7] The negated universal context in Table 3 should be understood as the scope, not the restriction, of a negated universal quantifier, as in *Not everybody need be present*. The possibility of negation is clearly due to the presence of negation, since the universal quantifier by itself does not license polarity items in its scope. The universal context in the table, however, has to be understood as the restriction of the quantifier, being a legitimate context for at least some polarity items. Compare examples (25–27) in the main text.

(21) Du brauchst nur eine Sache mitzubringen.
You need only one thing along-to-bring
'You need bring only one thing'

For most negative polarity items, restrictive adverbs are not a particularly common type of trigger. So why are these modal verbs different in that respect? It seems that this may be due to the general requirement on the use of the modals: while their lexical semantics is that of a deontic necessity operator, their global requirement appears to be that they are used in the expression of a *weak requirement*.[8] Negation, of course, yields the weakest requirement. If you are told 'You need not come', then what you are required to do is the absolute minimum: nothing at all. If you are told 'You need do very little', the requirement is still rather minimal. The same is true, arguably, if you are told 'You need bring only one thing'. Here, there is an implicature that you don't have to do much. Various collocations reinforce this point. A common combination in English and Dutch is 'one need not look far'/ 'niet ver hoeven te zoeken.' Not having to look far leaves only your own vicinity to explore, which is clearly a weaker requirement than having to look far. The notion of a minimal requirement may also be used to explain an otherwise odd fact about universal quantifiers. Ordinarily, *need – hoeven – brauchen* are not acceptable in the restriction of a universal quantifier:

(22) *Every student that need worry will be informed.

(23) *Elke student die zich zorgen hoeft te maken, wordt geïnformeerd.
Every student who self worries need to make, become informed
'Every student that need worry will be informed'

(24) *Jeder Student der sich Sorgen zu machen braucht, wird informiert.
Every student that self worries to make needs, becomes informed
'Every student that need worry will be informed'

However, when the quantifier is *all* or its Dutch or German counterpart, the result is fine:

---

[8] As before (cf. footnote 2), the global condition is not required to hold of the entire utterance, but may also just characterize a subformula of the utterance, e.g. in *Don't think you need not worry*. Here the clause *you need not worry* expresses a weak requirement, but the same is not true, necessarily, of the entire utterance.

(25) All you need know is in this booklet.

(26) *Alles wat je hoeft te weten staat in dit boekje.*
All that you need to know stands in this booklet
'All you need know is in this booklet'

(27) *Alles was du zu wissen brauchst, steht in diesem Büchlein.*
All that you to know need stands in this booklet
'All you need know is in this booklet'

Sentences such as (25)–(27) carry an implicature that is absent in (22)–(24), namely that nothing else is required. Hence the requirement might be considered light. When we add material to eradicate this implicature, the acceptability of (25)–(27) disappears:

(28) a. John told me all I have to know, and it's a lot!
b. John told me all I need know, #and it's a lot!

It remains to be seen why negated universal quantifiers may trigger *need* and its counterparts. Clearly a statement such as *Not everybody need come* expresses a requirement which is not necessarily small. However, modal commitments are often given with respect to a background set of assumptions. A statement such as *Not everybody need come* is most natural when content of the statement *Everybody has to come* is under discussion or somehow assumed. Compared to that strong requirement, 'not everybody need come' may be interpreted as relatively mild.

Questions involving *need* often involve some rhetorical interpretation involving, once more, some implicature of minimality:

(29) Need I say more? [implicature: nothing more need be said]

By contrast, a statement such as (30) is decidedly odd:

(30) Need I speak to all the delegates separately?

Here, there is no suggestion of a minimal requirement, and the more natural way of expressing the meaning of (30) would be something like

(31) Do I need to speak to all the delegates separately?

3.3. Mouse, chicken and dog in Dutch

Many European languages have polarity-sensitive animal names that refer to humans. For example, French *pas un chat* (lit. 'not a cat') means 'not a living soul, nobody' (von Bremen 1986), likewise the standard Dutch of Flanders *geen kat* 'no cat', standard (northern) Dutch *geen hond* 'no dog', German *kein Schwein* 'no pig' and Danish *ikke en kat* 'not a cat' (Jespersen 1917).

Particularly interesting is the situation in standard Dutch, where besides *geen hond* 'no dog', there are two other expressions in use: *geen kip* 'no chicken' (cf. the English "Just us (little) chickens" as in a response to "Who's there?") and *geen muis* 'no mouse'. The three expressions are not used interchangeably, but have become, to a very high degree, differentiated. The expression *geen kip* is typically used to indicate the absence of humans from a scene. *Geen muis* has a scalar use in contexts where size plays a role. *Geen hond*, finally, seems to be the default case, for use elsewhere. It is possible, in connection with this three-way division, to distinguish three general kinds of contexts: (1) existential contexts and complements of *see*; (2) scalar contexts (often involving the use of the modal *kunnen* 'can', which often adds a scalar flavor to a predicate; and (3) all other contexts.

Existential sentences and complements of *see* have in common that they describe some scene. Typical examples involving *geen kip* 'no chicken' are:

(32) a. *Er   was geen kip   op straat.*
There was no   chicken on street
'There wasn't a soul on the street'

b. *Er   was geen kip   te bekennen.*
There was no   chicken to discern
'There wasn't a soul to be seen'

c. *Ik heb geen kip   gezien.*
I have no   chicken seen
'I haven't seen a soul'

At the lexical level, *geen kip* simply means 'nobody', but its global requirement forces it to appear in contexts where an entire scene is characterized by the absence of humans. Sentences such as (33), where this is not the case, while not entirely impossible for all speakers, are quite unusual, as we will see:

(33) %*Geen kip sprak met de kinderen.*
No chicken spoke with the children
'Not a soul spoke to the children'

Scalar uses typical of *geen muis* involve situations where either a place is so packed, nobody could get in, not even a mouse, or where a barrier is so impenetrable, that nobody could get through, not even a mouse. A typical example would be

(34) *De politie laat vanavond geen muis meer door.*
The police let tonight no mouse anymore through
'The police won't let anybody through tonight'

Finally, an example involving *geen hond* 'no dog':

(35) *Geen hond gelooft dat jij onschuldig bent*
no dog believes that you innocent are
'No one believes that you are innocent'

In the following table, some corpus evidence is presented to show the contextual specialization of the three polarity items:

Table 4. Mouse, Chicken and Dog

TERM	N	% SCALAR	% EXISTENTIAL	% OTHER
muis	30	94%	3%	3%
kip	109	5%	78%	17%
hond	206	–	33%	67%

The distribution of *geen kip* is strongly reminiscent of that of *pas un chat* in French, as described in von Bremen (1986: 242–243). Von Bremen offers the following examples plus judgments:

(36) a. *Il n'y a pas un chat ici.*
There is not a cat here
'There's not a soul here'

b. *Je n'ai pas vu un chat dans le magasin.*
I not have seen a cat in the store
'I haven't seen a soul in the store'

    c. *?Pierre est très gentil. Il ne heurtera pas un chat.*
       Pierre is very kind. He not-would-hurt a cat
       'Pierre is very kind. He would not hurt a soul.'
    d. **Jean est très égoïste.* Il n'aide pas un chat.*
       Jean is very egotistical. He not-would-help a cat
       'Jean is very much an egotist. He would not help a living soul.'
    e. **Il n'a pas parlé à un chat.*
       He not-has spoken to a cat
       'He has not spoken to a cat'

As von Bremen concludes: "L'emploi le plus naturel d'*un chat* est dans les *propositions existentielles.*"

3.4. The likes of which

One of the most unusual polarity items is the English expression **the likes of which.** This expression heads relative clauses which express some kind of superlative quality by stating that something or somebody does not have an equal. Some typical examples are given in (37):

(37) a. a war, the likes of which the world has never known
      b. a genius, the likes of which we will never see again
      c. a linguistic oddity, the likes of which we are unlikely to find in any other language

In Table 5 below, my corpus data are presented. As one can see, it looks as if the expression has not yet fully grammaticalized into a polarity item, given the 6% positive occurrences. However, the predominance of negative environments is strong enough to convince us that we are dealing with a polarity-sensitive core in the use of this expression. In the right hand part of the table, the various predicates that cooccur with *the likes of which* are listed. Notice the predominance of verbs of experience, especially verbs of perception (which I take to indicate a subtype of experience) [cf. also "NP's like(s)", as in *never saw his like (again),* also an NPI, usually with *see* or *know*].

Table 5. The likes of which [n=72]

TRIGGERS		PREDICATES	
not	32%	encounter	3%
n-word	56%	experience	7%
negative predicate	1%	hear	4%
restrictive adverb	1%	imagine	3%
rarely/seldom	3%	see	60%
positive occurrences	6%	other	23%

There is some overlap here with the use of *kip* and *chat* discussed in the previous section, but there is also a major difference: while the latter expressions are almost always used to characterize some scene, the use of *the likes of which* is typically for the characterization of larger settings, such as world history, or the combined experience of an entire group of people, etc., in line with its superlative nature. So our general characterization of this expression is that a maximal degree is indicated indirectly through *nonexistence* of an equal. Nonexistence in turn is expressed indirectly by means of predicates of perception/experience. This might be viewed as a collocational effect. Only occasionally did I find existence predicates with this construction, such as *exist* or its fancier counterpart *walk the earth*.

Other examples where superlative degree is expressed by the lack of an equal abound, but usually, these involve some light verb such as *have* or *know*, such as Dutch *zijns gelijke kennen* 'know one's equal', *zijn weerga hebben* 'have one's counterpart', German *seines-gleichen haben* 'have one's equal' or *seinesgleichen suchen* 'seek one's equal = not to have an equal'. Notice, by the way, that *seinesgleichen haben* is a negative polarity item, whereas *seinesgleichen suchen* is a positive polarity item. Both items have the same global property of expressing a superlative degree by denying the existence of an equal.

3.5. English *in sight*

In section 3.3. we saw some expressions that serve to characterize absence of humans from a scene. Another scene-characterizing expression is English *in sight*. While this prepositional phrase does not appear to be a polarity-item if one looks at the entirety of its uses, some striking correlations emerge when we distinguish three subcases: sentences with definite subjects (not polarity sensitive), sentences with indefinite subjects (polarity sensitive) and use of *in sight* as a nominal modifier. These use are illustrated by the following examples:

(38) a. The end is (nowhere) in sight.
b. There was *(not) a student in sight.
c. Shoot every/*some opponent in sight.

Besides such sentences, *in sight* is very often used in absolute constructions involving the prepositions *with* and *without*:

(39) a. The war just went on and on with no end in sight.
b. Without an end in sight, the war just went on and on.

Here, too, indefinite subjects are typically associated with negation, but definite subjects are not:

(40) With the end of the war in sight, it was time to divert the public attention to other matters.

In Table 6, some newspaper data are presented.[9] Sentences with occurrences of *in sight* were classified according to subject type and adjunct status. Indefinite subjects were strongly associated with negation, adjuncts with universal quantifiers.[10] Definite subjects were not associated with negation in a strong way, although we may find some polarity sensitivity here, too, if

---

[9] The data in Table 6 are from LexisNexis, from the newspapers Los Angeles Times, the Observer, the Sydney Morning Herald and the Toronto Star. I searched for the string "in sight" in all articles from July 20, 2008 – January 20, 2009 (a 6-month period). From the original 406 hits, a fair number had to be discarded, as they contained occurrences of the light-verb combinations *keep in sight*, *have in sight*, or the string "in sight of" (which behaves differently from *in sight* as in (39)), as well as a few hard-to-classify cases in headlines.

[10] The fact that adnominal *in sight* occurs with universally quantified noun phrases might be seen as a consequence of its negative-polarity status. The restriction of a universal quantifier is, after all, one of the familiar contexts of polarity items. Note, however, that adnominal *in sight* appears to be less felicitous in other types of triggering environments:
(i) *Jones did not want to shoot a student in sight.
(ii) *Did Jones shoot a student in sight?
(iii) *If Jones shot at a student in sight, he would not have missed him.
This suggests that adnominal *in sight* cannot be simply regarded as a straightforward polarity item.

we distinguish further among subcases. Positive definite subjects are often instantiations of a few highly frequent combinations: the end is in sight, the goal is in sight, the final solution is in sight, etc. With other types of subject, it appears that negation is preferred:

(41) a. Jones was nowhere in sight.
     b. ??Jones was in sight.

As for the indefinite subjects, many appeared in absolute constructions such as *with no end in sight; with not a living soul in sight; without a solution in sight* etc. Especially common is *with no end(ing) in sight*.[11]

Table 6. *in sight* (raw numbers, not percentages)

	POSITIVE	NEGATIVE	UNIVERSAL	THE ONLY/ SUPERLATIVE
DEFINITE SUBJECT	36	15	–	–
INDEFINITE SUBJECT	20	191	–	–
ADJUNCT	–	–	19	4

It appears from these data, therefore, that *in sight* should be viewed in the context of the larger constructions it partakes in, before one can establish its polarity-sensitive nature in some of these contexts.

The difference between definite subjects and indefinite subjects may well be linked to veridicality and reference (cf. the work of Giannakidou 1998, Zwarts 1995 for discussion). A sentence like *A solution is in sight* entails that a solution does not (yet) exist. We might render it in predicate

---

[11] It is perhaps interesting to note that *without an end in sight* does not block *with no end in sight*, in spite of the fact that phrasal expression is often blocked by lexical expression (cf. Poser 1992). See Ackema and Neeleman (2001) for a different account of blocking, which states that syntax trumps morphology: if all things are equal, use a syntactic construction over a morphological one. What predictions this latter account makes for cases of complex lexical items such as *without*, which are not morphologically regular, versus *with no*, is unclear. The prospects for a general theoretical account of blocking are a bit bleak anyway, given that *with no* is not blocked by *without* (cf. *with no money down* alongside *without any money down*), whereas its Dutch counterpart *met geen* 'with no' is blocked by *zonder* 'without': cf. **met geen eind* 'with no end' versus *zonder eind* 'without end.'

logic as Op [∃x [solution(x)]], where Op is some kind of nonveridical operator such as "it is to be expected that"/"it is likely that it will be the case that. " The 20 positive occurrences of *in sight* consist of 7 cases of absolute constructions involving *with* ("with an end in sight, ..."), and 13 finite clauses. Of the 13 finite clauses, 10 had additional nonveridical operators, such *perhaps, hope, if, appear, confident that*, etc. This seems rather a lot, and may be a sign that the nonveridicality associated with *in sight* has to be bolstered by the presence of these additional elements. At the same time, the three positive occurrences without additional nonveridical elements suggest that this is merely a tendency, not a rule of grammar.

3.6. Litotes constructions

Litotes constructions are a common source of negative polarity items (Horn 1991, van der Wouden 1996, 1997, Israel 1998). The Dutch examples in (42) show a typical case of a negative member of an antonym pair, employed in a litotes construction of double negation to express something mildly positive:

(42) a. *Hij schaakt niet onverdienstelijk.*
 He chesses not without-merit
 "He plays chess rather well"
 b. **Hij schaakt onverdienstelijk.*

The lexical meaning of *onverdienstelijk* is 'without merit, lacking value', and the global requirement on its use is that it should state something mildly positive. This is possible only in the context of negation. In other types of environment other polarity items may be just fine, but litotes expressions tend to shy away from them. They only appear in contexts that are narrowly negative. For instance, conditionals or questions are out of the question for *onverdienstelijk*. In a set of 82 occurrences of *onverdienstelijk*, 91% cooccurred with *niet* "not, " 9% with *geen* "no". That means that a full 100% occurred in a negative sentence. This type of distribution is typical for litotes constructions in general. Some English examples of litotes NPIs are *lose sleep over* (example: *Bush lost no sleep over dead soldiers*), *love lost between* (example: *There was no love lost between the generative and the interpretive semanticists*) or *say no to* (example: *I would not say no to a free trip to Murmansk*).

A slightly more complex type of litotic construction is exemplified by the examples in (43) below. The construction can be found, with only minor differences, in English, Dutch and German:

(43) English
   a. Not a day went by that I did not miss her.
   b. Not a week went by without new developments.

German
   c. *Kein Tag vergeht ohne neue Entwicklungen.*
      No day passes without new developments
      'Not a day passes without new developments'
   d. *Kaum ein Tag geht vorbei dass ich nicht an sie denke.*
      Hardly a day goes by that I not to her think
      'Hardly a day goes by, that I do not think of her'

Dutch
   e. *Er gaat geen dag voorbij zonder nieuwe ontwikkelingen.*
      There goes no day by without new developments
      'Not a day goes by without new developments'
   f. *Er gaat zelden een week voorbij dat het niet regent*
      there goes seldom a week by that it not rains
      'Rarely a week goes by in which it does not rain'

The construction consists of a temporal indefinite, usually of midsize (so *moments, seconds, minutes* are not commonly used, nor are *decades, centuries, eons*), a verb expressing the passing of time (*go by, pass*) and either a *without*-PP, or a negated relative clause. When this whole construction is negated, the construction assumes universal force: if not a week went by without new developments, then every week brought new developments. In Table 7 below, Dutch data are compared to English data (German data were too scant to consider). We see the same sets of triggers, but with interesting differences in prominence. English tends to prefer the weaker negative adverbs *barely/hardly/scarcely*, whereas Dutch strongly prefers regular negation. It would seem plausible that this reflects a stylistic difference between the two languages (or rather, their users), and not a difference of semantic nature. Occasionally, we find positive occurrences of the construction (*a year went by without any news from the missionaries*). Slightly different variants, such as *Another year went by without a major hurricane* or *Most of the year went by without any new accidents*, were not included in the data set, since the polarity-sensitive use does not show up with definite subjects, or subjects introduced by *another*.

Table 7. Dutch corpus data (N=116) for *not an X goes by* versus English (N=45)

TRIGGER	N	%	TRIGGER	N	%
Geen "no"	104	90	Barely/hardly/scarcely	22	49
Nauwelijks 'hardly'	7	6	Negation	20	44
Zelden 'seldom'	4	3	Rarely/seldom	2	4
Weinig 'few/little'	1	1	Positive	1	2

In Table 8, I summarize the various temporal nouns involved in the construction. As you can see, the similarities between Dutch and English are quite obvious.

Table 8. an X goes by – types of nouns

DUTCH	%	ENGLISH	%
Dag 'day'	73	Day	71
Week 'week'	14	Week	16
Nacht 'night'	4	Night	7
Jaar 'year'	3	Month	2
Avond 'evening'	3	Year	2
Maand 'month'	1	Winter	2
Winter 'winter'	1		
Zomer 'summer'	1		
Twee dagen 'two days'	1		

A notable property of litotes constructions in general is that the secondary negation is generally unable to trigger polarity items. Compare:

(44) a. *Jones can't deny he has ever been to the Canary Islands.[12]
 b. *Not a day passed by without so much as a wink from the master.

---

[12] As noted by Larry Horn (p.c.), the sentence

 (i) On the stand, Jones didn't deny he has ever been to the Canary Islands.

 is somewhat better than (44a). Actually, (44a) is only bad on the reading where Jones can't deny having ever been to the Canary Islands because that would involve a lie. There is another reading where Jones is for some other reason unable to deny, e.g. because he has lost his voice. This reading is fine for (44a):

 (ii) I am glad Jones lost his voice. Now he can't deny he has ever been to the Canary Islands.

In this respect, litotic constructions differ significantly from other types of double negation, which are generally acceptable with polarity items (cf. the discussion of example (9) above). It would make sense to view litotes as a constructional unit which acts as an upward entailing context, whereas other types of double negation should be treated compositionally, each negative element contributing negative force to its scopal domain.

3.7. Adverbs of degree

Adverbs of degree may have a special understating use (cf. Horn 1989). For instance *Jones is not very smart* may be used to convey the information that Jones is actually somewhat dumb. Therefore, this sentence may be used as an understatement, given that its literal meaning only excludes that Jones is very smart, so could well be compatible with a state of affairs in which Jones is smart, but not very smart (e.g., when Jones' IQ equals, say, 120). Some adverbs of degree seem to have specialized in this understating use, having become negative polarity items. Some cases in point are English *all that*, German *sonderlich*, and Dutch *gek*.

(45) a. Frank was not all that happy.
 b. The show was not all that bad.
 c. The fish did not bite all that often.

(46) a. *Er war nicht sonderlich geliebt* (German)
  He was not especially well-liked
  'He was not all that well-liked'
 b. *Hij wist niet zo gek veel.* (Dutch)
  He knew not so crazy much
  'He didn't know all that much'

The lexical semantics of these adverbs is that of *very*, indicating a high degree, but their global requirement, which makes them negative polarity items, is that they express an understatement, i.e. indicating a fairly low degree. The distribution of the adverbs shows a preponderance of purely negative environments, much like that of litotes constructions, and the rest is of very little quantitative significance.

Table 9. all that, sonderlich, gek

ALL THAT	N=159	SONDERLICH	N=97	GEK	N=209
not	91%	nicht 'not'	85%	niet 'not'	97%
n-word	5%	n-word	10%	n-word	1%
as-if	1%	kaum 'hardly'	1%	positive	1%
neg. predicate	1%	ohne 'without'	4%	zonder	1%
question	1%				
without	1%				

As Klein (1998), among others, has pointed out, a great many adverbs of degree are positive polarity items. Perhaps the most notable among them are adverbs denoting a middling degree, such as *rather, pretty, fairly, somewhat,* and their counterparts in Dutch *nogal, tamelijk, vrij* 'rather, fairly' and in German *ziemlich*. Cf. also French *assez,* which like *enough* is a PPI when used as as a scalar adverb and not when it's used as a true comparative with an explicit or implicit complement. *Rather* is one of the oldest-known positive polarity items, already showing up in seminal works such as Baker (1970). Since midlevel degree adverbs appear to be generally averse to negation, it seems likely that there is some general semantic reason for it. Here is one possibility: let's assume that *rather smart* means something like: *smart, but not very smart*. When I assert that X is rather smart, and you disagree, I assume you disagree that the person is smart. Only with the use of metalinguistic negation (cf. Horn 1989: 400ff.) can we arrive at another possibility, where not the property *smart* but the property *not very smart* is negated, as in: *Jones is not RATHER SMART, he is EXTREMELY smart*. So I take it that *not rather smart* boils down, in the absence of metalinguistic negation, to *not smart*. This might be the reason why *rather* and similar expressions have turned into positive polarity items, because they are equivalent, under negation, to a shorter, and hence preferable combination (cf. also Grice's 1975 Maxim of Manner, in particular its submaxim 'Be brief'). If correct, this would be a general semantic explanation for the fact that cross-linguistically midlevel degree adverbs shy away from negation (excepting, as always, double negation, cf. Baker 1970, Horn 1989, Szabolci 2004, and metalinguistic or 'radical' negation, cf. Seuren 1985, Horn 1989).

For high degree modifiers, some of which are also positive polarity items, the above account won't work. But here, it is important to note that not all high degree modifiers are positive polarity items. Some are, some are not, and some are the opposite: negative polarity items, like *all that*.[13] Consequently, there cannot be a general semantic reason of the positive polarity status of such high degree modifiers as English *highly* or Dutch *hoogst*. Instead, I take it that their lexical semantics makes them markers of a high degree, and that the global condition on their use is simply that they express a high degree. This can be achieved in a positive context only. Compare for instance Dutch *hoogst* 'highly' and *erg* 'terribly, very'. While all 376 occurrences of *hoogst* in my database are positive, *erg* has 334 negative occurrences out of a total of 1776 (19%). One might state the difference between the two items in the following way: while both have the lexical semantics of high degree modifiers, only *hoogst* has developed an additional global condition on its use, namely that it always indicate a high degree within the larger sentential context. Hence any use with clause-mate regular negation needs to be ruled out. Of course, with echoic or metalinguistic negation, *hoogst* is fine:

(47) *Het is misschien een beetje ongewoon, het is zeker niet 'hoogst*
      It   is perhaps   a   bit   unusual   it is surely not 'highly
      ongewoon.'
      unusual'
     'It is perhaps a bit unusual, it is definitely not 'highly unusual.'

My main point here is that with minor differences in the global conditions on usage, we can characterize both positive and negative polarity items among the class of degree adverbs. In the following section, I will argue that many other types of positive and negative polarity items may be handled in like manner.

3.9. Positive polarity items with negative polarity counterparts

Often, positive polarity items resemble nothing so much as negative polarity items. Consider for instance the two Dutch expressions *op rozen zitten* 'sitting on roses' and *over rozen gaan* 'going over roses.' The former expression is a positive polarity item, the latter a negative polarity item. To be sure, there are more differences. The first item is static, and the second one dynamic. The first tends to be predicated of people, or groups of people,

---

[13] For English adverbs of degree, cf. also Bolinger (1972: 115–125) and Horn (1989: 353).

the latter is typically predicated of path-like expressions such as a life, a career, a journey, a quest, a marriage, all of which have in common that they involve a stretch of time. However, I don't think those selectional differences matter for the polarity status of the expressions.

(48) a. *Na   de 2-0 zat Feyenoord op rozen.*
After the 2-0 sat Feyenoord on roses
"After the 2-0, Feyenoord was doing just fine"

b. *De tocht   naar het kampioenschap ging niet over rozen.*
the journey to the championship   went not over roses
"The journey to the championship was hard"

In both cases, roses are indicative of a situation ranked highly on an evaluative scale. *Op rozen zitten* 'sitting on roses' requires a global context which maintains this high ranking (positive) whereas *over rozen gaan* 'going over roses' has the global property of an understater, hence it requires negative contexts. These differences cannot be explained from lexical semantics only (e.g. the distinction static/dynamic does not in any way explain them). Rather, I assume that we are dealing with arbitrary and conventional conditions associated with particular expressions.

A similar type of example is provided by Dutch *kwaad kersen eten* 'bad cherry-eating', exemplified in (49):

(49) *Met hem is het kwaad kersen eten*
with him, it is   bad   cherry-eating
'He is a tough person to deal with'

This expression is a positive polarity item, indicating the old-time habit of spitting out the pits of cherries. Somebody who will spit cherry pits in your face, is a tough bastard to deal with, is the idea underlying this expression. German has almost the same expression, but based on German *gut* 'good' rather than Dutch *kwaad* 'bad', and not surprisingly, it is a negative polarity item (cf. Van der Wouden 1997):

(50) *Mit ihm ist nicht gut   Kirschen essen.*
With him is not   good cherries eat
'He is a tough person to deal with'

In both cases, the general import is the same, but due to lexical differences, one case requires negation, while the other must shun it.

A final example of this kind can be found in English. In English, the construction consisting of *have* + *possessive pronoun* + *noun*, where the possessive pronoun co-refers to the subject of *have*, is a positive polarity item:

(51) a. Membership has its advantages.
 b. The Rolling Stones had their moments.
 c. Fred has his quirks.
 d. I have my doubts.

Presumably, the combination of the possessive pronoun and the possessive verb *have* is what makes this a positive polarity construction. Membership may or may not have advantages, but when you add *its* to *advantages*, clearly you are already presupposing the existence of such advantages. Hence it seems natural to use *have its advantages* only in positive contexts:

(52) a. *Membership does not have its advantages.
 b. *The Rolling Stones did not have their moments.
 c. *Fred did not have his quirks.
 d. *I don't have my doubts.

As it happens, there is also a related construction in English, employing not the verb *have*, but the preposition *without*, in combination with the verb *be*:

(53) a. Membership is not without its privileges.
 b. The Rolling Stones were not without their moments.
 c. Fred is not without his quirks.
 d. I am not without my doubts.

It has been noted many times that the verbs *have* and *be* share many properties (Benveniste 1966, Kayne 1993, among others). As Benveniste noted, sentences such as English *John has a car* are rendered in many languages as *To John is a car*, involving a dative or dative-like preposition to mark the possessor, with the subject corresponding to the object possessed. Copula-constructions using *with* can be viewed denoting the converse relation, where the subject is the possessor and the object of the preposition is the possessed item. Hence *Peace be with you* is equivalent to the wish *You have peace*. (Of course, by pointing out these similarities, I do not want to suggest that any sentence involving the main verb *have* has an equally acceptable counterpart involving *with*.[14]) Given these deep semantic similarities, the existence of two related constructions in English, one involving, in part, *have*, the other involving *not be+without*, is not entirely unexpected.

Here, we are clearly dealing with a negative polarity construction. When negation is removed from the examples in (53), the result is decidedly odd:

(54) a. *Membership is without its privileges.
 b. *The Rolling Stones were without their moments.
 c. *Fred is without his quirks.

Compare also:

 d. *Fred is (not) with his quirks. (but OK: *Fred, with his quirks, ...*)

So the two constructions *have+possessive+N* and *be without+possessive+N* have the same global requirement: they must assert existence. In one case, this leads to positive-polarity status, in the other to negative-polarity status.

4. Conclusions

In this chapter, I have argued that polarity items are more than peculiar expressions with a special licensing requirement. Together with their triggers they play a role in a construction consisting of at least a polarity item and its licenser, but potentially also other material. This construction may have properties of its own, which do not arise from the requirements of the con-

---

[14] As a case in point, Dutch has the positive-polarity construction involving the verb *hebben* 'have', but not the negative-polarity counterpart involving *zijn* 'be' + *zonder* 'without' cf.:

(i) *Elk land heeft (*niet) zijn problemen.*
 Each country has (*not) its problems
 'Every country has its problems'

(ii) **Geen land is zonder zijn problemen.*
 No country is without its problems
 'No country is without its problems'

Occasionally, variants without the possessive pronoun are acceptable as litotes-combinations, though:

(iii) *De reis is niet zonder gevaar.*
 The trip is not without danger
 'The trip is not free of danger'

stituting words, but call for the kind of non-atomistic treatment to be found in construction grammar and related approaches. The atomistic outlook of compositional semantics tends to steer away from the kind of global conditions on the expression of meaning which I have argued are at stake here, and therefore needs to be supplemented with a subtler awareness of the constructional properties of both negative and positive polarity items.

Another point I have argued for above is that polarity items are not all alike, but differ greatly, both in their distributional and in their lexical-semantic properties. Current approaches to classification, such as Zwarts's (1998) Boolean approach, fail to provide a satisfactory framework, because they lack an account linking lexical semantics to distributional properties, in addition to various shortcomings vis-à-vis descriptive adequacy. It is my hope that the framework sketched here, which relies both on lexical semantics and various global conditions on usage, can be developed into a better tool for analyzing the messy facts of negative and positive polarity items.

Given the kind of account provided here, one may well ask the question whether notions such as 'negative polarity item' or 'positive polarity item' are more than descriptive terms for classes of expressions whose distribution is somehow affected by the presence of negation. In the preceding pages I have argued that a unified account of polarity licensing is not forthcoming. Rather, different individual expressions make different demands on their linguistic contexts, and it appears increasingly unlikely that there will ever be a theory that provides a uniform treatment of all polarity items. I believe that 'negative polarity item' may well be a grab bag, similar to, say, 'adverb', that does not directly play a role in the grammar, but serves as a convenient term to refer to a loosely knit group of expressions with overlapping distributional properties.

References

Ackema, Peter and Ad Neeleman
    2001   Competition between syntax and morphology. In *Optimality theoretic syntax*, Geraldine Legendre, Jane Grimshaw and Sten Vikner (eds.), 29–60. Cambridge, MA: MIT Press.
Auwera, Johan van der and Martine Taeymans
    to appear   The *need* modals and their polarity.
Baker, C. Lee
    1970   Double Negatives. *Linguistic Inquiry* 1: 169–186.

Benveniste, Émile
 1966 *Etre* et *avoir* dans leurs fonctions linguististiques. In *Problèmes de linguistique générale*, E. Benveniste (ed.), 187–207. Paris: Gallimard.
Blutner, Reinhard
 2002 Pragmatics and the lexicon. In *Handbook of Pragmatics*, Laurence R. Horn and Gregory Ward (eds.), 488–519. Oxford: Blackwell.
Bolinger, Dwight
 1972 *Degree words*. The Hague: Mouton.
Bremen, Klaus von
 1986 Le problème des forclusifs romans. *Lingvisticae Investigationes* X: 223–265.
Dayal, Veneeta
 1998 ANY as inherently modal. *Linguistics and Philosophy* 21: 433–476.
Fauconnier, Gilles
 1975a Polarity and the Scale Principle. *CLS* 14, Chicago.
Fauconnier, Gilles
 1975b Pragmatic Scales and Logical Structure. *Linguistic Inquiry* 6: 353–375.
Fauconnier, Gilles
 1978 Implication reversal in a natural language. In *Formal semantics and pragmatics for natural languages*, F. Guenthner and S. J. Schmidt (eds.), 289–301. Dordrecht: Reidel.
Fischer, Olga, Muriel Norde and Harry Perridon
 2004 *Up and down the cline – the nature of grammaticalization*. Amsterdam/Philadelphia: John Benjamins.
Geach, Peter T.
 1962 *Reference and Generality*. Cornell University Press, Ithaca.
Giannakidou, Anastasia
 1997 *The Landscape of Polarity Items*. Dissertation. University of Groningen.
Giannakidou, Anastasia
 1998 *Polarity Sensitivity as (Non)veridical Dependency*. John Benjamins, Amsterdam-Philadelphia.
Giannakidou, Anastasia
 1999 Affective depencencies. *Linguistics and Philosophy* 22 (4): 367–421.
Giannakidou, Anastasia
 2001 The meaning of free choice. *Linguistics and Philosophy* 24 (6): 659–735.
Giannakidou, Anastasia
 2009 Negative and positive polarity items: licensing, compositionality and variation. To appear in *Semantics: An International Handbook of Natural Language*, Meaning Maienborn, Claudia, Klaus von Heusinger, and Paul Portner (eds). Berlin/New York: Mouton de Gruyter.

Grice, H. Paul
    1975   Logic and conversation. In *Syntax and semantics*, Vol 3., Cole, P. and Morgan, J. (eds.), 41–58 . New York: Academic Press.

Hoeksema, Jack
    1986   Monotonicity Phenomena in Natural Language. *Linguistic Analysis* 16 (1/2): 235–250. [Reprinted 2004 in *Semantics – Critical concepts*, Javier Gutierrez-Rexach (ed.). London: Routledge.]

Hoeksema, Jack
    1994   On the grammaticalization of negative polarity items. *BLS* 20: 273–282.

Hoeksema, Jack
    1998a   On the (non)loss of polarity sensitivity: Dutch *ooit*. In *Historical Linguistics 1995. Vol. 2: Germanic Linguistics*, R. M. Hogg and L. van Bergen (eds.) 101–114. Amsterdam: John Benjamins.

Hoeksema, Jack
    1998b   Corpus Study of Negative Polarity Items. In *IV–V Jornades de corpus lingüístics 1996–1997*, IULA, Universitat Pompeu Fabra, Barcelona, 67–86.

Hoeksema, Jack
    1999   Blocking Effects in the Expression of Negation. *Leuvense Bijdragen/Leuven Contributions in Linguistics and Philology* 88 (3/4): 403–423.

Hoeksema, Jack
    2000   Negative Polarity Items: Triggering, Scope and C-Command. In Horn and Kato (2000), 115–146.

Hoeksema, Jack and Donna-Jo Napoli
    2008   Just for the hell of it: A comparison of two taboo-term constructions. *Journal of Linguistics* 44 (2): 347–378.

Horn, Laurence R.
    1989   *A Natural History of Negation*. Chicago: University of Chicago Press.

Horn, Laurence R.
    1991   Duplex negation affirmat: The economy of double negation. In *Papers from the 27th regional meeting of the Chicago Linguistic Society. Part 2: The parasession on negation*, L. M. Dobrin et al (eds.), 180–196, Chicago: Chicago Linguistic Society.

Horn, Laurence R.
    2000a   Pick a theory (not just *any* theory). Indiscriminatives and the free-choice indefinite. In Horn and Kato (2000), 147–192.

Horn, Laurence R.
    2000b   *any* and *ever*: Free choice and free relatives. In *IATL 7. The Proceedings of the 15th Annual Conference*, Adam Zachary Wyner (ed.), 71–111, University of Haifa.

Horn, Laurence R.
2005   Airport '86 revisited: Toward a unified indefinite *any*. In *Reference and Quantification: The Partee Effect*, Gregory N. Carlson and Francis Jeffry Pelletier (eds.), 179–205. Stanford: CSLI.

Horn, Laurence R. and Yasuhiko Kato, eds.,
2000   *Negation and Polarity. Semantic and Syntactic Perspectives.* Oxford: Oxford University Press.

Israel, Michael
1996   Polarity Sensitivity as Lexical Semantics. *Linguistics and Philosophy* 19: 619–666.

Israel, Michael
1998a   *The Rhetoric of Grammar: Scalar Reasoning and Polarity Sensitivity.* Dissertation, UC San Diego.

Israel, Michael
1998b   Ever: Polysemy and Polarity Sensitivity. *Linguistic Notes from La Jolla* 19: 29–45.

Israel, Michael
2001   Minimizers, Maximizers, and the Rhetoric of Scalar Reasoning. *Journal of Semantics* 18 (4): 297–331.

Israel, Michael
2004   The Pragmatics of Polarity. In *The Handbook of Pragmatics*, Laurence R. Horn and Gregory Ward (eds.), 701–723. Oxford: Blackwell.

Israel, Michael
2008   *The Least Bits of Grammar: Pragmatics, Polarity, and the Logic of Scales.* Ms., to be published by Cambridge University Press, Cambridge.

Jackson, Eric
1995   Weak and Strong Negative Polarity Items: Licensing and Intervention. *Linguistic Analysis* 25: 181–208.

Janda, Richard D.
2000   Beyond 'pathways' and 'unidirectionality': on the discontinuity of language transmission and the counterability of grammaticalization. *Language Sciences* 23 (2/3): 265–340.

Jennings, R. E.
1994   *The Genealogy of Disjunction.* New York/Oxford: Oxford University Press.

Kadmon, Nirit and Fred Landman
1993   Any. *Linguistics and Philosophy* 16: 353–422.

Kayne, Richard S.
1993   Toward a Modular Theory of Auxiliary Selection. *Studia Linguistica* 47: 3–31.

Kiparsky, Paul
1973 'Elsewhere' in Phonology. In *A Festschrift for Morris Halle*, S. Anderson and P. Kiparsky, eds., 93–106. New York: Holt, Rinehart and Winston.
Klein, Henny
1998 *Adverbs of Degree*. Amsterdam/Philadelphia: John Benjamins.
Klein, Henny and Jack Hoeksema
1994 *Bar* en *Bijster*: Een onderzoek naar twee polariteitsgevoelige adverbia. *Gramma/TTT* 3 (2): 75–88.
König, Ekkehard
1981 The meaning of Scalar Particles in German. In *Words, Worlds and Context*, H.-J. Eikmeyer and H. Rieser (eds.), 107–132. Berlin/New York: Mouton de Gruyter.
Krifka, Manfred
1995 The Semantics and Pragmatics of Polarity Items. *Linguistic Analysis* 25 (3/4): 209–257.
Ladusaw, William A.
1979 *Polarity Sensitivity as Inherent Scope Relations*. Dissertation, University of Texas.
Levinson, Stephen C.
2000 *Presumptive Meaning*. Cambridge, MA: MIT Press.
Lichte, Timm and Jan-Philipp Soehn
2007 The Retrieval and Classification of Negative Polarity Items using Statistical Profiles. In *Roots: Linguistics in search of its evidential base*, S. Featherston and W. Sternefeld (eds.), 249–266. Berlin/New York: Mouton de Gruyter.
Lichte, Timm and Manfred Sailer
2004 Extracting Negative Polarity Items from a Partially Parsed Corpus. In *Proceedings of the Third Workshop on Treebanks and Linguistic Theory (TLT2004)*, S. Kübler, J. Nivre, E. Hinrichs and H. Wunsch (eds.), 89–101. Seminar für Sprachwissenschaft, Universität Tübingen.
Linebarger, Marcia
1981 *The Grammar of Negative Polarity*. Dissertation, MIT. Distributed 1981 by the Indiana University Linguistics Club.
Linebarger, Marcia
1987 Negative polarity and grammatical representation. *Linguistics and Philosophy* 10 (3): 325–387.
Partee, Barbara H.
1986 *Any, almost* and superlatives. Manuscript, published 2004 in *Formal Semantics: Selected Papers by Barbara H. Partee*, Barbara H. Partee (ed.), 31–40.

Pereltsvaig, Asya
: 2004 : Negative Polarity Items in Russian and the 'Bagel Problem'. In *Negation in Slavic*, A. Przepiorkowski and S. Brown (eds.). Slavica Publishers, Bloomington.

Poser, William
: 1992 : Blocking of Phrasal Constructions by Lexical Items. In *Lexical Matters*, Ivan Sag and Anna Szabolsci (eds.), 111–130. Stanford: Center for the Study of Language and Information.

Quine, W. V. O.
: 1960 : *Word and Object*. Cambridge, MA: MIT Press.

Rooij, Robert van
: 2008 : Towards a uniform analysis of *any*. *Natural Language Semantics* 16 (4): 297–315.

Sahlin, Elisabeth
: 1979 : Some *and* Any *in Spoken and Written English*. Uppsala: Almqvist and Wiksell.

Sailer, Manfred and Frank Richter
: 2002 : *Not for Love or Money: Collocations!* In *Proceedings of Formal Grammar 2002*, Gerhard Jäger, Paola Monachesi, Gerald Penn and Shuly Wintner (eds), 149–160.

Seuren, Pieter A. M.
: 1976 : Echo: een studie in negatie. In *Lijnen van taaltheoretisch onderzoek. Een bundel oorspronkelijke artikelen aangeboden aan prof. dr. H. Schultink*, Geert Koefoed and Arnold Evers (eds.), 160–184. Groningen: Tjeenk Willink.

Seuren, Pieter A. M.
: 1985 : *Discourse Semantics*. Oxford: Blackwell.

Szabolcsi, Anna
: 2004 : Positive polarity – negative polarity. *Natural Language and Linguistic Theory* 22 (2): 409–452.

Szabolcsi, Anna and Frans Zwarts
: 1993 : Weak islands and an algebraic semantics for scope taking. *Natural Language Semantics* 1: 235–285.

Swart, Henriëtte de
: 2006 : Expression and interpretation of negation. MS, University of Utrecht.

Traugott, Elisabeth Closs
: 1990 : From less to more situated in language: the unidirectionality of semantic change. *Papers from the 5th International Conference on English Historical Linguistics*, Sylvia Adamson, Vivien Law, Nigel Vincent and Susan Wright (eds.), 497–517. Amsterdam: John Benjamins.

Traugott, Elisabeth Closs and Berndt Heine
  1991   *Approaches to Grammaticalization.* Vol. 1, Amsterdam: John Benjamins.
Vendler, Zeno
  1967   *Each* and *Every, Any* and *All.* In *Linguistics in Philosophy.* Ithaca: Cornell University Press.
Westerståhl, Dag
  1985   Determiners and context sets. In *Generalized Quantifiers in Natural Language*, J. van Benthem and A. ter Meulen (eds.), 45–71. Dordrecht: Foris.
Wouden, Ton van der
  1996   Litotes and Downward Monotonicity. In *Negation: a notion in focus*, Heinrich Wansing (ed.), 145–167. Berlin/New York: Mouton de Gruyter.
Wouden, Ton van der
  1997   *Negative Contexts: Collocation, Polarity and Multiple Negation.* London: Routledge.
Wouden, Ton van der
  2001   Three modal verbs. In *Zur Verbmorphologie germanischer Sprachen*, Sheila Watts, Jonathan West and Hans-Joachim Solms (eds.), 189–210. Tübingen: Niemeyer.
Zwarts, Frans
  1981   Negatief Polaire Uitdrukkingen I. *GLOT* 4 (1): 35–132.
Zwarts, Frans
  1986   *Categoriale Grammatica en Algebraïsche Semantiek.* PhD dissertation, University of Groningen.
Zwarts, Frans
  1995   Nonveridical Contexts. Linguistic Analysis 25 (3/4): 286–312.
Zwarts, Frans
  1998   Three types of polarity. In *Plurality and Quantification*, F. Hamm and E. Hinrichs (eds.), 177–238. Dordrecht: Kluwer.

Negation as a metaphor-inducing operator

Rachel Giora, Ofer Fein, Nili Metuki and Pnina Stern

> This paper is dedicated to the White Rose resistance whose motto was: *We will not be silent. We are your bad conscience. The White Rose will not leave you in peace!*[1]

1. Introduction: Processing Negation

Consider the following cartoon (Babin 2008):

(1)

How do we go about processing it? How do we make sense of it? Suppose we start by trying to make sense of the negative statement which cites McCain's in the third presidential debate (York 2008): *I am not President Bush*. Taken at face value, this statement is literally true but redundant. Even replacing the negated concept – *Bush* – with an available alternative – *McCain* – is literally true but similarly uninformative. This, then, is most probably not the way we represent the statement. Suppose then that we attempt an alternative nonliteral interpretation such as 'I am not like/similar

---

[1] http://en.wikipedia.org/wiki/White_Rose#Quotes (Retrieved 4 October, 2008).

to President Bush'; 'Do not compare me to President Bush'. What processes do these more informative interpretations involve? By rejecting a similarity to Bush, McCain allows an instant activation of an affirmative metaphoric comparison "I am Bush", which highlights the affinities he intends to deny. *I am not President Bush* is thus represented as "I am President Bush" (alongside its rejection), as has also been ironically interpreted by one of the postings on YouTube which records Bush and McCain's closeness and similarity[2] (On negative comparisons as comparisons see Giora, Zimmerman and Fein 2008).

Moreover, taken as a whole, the cartoon invokes yet another comparison, which also results in an ironic reading of the statement. By giving the image of McCain a Nixonian hunch and the double-V sign Nixon was famous for, the cartoonist invites comprehenders to access Nixon's infamous *I am not a crook* (Kilpatrick 1973), which was also subjected to the same interpretation processes as McCain's statement: It evoked a public perception of Nixon as a crook, in spite of the use of negation (and despite the availability of an alternative opposite such as "honest").[3]

These (possible) interpretation processes must rely heavily on the accessibility of information within the scope of negation. They could not have emerged had negated information been either initially inaccessible or initially accessible but rendered inaccessible later on due to suppression processes assumed obligatory following negation (MacDonald and Just 1989). Can studies into the online processes involved in interpreting negated concepts and utterances shed light on the interpretive insights exemplified above? Will negation allow comprehenders access to what is within the scope of negation, as might be deduced from the example above, or will that concept be blocked, either initially or later on, given the negative operator?

Recent findings in psycholinguistics show that, across the board, when negated and nonnegated concepts (*rocket* in The train to Boston was/was **not a rocket**; *open* in The door is/is **not open**) are encountered, they are accessed immediately, regardless of whether they are negated or not.

---

[2] http://www.youtube.com/watch?v=Oq7yJh08VHU (retrieved on October 20, 2008).

[3] "Note that the cartoon as displayed on that site is in fact "animated", and the Nixonian double-V salute pops up only after McCain denies he's Bush, as if it's the inevitable subtext. What it especially reminds me of is the Dr. Strangelove character (the ex-Nazi rocket-scientist-cum-mad-bomber working for the U. S. nuclear program) in the eponymous film whose right hand, as if it had a will of its own, would suddenly extend into the Heil Hitler salute" (Laurence R. Horn, email communication, 20.11.08).

Initially, then, processing negated and nonnegated concepts follow similar processing routes and exhibit no asymmetry: Both make available the affirmative (salient[4]) meaning of the concept, although in the negative condition this meaning is contextually incompatible (Ferguson and Sanford 2008; Ferguson, Sanford and Leuthold, 2008; Fischler, Bloom, Childers, Roucos and Perry 1983; Giora, Balaban, Fein and Alkabets 2005; Hasson and Glucksberg 2006; Kaup, Lüdtke and Zwaan 2006; MacDonald and Just 1989; for a review, see Giora 2006).

Will that meaning resist negation effects even when extra processing time is allowed? Theoretically, later processes might be susceptible to a number of effects following negation: they might result in reducing the accessibility of the negated concept, either fully or partially, or keep it intact. Whereas the various theories in psycholinguistics converge on the initial access phase, they tend to disagree on the effects of negation occurring at the later processing stages.

The prevailing assumption in psycholinguistics is that, once enough processing time is allowed, negation affects suppression of negated concepts unconditionally so that they are (i) discarded from the mental representation altogether and (ii) replaced by an available opposite (The suppression hypothesis). Findings indeed show that when presented in isolation, negated concepts are eliminated from the mental representation about 500–1000 msec following their offset, at which stage, initial levels of activation are reduced to base line levels (Hasson and Glucksberg 2006; Kaup, Lüdtke and Zwaan 2006; MacDonald and Just 1989). Findings also show that when sufficient processing time (1500 msec) is allowed, negated concepts are often replaced by an alternative opposite (Kaup, Lüdtke and Zwaan 2006), should this be available (Mayo, Schul and Burnstein 2004, but see Prado and Noveck 2006). Thus, while 750 msec following its offset *open* in *not open* lost initial levels of activations, another 750 msec later, it was replaced by an opposite – "closed" (Kaup et al. 2006). Similarly, between 150–500 msec following their offset, negated and nonnegated concepts (*not a rocket/a rocket*) were both represented as "fast". However, 1000 msec following their offset, their initial levels of activations were preserved only following affirmative contexts (*The train to Boston was a rocket*), in which this meaning was contextually compatible.

---

[4] A meaning is *salient* if it is coded and foremost on our mind due to factors such as experiential familiarity, frequency, conventionality, prototypicality etc. (see Giora 1997, 2003).

Outside a specific context, then, negated concepts are eventually suppressed. Indeed, it has been widely acknowledged that negation (mainly 'No' and 'Not') reduces the accessibility of the affirmative meaning of the concept within its scope so that it can deny, reject, convey disagreement, or correct this information by activating an alternative replacement (for reviews, see Ferguson, Sanford and Leuthold 2008, Giora 2006, Horn 2001, Israel 2004, Jespersen 1924, Pearce and Rautenberg 1987 inter alia).

An alternative view to the suppression hypothesis – the SUPPRESSION/RETENTION HYPOTHESIS – has been proposed by Giora and colleagues, which argues that suppression following negation is not obligatory but sensitive to discourse goals and requirements. Information will be disposed of if it is deemed unnecessary or obstructing, regardless of negation. (On suppression following affirmative concepts and on suppression as a context driven mechanism, see Gernsbacher 1990). In this respect, negation is not different from affirmation – both might lead to suppression or retention of concepts depending on specific contextual information and speaker's intent (Giora 2006).

According to Giora and colleagues (Giora 2006; Giora, Fein et al. 2007), then, both negated and nonnegated concepts can either maintain their initial levels of activation or allow their gradual reduction up to base line levels and below, depending on discoursal factors (the SUPPRESSION/RETENTION HYPOTHESIS, Giora 2003). Contra the received view, then, in this respect, negation and affirmation are not different; they do not exhibit asymmetric effects even when later processes are concerned (Giora 2006, 2007; Giora, Balaban et al. 2005; Giora, Fein et al. 2005).

Indeed more recent studies have shown that when negated concepts are not presented in isolation but instead are embedded in a supportive context, they need not be suppressed and replaced by an alternative. Instead they can be retained if deemed useful for the unfolding context. Thus, when negated concepts (*The train to Boston was **not a rocket***) were furnished with a relevant late context discussing the same discourse topic (*The trip to the city was fast, though*), their so-called contextually inappropriate interpretation (*fast*) was not discarded from the mental representation but instead remained accessible at least as long as 1000 msec following their offset. In contrast, when followed by an irrelevant context, these interpretations were dampened. Similarly, when embedded in a supportive prior context (*millionaires* in *I live in the neighborhood of **millionaires** who like only their own kind. Nonetheless on Saturday night, I also invited to the party at my place a woman who is not **wealthy***) negated concepts (*wealthy*) preserved their accessibility as long as 750 msec following their offset (Giora, Fein, Aschkenazi and Alkabets-Zlozover 2007).

It is precisely this persisting accessibility of negated information that allows negation to affect its representation in various ways. For instance, negated concepts have been shown to induce mitigation of their interpretations so that "not pretty", for instance, was represented as "less than pretty" rather than as "ugly" (Giora, Balaban et al. 2005; Horn 1989, 2001; Jespersen 1917, 1924; Paradis and Willners 2006). In addition, compared to affirmative modifiers *(almost)* negation is a rather strong mitigator, representing a weaker or more hedged version of the affirmative (Giora, Balaban et al. 2005). Negated concepts have also been shown to be represented as a mitigated version of their alternative opposites, so that "not pretty" was represented as a hedged version of "ugly" (Fraenkel and Schul 2008). However, when negating an end of the scale member of the set ("not very pretty"), mitigation via negation invited an ironic interpretation even outside a specific context (Giora, Fein, Ganzi, Alkeslassy Levi and Sabah 2005; Horn 2001: Chapter 5).

Along the same lines, it is this accessibility of negated information that allows negative comparisons to come across as comparisons, maintaining the prototypical features of the source domains (*Hitler* in *Bush isn't Hitler*) as shown by both lab results and natural data (Giora 2007; Giora, Zimmerman and Fein 2008; see also Ward 1983). Specifically, in Giora, Zimmerman and Fein (2008), negated comparisons and their affirmative counterparts (*Saddam Hussein was/wasn't like Hitler*), came across as similarly appropriate. In addition, reference to a salient, prototypical feature of a negated concept (*a well known masterpiece* in *Susie's drawing is **not the Mona Lisa**. Susie's drawing is not **a well known masterpiece***) elicited higher appropriateness ratings and faster reading times compared to a less prototypical feature (*Susie's drawing didn't warrant **a parody by Marcel Duchamp***). Their respective controls, however, in which the context sentence was unrelated (*Susie's drawing is not the armored corps*), were rated as least appropriate and took longest to read.

Similarly, it is this accessibility that allows negative categorizations to come across as affirmative categorizations, obeying same categorization constraints. Thus, like affirmative conjunctions, negative ones (*What I bought yesterday was not a bottle but a jug*), which obey categorization principles, were found acceptable. In contrast, negative conjunctions which do not (*What I bought yesterday was not a bottle but a closet*) were unacceptable (Giora, Balaban et al. 2005: Ex. 2). By the same token, it is this accessibility that allows negated concepts (*apple* in *Justin bought a mango but not an apple. He ate the fruit*) to interfere with anaphor resolution (*fruit*) when the antecedent (*mango*) was not a prototypical member of the set (Levine and Hagaman 2008; Shuval and Hemforth 2008); it is this accessibility that allows negation to also serve as an enhancer, highlighting information within its scope (*Who wasn't there in the coronation balls in Little Rock and Washington?* ***All the who's and who's*** *in the entertainment industry*) as shown by studies of Hebrew expletive negation (Eilam 2009, Giora 2006: 993).

This retention of negated concepts applies equally well to visual negation markers (e.g., a cross or a line) superimposed on visual percepts. In Giora, Heruti, Metuki and Fein (2009), we show that when presented with a crossed over image (an open door with a cross superimposed on it) and asked to select the appropriate interpretation, participants did not select the one that manifested an alternative opposite (***Close*** *the door!*). Instead, they opted for the interpretation which included mention of the negated concept (*Don't leave the door* ***open****!*). Visual negation, then, is processed along the same lines as verbal negation. It does not unconditionally invoke suppression of negated concepts even when an alternative opposite is available. Rather it retains the concept within its scope which partakes in the representation of the visually negated stimulus. This study into the processes involved in interpreting visually negated percepts suggests that both suppression and negation are general cognitive processes not specific to linguistic systems only.

It is this accessibility of negated concepts that also allows them to "resonate" with related concepts in their environment, that is, to activate affinities across utterances (Du Bois 2001), as shown for affirmative concepts (see Du Bois 2001); it is this accessibility of negated concepts that allows activating an array of linguistic and conceptual elements in one speaker's utterance which "resonate" with elements in hers or another's in both prior and late context, as shown by both studies of natural discourses (Giora 2007) and lab results (Giora et al. 2007).

To illustrate, consider the following example which features a concept within the scope of negation (*don't give away even a slight* ***quiver*** *of a [combat aircraft's] wing*)[5], which nonetheless resonates with prior context (*earth****quake****):

---

[5] This is a reference to Dan Haluz's admission (following the killing of 14 Palestinian civilians by the Israeli Air Force, which he headed at the time) that when he drops a bomb all he feels is a slight quiver of the aircraft's wing.

(2) For months they tell us about an "earth**quake**" [the 2nd Lebanon war]. But the memorials have a mind of their own as if bereavement is a natural disaster or fate, and they do**n't** give away even a slight **quiver** of a [combat aircraft's] wing. (Misgav 2007)

As for forward resonance, consider the title of the article of the example cited above. This title – *Not a quake and not a quiver* – demonstrates forward resonance in which a given negated concept (*Not a quake*) resonates with, that is, activates affinities with the next negated concept, appearing in its late context (*not a quiver*).

Forward resonance allowed by negated concepts (*no monument and no memorial*) can also make accessible an affirmative (related) concept (*grave*). This is afforded only by the retainability of the concept within the scope of negation:

(3) [T]he time has come to ask Kastner's forgiveness. Perhaps this important film [*Killing Kastner* by Gaylen Ross] will carry out the historical task, in a place where Kastner has **no monument** and **no memorial**, except for his **grave**. (Levy 2008)

A recent event-related fMRI study further demonstrates that negation and affirmation need not exhibit asymmetrical behavior, since how they are processed often depends on their context (task included). Looking into the neural substrates of making negative and affirmative decisions about semantic relatedness, Stringaris, Medford, Giora, Giampietro, Brammer and David (2006) show that rejecting and endorsing semantic relatedness activates similar brain areas when related (*honesty*) and unrelated (*meetings*) probes are presented following (conventional) metaphors (*Some answers are* ***straight***). This, however, is not the case when related (*passion*) and unrelated (*meetings*) probes are presented following literals (*Some answers are* ***emotional***). More specifically, findings show that both rejecting a relation (saying "no" to *meetings*) and endorsing it (saying "yes" to *honesty*) following open-ended polysemous words (*straight*) invoke a similar search for (a wide range of) associations, activating the right ventrolateral prefrontal cortex (see Figure 1, images a and c). However, a non-open-ended, non-polysemous (literal) context (*emotional*) exhibits a neural asymmetry between negative and affirmative responses, activating different brain areas (see Figure 1 images b and d), showing that only endorsement involves a search for associations:

"No" "Yes"
Metaphor   Literal   Metaphor   Literal

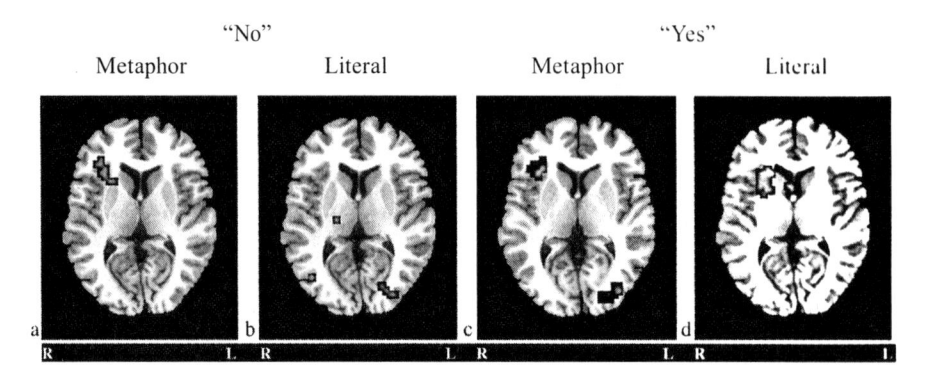

Figure 1. Right Frontal Cortex activation following metaphor: (a) Rejection ("No"); (c) Endorsement ("Yes")

Most of our previous studies, then, have argued that, contra the received view, negation and affirmation are functionally equivalent, because negation allows activation and retention of information within its scope (Giora 2006). In contrast to our previous research, however, this chapter focuses on one of the most intriguing asymmetries between negatives and affirmatives. Still, even this asymmetry relies heavily on the accessibility of information within the scope of negation.

In what way is this asymmetry unique? Although pragmatic and psycholinguistic studies of negation have demonstrated that negation might have a great number of effects (see Horn 2001), they all, however, have been shown to share a common feature. They all seem to mostly operate on an affirmative concept so that that concept ("X" in "Not X") undergoes some modification while being negated. This simply amounts to saying that negative constituents have been often shown to be semantically and pragmatically a derivative – the consequence of an operation on the (more basic, unmarked) affirmative. As shown earlier, "not clear", for instance, might mean any of the following: 'less than clear', 'not clear enough', 'should be clearer', 'a disagreement that it is clear', 'kind of vague', 'vague', etc.. In such cases, negation induces a variety of weakening effects of the affirmative X, ranging between slightly (less than 'clear') to wholly mitigating X to the extent that X is suppressed and replaced by an alternative opposite Y ('vague'). This is particularly true of scalar adjectives and predicates (Fraenkel and Schul 2008; Giora 2006; Giora, Balaban, Fein and Alkabets 2005; Hasson and Glucksberg 2006; Horn 1989/2001; Paradis and Willners 2006, inter alia).

The kind of negative utterance we focus on here is, however, different. It includes a set of negative utterances of the form "X is not Y" (*I am not your maid; This is not food*) that pragmatically are not derivable from their affirmatives. That is, their negative interpretation is radically different from their affirmative interpretation. Whereas the affirmative version of these utterances mostly gives rise to non-metaphoric interpretations, the negative versions mostly induce metaphoric interpretations.

How do people go about processing such negative utterances? What is the default context they activate to render such negative statements plausible? As will be seen later, to render such statements meaningful, speakers often activate a context in which the predicate (*not your maid*; *not food*) is related to the topic (*I, This*) in a nonliteral way. For instance, in the following examples (4)–(5), what the speaker means by the negated utterances (with bold highlighting added) is fleshed out later on (italicized, for convenience), making clear that, in both cases, the information within the scope of negation is intended metaphorically. That is, it is not the literal interpretation of either *your maid* ('an employed woman hired to do her job') or *food* ('foodstuff') that is dismissed here, but rather the nonliteral interpretation of these concepts ('someone that you can lay your demands [on] all of [the] time'; 'foodstuff fit for human consumption'). Put differently, it is not metaphor-irrelevant ('foodstuff') meanings that are rejected here but rather salient (Giora 1997, 2003; Ortony, Vondruska, Foss and Jones 1985), metaphor-relevant features ('fit for human consumption') that are dismissed, yet not at the cost of being dispelled from the mental representation:

(4) You tell me what to do all of the time, what to say, where to hide, and what to do. *I am not your wife* **I am not your maid**, *I'm not someone that you can lay your demands [on] all of [the] time, I'm sick of this it's going to stop!* (Blige 2007)

(5) "Tell TBS **this is not food**. They should concentrate on checking upon foodstuff imports many of which are *expired or sub-standard or unfit for human consumption*," said stall holder Saidi Abdallah Umbe. (BBC News 2003)

How do people make sense of the affirmative counterparts of these statements? What is the default context affirmative utterances such as *I am your maid, This is food* activate to render these statements plausible? As will be seen later, to render such statements meaningful, speakers often activate a context in which the predicate (*your maid; food*) is related to the topic (*I, This*) in a literal way, that is, in a way that also communicates or assumes metaphor-irrelevant meanings ('a woman employed to do certain jobs'; 'foodstuff to be eaten'; italicized, for convenience):

(6) "No, mum. **I am your maid**. It is *you, who picked me*. It is *my job to attend to you*, mum." (Summerfield 1998)

(7) **This is food,** and this is how you *eat* it. (Chamberlain 2005)

In this chapter we focus on this set of negative utterances which we term "negation-induced metaphors" (4, 5), and their affirmative counterparts (6, 7). Unlike regular metaphors, which communicate metaphoricity in both their affirmative and negative versions (Hasson and Glucksberg 2006, Keil 1979), this set of metaphors is unique. Pragmatically, negation-induced metaphors are not derived from their affirmative versions. Whereas the default interpretation of the affirmative versions is literal, the negative versions are by and large metaphoric. Like examples (4–7) above, the following examples (8–9) are illustrative: their (a) versions feature negation-induced metaphors (in bold) and instantiations/explications of their metaphoric interpretations (in italics); their (b) versions feature equivalent affirmative versions, whose interpretation is literal (in bold) accompanied by instantiations/explications of their literal interpretations (in italics):

(8) a. If you do not want to attend the class please drop it or let yourself get an 'F'. **I am not your secretary** *to file all the documents and keep track of the learning materials for you.* (Student 2008)

b. Hi everyone! My name is Stephanie Zguris and **I am your secretary**! *If you ever miss a meeting or want to know about upcoming meetings or events, I am the one to talk to!* (Zguris 2008)

(9) a. *Don't ever tell me that "I better do something on my blog."* **You are not my boss** *so don't tell me what to write.* (Joan 2008)

b. No keeping someone on staff. No extra payroll costs. No third party human resource company. This means *I work for you*, and **you are my boss**. (Banda 2008)

At first glance, one might suspect that negation-induced metaphors are negative polarity items (NPIs), items exhibiting asymmetric behavior in minimal pairs of negative and affirmative sentences (Israel 2004). Admittedly, on the face of it, there is some striking resemblance. Like polarity items, they do exhibit asymmetric behavior in minimal pairs of negative and affirmative sentences: Whereas the negative utterances are primarily metaphoric, their affirmative versions are primarily literal.

However, despite this superficial similarity, negation-induced metaphors differ from NPIs in various respects. First, NPIs are typically highly conventional/fossilized and appropriate whereas their affirmatives are not and are often nonexistent (Horn 1989: 49). By contrast, negation-induced metaphors need not be conventional. Instead, they can be entirely novel, their metaphoric interpretation constructed on the fly (see Experiment 3 below). Additionally, their affirmative counterparts are prevalent and appropriate, only intended to convey a different (literal) sense.

What, then, allows negation to generate figurativeness? Using affirmative and negative statements (such as 4–9 above), the present study tests the hypothesis that, among other things, negation generates figurativeness via highlighting metaphor-related features of the affirmative concept within its scope, while rendering its metaphor-unrelated (literal) features pragmatically irrelevant, regardless of whether they are true or false. Negating an affirmative concept, then, may enhance its salient/distinctive properties which may then be attributed to the topic of the (negative) statements.[6]

For example, by negating **food, wife, maid, secretary,** or **boss** etc. (see 4–5; 8a–9a above), the speaker enhances metaphor-relevant properties (italicized) where, for example, **not food** means [*food*] *unfit for human consumption;* **not your wife, not your maid** means *not someone that you can lay your demands[on] all of [the] time;* **not your secretary** means [*will not*] *file all the documents and keep track of the learning materials for you;* **not my boss** means *don't tell me what to write* etc. By bringing out the features of the source domain (*food, wife, maid, secretary, boss*), whether they are made explicit or need to be inferred, negation allows their attribution to the target domain (*this, I, you*) while rejecting its applicability. (On metaphor residing in attributing features of the source domain to the target domain, see e.g., Glucksberg 1995, Glucksberg and Keysar 1990).

It should be noted that, so far, the present hypothesis relates to utterances of the form "X is not Y" where X is a high accessibility referring expression (a pronoun), and hence hardly informative (Ariel 1990), and Y is a noun phrase.

It should be noted further that the source domain features attributable to the target domain need not be of superordinate abstraction level (as assumed by Glucksberg and Keysar 1990).[7] They should, however, share a

---

[6] On the process involved in ordinary metaphors whereby metaphor-related features are attributed to the topic (or target) of the metaphoric statement, see Glucksberg (1995).

[7] On metaphor-irrelevant meanings as pertaining to superordinate abstractions and metaphor-relevant meanings as pertaining to lower level abstractions, see e.g. Gernsbacher, Keysar, Robertson and Werner (2001), Rubio Fernández (2007).

common feature and be classifiable under a common superordinate category. For instance, while **not your secretary** (8a) could mean [*will not*] *file all the documents and keep track of the learning materials for you* or *will not print out and staple your work*,[8] these features should be categorizable as, for example, instances of 'servility' typical of secretarial assistance.

In three experiments involving native speakers of Hebrew (Section 2) and corpus-based studies examining equivalent English, German, and Russian utterances (Section 3), we test the hypothesis that some negative utterances (see 4, 5, 8a, 9a for typical constructions) tend to be interpreted nonliterally, even when no specific context is provided. Specifically, we aim to show that negation functions as a metaphor-inducing operator – a device that enhances the figurative interpretation of the concept it rejects, while rendering its literal interpretation pragmatically irrelevant to the interpretation process.

In Experiment 1, participants were instructed to decide whether contextless affirmatives (*I am your maid; this is food*) and their negative counterparts (*I am not your maid; this is not food*) communicate either a literal or a metaphoric interpretation. Experiment 2 compared affirmative (*almost*) and negative (*not*) modifiers in order to demonstrate that a negative but not an affirmative modifier is a metaphor-inducing device. Participants were presented affirmative statements (*I am almost your maid; this is almost food*) and their negative counterparts (*I am not your maid; this is not food*) and were asked to rate the extent to which they were (non)literal.

Because many of the negative items of Experiments 1–2 could be rather familiar, Experiment 3 used only highly novel negative statements (**This is not Memorial Day; I am not your doctor**) and their equally novel affirmatives (**This is Memorial Day; I am your doctor**). Design and procedure were the same as in Experiment 2.

Findings, demonstrating the prevalence of figurative meanings in negative but not in affirmative constructions, are then corroborated by naturally occurring uses which also demonstrate the way in which the discourse environment of these negative items resonates with their metaphoric interpretation (Section 3).

On metaphor vehicles representing superordinate categories, see Glucksberg and Keysar (1990), Shen (1992).

[8] http://iweb.tntech.edu/kosburn/history-202/Cause%20and%20effect.htm (retrieved July 27, 2008)

2. Experimental data

Experiment 1

The aim of Experiment 1 was to test the hypothesis that negation enhances metaphor-related properties. Specifically, it tests the prediction that, when having to decide whether a statement is intended either literally or metaphorically, participants will opt for the metaphoric interpretation when encountering a negative statement but significantly less so when encountering its affirmative counterpart.

Method

Participants. Forty-eight students of linguistics at Tel Aviv University (33 women, 15 men), mean age 24.4 years old, volunteered to participate in the experiment.

Materials. Materials included 24 context-less affirmatives (***I am your maid; this is food***) and their negative counterparts (***I am not your maid; this is not food***). Two booklets were prepared so that each participant would be presented with only one item of a pair. Each booklet included 12 affirmative items, 12 negative items and 17 filler items, about half of which were negative (***I am not hungry now***).

Procedure. Participants were instructed to decide whether each of the items either communicates a literal or a metaphoric interpretation. No participant saw more than one version of each item.

Results and Discussion

As predicted, comprehenders opted for the metaphoric interpretation when they were judging the negative items but significantly less so when they were judging the affirmative items, which tended to be interpreted literally. Specifically, the mean probability of negative items to be judged as metaphoric was higher (68%) than the mean probability of their affirmative versions (43%). The difference was significant in both subject (t_1) and item (t_2) analyses, ($t_1(47)=7.09, p<.0001; t_2(23)=7.19, p<.0001$). Results thus support the view that negation, can, indeed, function as a metaphor inducing operator.

Since we forced comprehenders to choose between two alternatives and did not allow them a chance to grade their responses or even decide on another response, we ran another experiment. In Experiment 2, participants were asked to rate the interpretation of the targets on a 7 point scale ranging between two specific (either literal or metaphoric) interpretations. In addition to these interpretations or as an alternative, they were allowed to come up with an interpretation of their own. This experiment focused on comparing negative items (involving the negative modifier "not") and their affirmative alternatives (involving the affirmative modifier "almost"[9]). This experiment thus allows us to compare negatively and affirmatively marked versions of utterances.

Experiment 2

The aim of Experiment 2 was to show that even when allowed a wider range of choices, comprehenders find the negatively marked items more metaphoric than their affirmative counterparts.

Method

Participants. Participants were 24 students at Tel Aviv University (9 women and 15 men), mean age 25.3 years old who volunteered to participate in the experiment.

Materials. Materials included 16 items involving a negative modifier (***You are not my boss***) and their counterparts including an affirmative modifier – *almost* (***You are almost my boss***). Two booklets were prepared so that each participant would be presented with only one item of a pair. Each booklet included about 8 affirmative items, 8 negative items and 7 similar filler items.

Each item was followed by a 7 point scale which featured two different interpretations – either literal or metaphoric – presented randomly at each end of the scale:

(10) **You are not my maid**

☐——☐——☐——☐——☐——☐——☐

Don't serve me   You are not the
                                                                                  person who cleans my
                                                                                  place for a living

(11) **You are almost my maid**

☐——☐——☐——☐——☐——☐——☐

You help me a lot   You are about to
with the house-   get the job as a
keeping chores   maid in my house

---

[9] Although *almost* entails negation (see Horn, 2002, 2009 and references therein), it is considered here an affirmative modifier, because it is not overtly marked for negation.

Procedure. Participants who agreed to take part in the experiment were sent an electronic booklet. They were asked to rate, on a 7 point scale, whose ends instantiated either a literal (=1) or a metaphoric (=7) interpretation of each item, the proximity of the interpretation of the item to any of those instantiations at the scale's ends. In case they did not agree with both interpretations, they were allowed to come up with an interpretation of their own.

Results and Discussion

Since there were only 16 cases out of 384 (4%) in which participant offered their own interpretations, we did not include them in the analysis. Negative statements were rated as more metaphoric (M=6.02, SD=0.65) than their affirmatives counterparts (M=5.59, SD=0.70). The difference was significant in the subject (t_1) analysis, and marginally significant in the item (t_2) analysis ($t_1(23)=2.50, p<.01$; $t_2(15)=1.56, p=.07$).

In all, results of Experiments 1–2 show that, as assumed, negation generates metaphoricity. When faced with an either/or choice, participants decided on a metaphoric interpretation for the negative but not for the affirmative items (Experiment 1). When allowed a graded choice, they attributed a metaphoric interpretation to the negative rather than to the affirmative items (Experiment 2).

Because many of the negative items of Experiments 1–2 are used metaphorically quite frequently, we designed Experiment 3, in which novel negative statements (***This is not Memorial Day; I am not your doctor***) and their equally novel affirmatives (***This is Memorial Day; I am your doctor***) were tested.

Experiment 3

Method

Participants. Participants were 48 students and high-school graduates (31 women and 17 men), mean age 25.6 years old who either volunteered to participate in the experiment or were paid 15 Israeli shekels (about $4).

Materials. To ensure that we use only novel metaphors, we ran a pretest, in which 31 affirmative utterances and their negative counterparts were rated for familiarity by 50 participants (students and high-school graduates). Two booklets were prepared so that each participant would be presented only one item of a pair. Each booklet included about 15 affirmative items, 15 negative items, and 15 filler items (familiar metaphors, half of which were negated) which were the same for both booklets.

Participants were asked to rate the items' familiarity on a 7 point familiarity scale ranging from 1 ("Not familiar at all; Never heard it") to 7 ("Highly familiar; I hear it all the time"). For the actual experiment, 15 items were selected, those that scored below 4 (in both the affirmative and negative versions) and which, in addition, had similar familiarity ratings for the affirmative and negative versions, as shown by t-tests, which did not reveal any significant differences (p value was always above .20).

Materials for the actual experiment, then, were the 15 novel affirmatives and their (equally novel) negative counterparts, selected on the basis of the pretest's results described above. Two booklets were prepared so that each participant would be presented only one item of a pair. Each booklet included 7 or 8 affirmative items and 7 or 8 negative items, modeled after the presentation of items in Experiment 2 (see 12–13), and 15 filler items (familiar metaphors, half of which were negated):

(12) **This is not Memorial Day**

☐————☐————☐————☐————☐————☐————☐

No need to   We are not cele-
be so sad   brating Memorial
                                                        Day today

(13) **This is Memorial Day**

☐————☐————☐————☐————☐————☐————☐

Everybody   We are celebrating
is sad today   Memorial Day
                                                    today

Procedure. As in Experiment 2.

Results and Discussion

Since there were only 27 cases out of 720 (3.8%) in which participants offered their own interpretations, we did not include them in the analysis. Results show that novel negative statements were rated as more metaphoric (M=5.50, SD=0.96) than their affirmative counterparts (M=3.48, SD=1.27). The difference was significant in both subject (t_1) and item (t_2) analyses ($t1(47)=10.17, p<.0001; t2(14)=4.36, p<.0005$).

Overall, results from novel and familiar utterances (of the form "This is not..."; "I am not..."; You are not..."), where the topic is a pronoun hardly informative about the specific nature of the referent and the predicate includes a noun phrase, support the view that negation generates figurativeness as a default interpretation.

3. Corpus data

If negation indeed generates figurativeness as a default interpretation, we should be able to show that natural instances of some negative items (of the form "This is not..."; "I am not..."; "You are not...") are used figuratively more often than their affirmative counterparts. Since Experiments 1-3 providing support for this hypothesis were run in Hebrew, we tested its predictions on other languages such as English, German, and Russian. To do that, we selected a few examples and searched their first ~50 occurrences in both their affirmative and negative versions, using engines such as Google, Yahoo, Start, MSN, and Netex. We first studied their interpretations: On the basis of their context, 2 judges (a research assistant and the first author) decided whether each utterance was used either figuratively or literally. Agreement between judges was high overall, and all differences were resolved after a discussion. We expected these negative items to be considered figurative more often than their affirmative counterparts (3.1).

Second, we studied their context and how it reflects their interpretations: the same judges looked at the negative items' environment to see the extent to which contextual information resonates with either the metaphoric or the literal interpretation of the negative items. Again, agreement between judges was high overall, and all differences were resolved after a discussion. We expected the environment of the negative items to reflect their metaphoric interpretation to a greater extent than their literal interpretation (3.2).

3.1. Distribution of metaphoric and literal interpretations

Corpus-based studies were run in 3 languages: English, German, and Russian. Findings of these studies are presented in Table 1 (English), Table 2 (German), and Table 3 (Russian), and in Figures 2–4 accordingly. As predicted, they demonstrate that, invariably, the negative versions are more metaphoric than their affirmative counterparts, as shown by z-ratio tests for the difference between two independent proportions (proportion of metaphoric interpretations of Negative vs. Affirmative statements). They further show that the negative versions are primarily metaphoric, that is, used metaphorically more often (i.e., in more than 50% of the cases) than literally and that, by the same token, the affirmative versions are primarily literal (i.e., used metaphorically in less than 50% of the cases). This is strikingly true of almost all items:

Table 1. Proportions of metaphoric interpretations of negative vs. affirmative utterances in *English* and results of z-ratio tests for the difference between them

	Negative	Affirmative	z-ratio, significance
I am not your maid / I am your maid	90.4% (47/52)	30% (15/50)	6.24 **
I am not your secretary / I am your secretary	95.7% (44/46)	12% (6/50)	8.20 **
You are not my mom / You are my mom	36% (18/50)	6% (3/50)	3.68 **
I am not your mom / I am your mom	50% (25/50)	16% (8/50)	3.62 *

* p<.0005, ** p<.0001

Table 2. Proportions of metaphoric interpretations of negative vs. affirmative utterances in *German* and results of z-ratio tests for the difference between them

	Negative	Affirmative	z-ratio, significance
Das ist kein Essen (This is not food) / Das ist Essen (This is food)	80% (20/25)	12.8% (6/47)	5.66 **
Das ist kein Spiel (This is not a game) / Das ist ein Spiel (This is a game)	66% (33/50)	22% (11/50)	4.43 **
Du bist nicht meine Mutter (You are not my mom) / Du bist meine Mutter (You are my mom)	82% (41/50)	20% (10/50)	6.20 **
Ich bin nicht deine Mutter (I am not your mom) / Ich bin deine Mutter (I am your mom)	65.9% (29/44)	12% (6/50)	5.40 **

* p<.0005, ** p<.0001

Table 3. Proportions of metaphoric interpretations of negative vs. affirmative utterances in *Russian* and results of z-ratio tests for the difference between them

	Negative	Affirmative	z-ratio, significance
Я не твоя секретарша (I am not your secretary) / Я твоя секретарша (I am your secretary)	85% (17/20)	20% (6/30)	4.52 **
Он не мой сын (He is not my son) / Он мой сын (He is my son)	80% (40/50)	2% (1/50)	7.93 **
Ты не моя мама (You are not my mom) / Ты моя мама (You are my mom)	24% (12/50)	0% (0/50)	3.69 **
Я не твоя мама (I am not your mom) / Я твоя мама (I am your mom)	72% (36/50)	10% (5/50)	6.30 **
Это не моё тело (This is not my body) / Это моё тело (This is my body)	80% (40/50)	12% (6/50)	6.82 **

* p<.0005. ** p<.0001

English Data

Bar chart showing % of Metaphoric Interpretations (0–100) for Affirmative vs. Negative:
- I am (not) your maid
- I am (not) your secretary
- You are (not) my mom
- I am (not) your mom

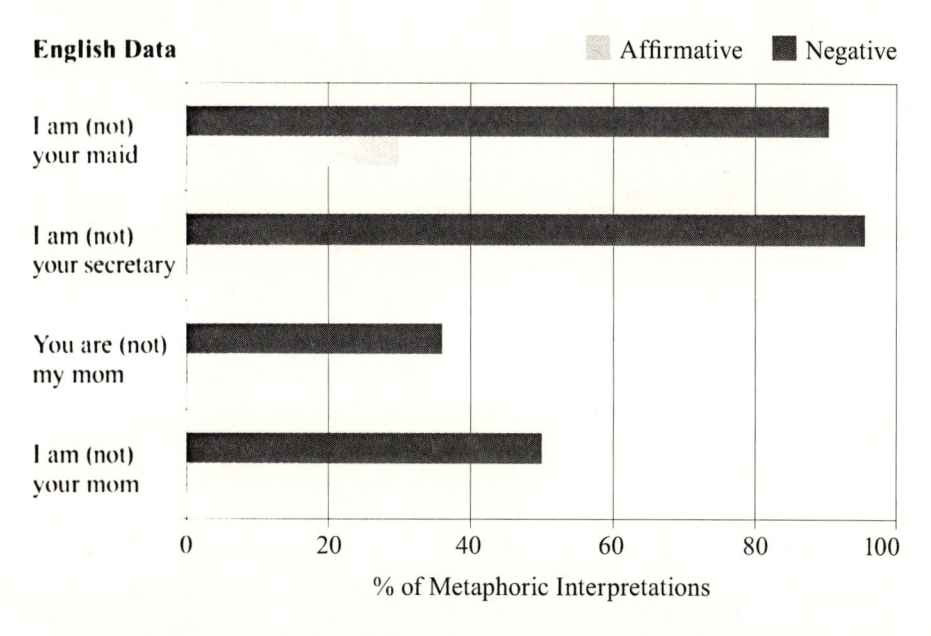

Figure 2. Percentage of Metaphoric interpretations of Affirmative vs. Negative Utterances – English

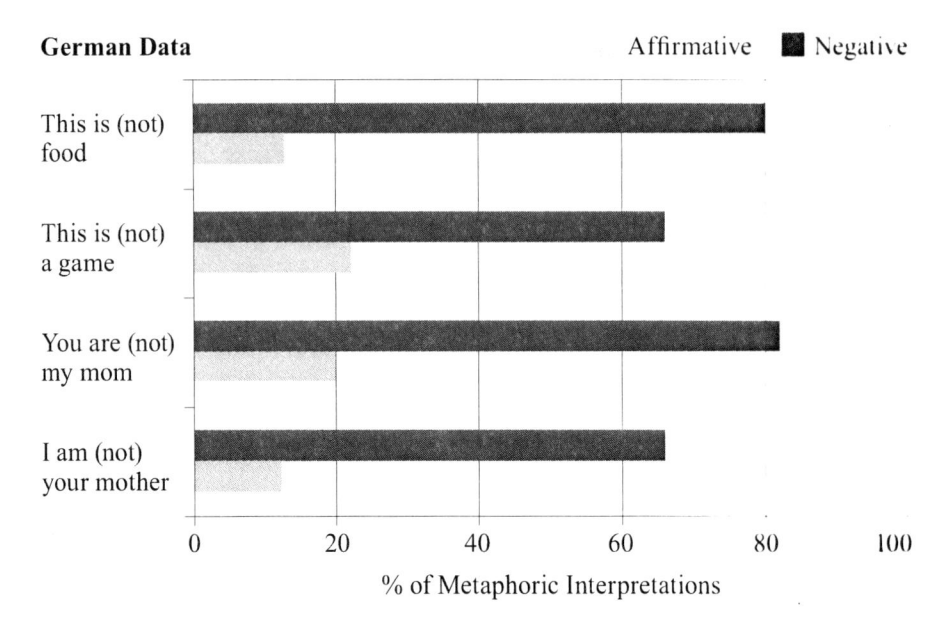

244   *Rachel Giora, Ofer Fein, Nili Metuki and Pnina Stern*

German Data

Figure 3. Percentage of Metaphoric interpretations of Affirmative vs. Negative Utterances – German

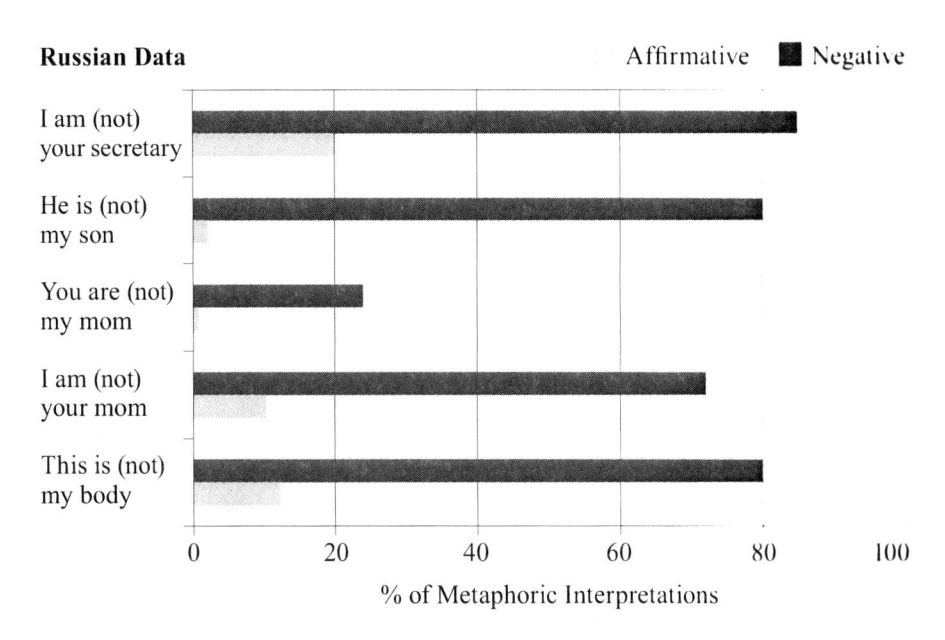

Russian Data

Figure 4. Percentage of Metaphoric interpretations of Affirmative vs. Negative Utterances – Russian

3.2. Environment of negative items – findings

Given that negative utterances of the form X is not Y (where X is a pronoun and Y is a noun phrase) are primarily metaphoric, we expect linguistic elements in their environment to resonate (à la Du Bois 2001) with their metaphoric rather than literal interpretation. Moreover, since these utterances often express a kind of a complaint (see 14) afforded by the negation, we expect their resonance effect to be rather intense. This can be achieved either by generating similar metaphors expressing the same negative stance or by explicating what is communicated by them:

(14) She was the principal in a high school; I feared her and respected her for her cleverness and professionalism. She brought me to the elementary school, and trained me to be independent, and punished me when I did something wrong as well. Often I **complained** from my heart, "**you are not my mom.**" (Irene 2000)

An instance of the environment of negative items can be found in example (4) above, which features the metaphoric *I am not your maid* (repeated in 15, for convenience). It is illustrative of both, the way the context resonates with the (target) metaphor (*I am not your maid*) via a prior negative metaphor (*I am not your wife*), conveying the same meaning (bolded and italicized) and by what it communicates, which is made explicit later on (*italicized*):

(15) You tell me what to do all of the time, what to say, where to hide, and what to do. ***I am not your wife* I am not your maid**, *I'm not someone that you can lay your demands [on] all of [the] time, I'm sick of this it's going to stop!* (Blige 2007)

Example (16) is also a case in point. It features a great number of negative metaphors (in bold) all resonating with each other, intending to convey the same metaphoric interpretation, which is then made explicit (*does not mean I am yours*):

(16) **I am not your therapist.**
I am not your savior.
I am not your way out.
I am not your servant.
I am not your secretary.
I am not your mother.
I am not your whore.

I am not your entertainment.
I am not your encyclopedia.
I am not your babysitter.

Just because I love you *does not mean I am yours.* (GUCK 2001)

Having shown that the negative items we studied are metaphoric (Sections 2, 3.1), we expect, then, that the context of these negative metaphors will abound in either or both: (a) uses of similar metaphors which communicate the same metaphoric interpretation; (b) explications of the metaphoric interpretation. Crucially, we expect metaphoric resonance to outweigh literal resonance (examples of the latter can be found in (17)–(18), italicized for convenience):

(17) User: Umm. I may have to go soon…
Jabberwacky: Oh, you're not going anywhere.
User: In a minute I'll scream.
Jabberwacky: **I'm not your maid.**
User: *No, but I'm sure you'd make a great maid!??*

(thisfred 2004)

(18) 1. **You are not my boss**.
2. *If you were my boss, we would have a serious discussion about THE MONEY.*[10]

Corpus-based findings concerning the kind of contextual resonance with negative metaphors are presented in Table 4 (English), Table 5 (German) and Table 6 (Russian), and in Figure 5 (English) and Figure 6 (German and Russian). They are divided between "Only metaphoric resonance" results, where context resonated only with the metaphoric interpretation of the target negative statement; "Only literal resonance" results, where context resonated only with the literal interpretation of the target negative statement; "Both metaphoric and literal resonance" results, where context resonated with both the metaphoric and literal interpretations of the target negative statement; and "No resonance" results, where context did not resonate with any interpretation.

---

[10] http://multiplex.integralinstitute.org/Public/cs/forums/11/54536/ShowThread.aspx (Retrieved June 9, 2008).

Findings show that, overall, the environment of negative metaphors reflects their metaphoric interpretation rather than their literal interpretation. For instance, the environment of *I am not your maid* (first raw of Table 4) resonates with its metaphoric interpretation in 74.5% (35/47) of the cases, while its literal interpretation is resonated with in only 25.5% (12/47).

However, *p*-values presented in the tables, are the results of Exact Binominal Probability tests, performed in each case for "Only metaphoric resonance" against "Only literal resonance". They test whether the probability of getting metaphoric resonance is significantly higher than chance level (50%). For example, in "I am not your maid", from the 35 occurrences with "Only metaphoric" and "Only literal" resonance, "Only metaphoric" resonance occurred 29 times (82.9%), which is significantly higher than 50% (p<.005). In all the cases, the superiority of the metaphoric resonance was evident, and only in one case ("I am not your maid" in Russian) it was not significant.

The figures present occurrences in which metaphoric or literal resonance appeared. That is, "metaphoric resonance" is the sum of "Only metaphoric resonance" and "Both metaphoric and literal resonance". The same holds for "literal resonance". As shown by the figures, in all the cases, except for the two Russian examples (which, in all, exhibited poor resonance), the environment included metaphoric resonance in more than 50% of the cases.

Table 4. Distribution of different types of resonance in the environment of negative utterances in English and results of exact binominal probability test for the superiority of metaphoric resonance

	Only Metaphoric resonance	Only Literal resonance	Both Metaphoric and literal resonance	No resonance	p-values
I am not your maid	61.7% (29/47)	12.8% (6/47)	12.8% (6/47)	12.8% (6/47)	p<.0005
You are not my mom	55.6% (10/18)	5.6% (1/18)	27.8% (5/18)	11.1% (2/18)	p<.01
I am not your secretary	79.5% (35/44)	4.5% (2/44)	9.1% (4/44)	6.8% (3/44)	p<.0005

Table 5. Distribution of different types of resonance in the environment of negative utterances in German and results of exact binominal probability test for the superiority of metaphoric resonance

	Only Metaphoric resonance	Only Literal resonance	Both Metaphoric and literal resonance	No resonance	p-values
Ich bin nicht deine Mutter (I am not your mom)	58.6% (17/29)	3.5% (1/29)	13.8% (4/29)	24.1% (7/29)	p<.0005
Du bist nicht meine Mutter (You are not my mom)	63.4% (26/41)	4.9% (2/41)	17.1% (7/41)	14.6% (6/41)	p<.0005
Das ist kein Essen (This is not food)	40% (14/35)	5.7% (2/35)	14.3% (5/35)	40% (14/35)	p<.005
Das ist kein Spiel (This is not a game)	54.5% (18/33)	3% (1/33)	15.2% (5/33)	27.3% (9/33)	p<.0005

Table 6. Distribution of different types of resonance in the environment of negative utterances in Russian and results of exact binominal probability test for the superiority of metaphoric resonance

	Only Metaphoric resonance	Only Literal resonance	Both Metaphoric and literal resonance	No resonance	p-values
Я не твоя секретарша (I am not your secretary)	20% (4/20)	5% (1/20)	5% (1/20)	70% (14/20)	p=.19
Я не твоя мама (I am not your mom)	12% (6/50)	0% (0/50)	2% (1/50)	86% (43/50)	p<.05

Negation as a metaphor-inducing operator 249

English Data Metaphoric resonances ■ Literal resonances

- I am not your maid (English)
- You are not my mom (English)
- I am not your secretary (English)

% of Metaphoric Interpretations

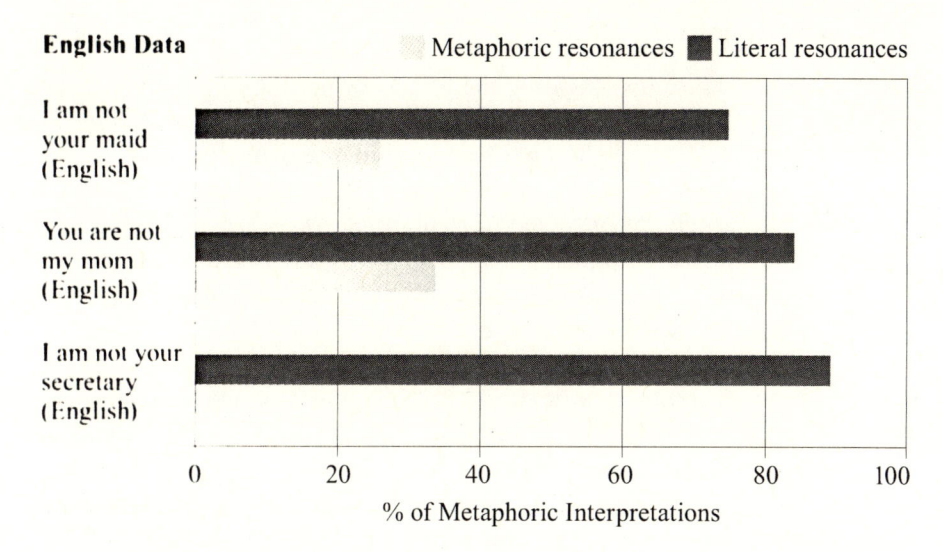

Figure 5. Percentage of Metaphoric vs. Literal Resonance in the Environment of Negative Utterances – English

German and Russian Data Metaphoric resonances ■ Literal resonances

- I am not your mom (German)
- You are not my mom (German)
- This is not food (German)
- This is not a game (German)
- I am not your secretary (Russian)
- I am not your mom (Russian)

% of Metaphoric Interpretations

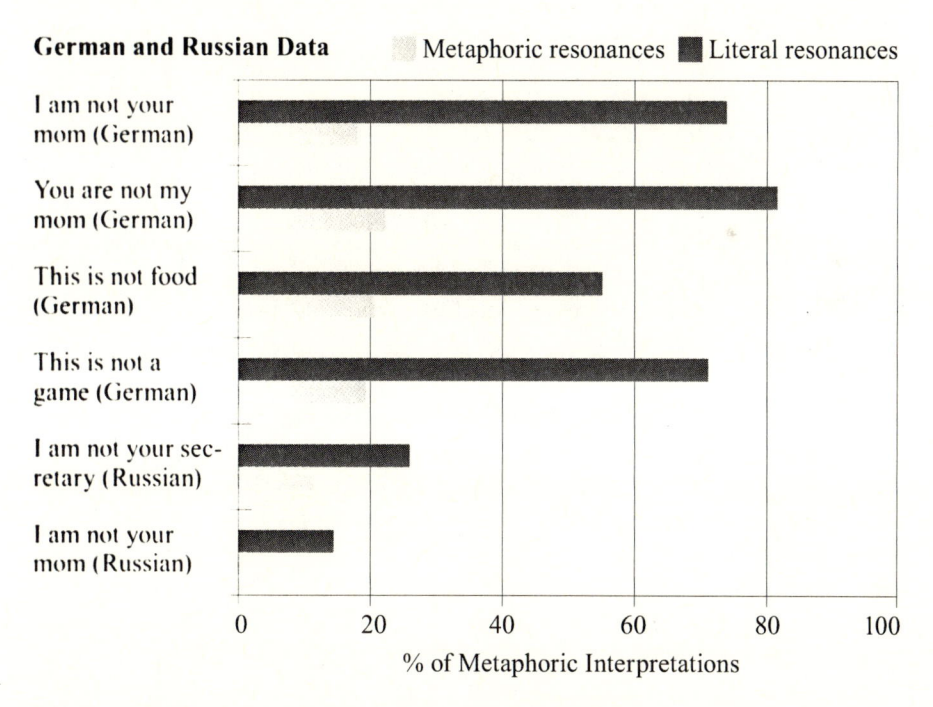

Figure 6. Percentage of Metaphoric vs. Literal Resonance in the Environment of Negative Utterances – German and Russian

4. General discussion

Most of the literature investigating the effects of negation on the concepts within its scope views negation as an accessibility-reducing operator. Its effects mostly range from slight modification (mitigation) to total suppression of the negated concept (for a review, see Giora 2006 and Section 1 above).

In three experiments conducted in Hebrew (Section 2), accompanied by corpus-based studies of English, German, and Russian (Section 3), we show that negation need not be a suppressor. Instead it can be an enhancer, inducing metaphoricity. Thus, in negative utterances of the form of "X is not Y", where X is a high accessibility referring expression (a pronoun) and Y a noun phrase, negation is a device that highlights metaphor-related properties of the source domain concept (Y). Their projection onto the target domain (X), however, is rejected as inapplicable. By contrast, affirmative counterparts come across as significantly less metaphorical since, in the absence of an enhancer, metaphor-related features are not brought out.

In Experiment 1, participants were instructed to decide whether context-less negatives (*I am not your maid*) and their affirmative counterparts (*I am your maid*) have either a literal or a metaphoric interpretation. Findings show that negative but not affirmative items were interpreted figuratively: The mean probability to be judged as metaphoric was higher for the negative utterances than for their affirmative versions. Experiment 2 compared affirmative (*almost*) and negative (*not*) modifiers and allowed a graded rather than a dichotomous response. Participants were presented negative statements (*I am not your maid*) and their affirmative counterparts (*I am almost your maid*). They were asked to rate the proximity of their interpretation of the items to those instantiations at a scale's ends. Results show that, compared to an affirmative modifier, a negative modifier is a stronger metaphorizing device, promoting metaphoric interpretations.

Given that experiments 1–2 might have included conventional (negative) items, Experiment 3 was designed to test the metaphoricity hypothesis with regard to novel items. Degree of novelty was established by a pretest. In this experiment, novel negative statements (*This is not Memorial Day*) and their equally novel affirmatives (*This is Memorial Day*) were tested in the same way previous items were (see Experiment 2). Results show that the novel negative statements were rated as significantly more metaphoric than their equally novel affirmative counterparts. Overall, results from the three experiments support the view that negation may generate figurativeness as a default interpretation (Section 2).

Corpus-based studies in three languages (English, German, and Russian) corroborate the experimental results (Section 3). They show, first, that in various languages speakers use negative statements (of the form mentioned above) metaphorically while their affirmative counterparts are used literally. An additional inspection of the environment of the negative statements further supports this asymmetry. It demonstrates that, as expected, in most of the cases studied, the environment of these utterances resonates with their metaphoric rather than with their literal interpretation. This provides further support for the view that negation can retain the concepts within its scope, which, under certain circumstances, allows negative utterances to come across as metaphoric.

A brief look at instances of implicit negation suggests that even when negation is implied, it has a similar effect (see also the figurativeness ratings following *almost* in Experiment 2). For instance, rhetorical questions, whose implication is negative, such as *What am I, your secretary?/Am I your secretary?* (19)–(20) are used metaphorically (94%, 17/18) rather than literally (6%, 1/18). This is further supported by their environment, which resonates with their metaphoric interpretation (italicized). Similarly, *Am I your mom?/What am I, your mom?* (21)–(22), are also used metaphorically (84%, 16/19) rather than literally (16%, 3/19), as also shown by their environment, which resonates with their metaphoric interpretation (italicized):

(19) I'm sorry? **Am I your secretary?** *Am I even a secretary? So stop handing off your work to me, like you always do, and do it your damn self.*
 (Seanzky 2008)

(20) **What am I your secretary?** *google [age of conan xbox 360] and see for yourself.* (Robusto 2008)

(21) You want a decsription? *Read the comic and write it yourself.* **What am I, your mom?** (Mann, retrieved 7 October, 2008)

(22) Oh, *I don't care what you say to who.* **What I am I, your Mom?**
 (Victoria "cloroxcowgirl" B., 2008)

Negation, then, induces metaphoricity by denying the attribution of metaphor-related properties to the topic of the negative statement. Rejecting via negation then need not reduce the accessibility of the negated concept, nor need it dispel that concept from the mental representation. Instead, negation may enhance information within its scope, which in turn, may effect metaphoricity.

Acknowledgments

This research was supported by The Israel Science Foundation grant (No. 652/07) to Rachel Giora. We thank John Du Bois for the cartoon example and its contextual background. We thank Arnon Kehat and Hagit Sadka for running the first experiment and Haim Dubossarsky, Kerstin Winter, and Polina Zozulinsky for their help in the corpus searches. Thanks are also extended to John Du Bois, Sam Glucksberg, Laurence R. Horn, Joseph Lubovsky, Aviah Morag, Eran Neufeld, Yeshayahu Shen, and Argyris Stringaris for very helpful comments and examples.

References

Ariel, Mira
    1990   *Accessing noun phrase antecedents.* London: Routledge.

Babin, Rex
    2008   I am not President Bush. The Sacramento Bee. (2008-10-17). http://www.sacbee.com/1088/rich_media/1322758.html.

Banda, Tracy
    2008   Tracy's Office Services. (2008-08-21). http://www.tracysofficeservices.com/rightbenefits.htm.

BBC News Africa
    2003   Used pants ban in Tanzania. (2003-10-23). http://news.bbc.co.uk/2/hi/africa/3198746.stm.

Blige, Nellie
    2007   http://www.streetpoetry.net/id12.html (2008-05-03).

Chamberlain, Larry
    2005   Why Does My Cat Bring Home Her Prey? www.articlealley.com/article_19599_54.html.

Du Bois, John W.
    2001   Towards a dialogic syntax. Unpublished ms., University of California, Santa Barbara.

Eilam, Aviad
    2009   The crosslinguistic realization of -ever: Evidence from Modern Hebrew. *Proceedings of the 43rd Annual Meeting of the Chicago Linguistic Society (CLS), Vol. 2*, Malcolm Elliott et al. (eds.), 39–53. Chicago: Chicago Linguistic Society.

Fischler, Ira, Paul A. Bloom, Donald G. Childers, Salim E. Roucos and Nathan W. Perry Jr.
    1983   Brain potentials related to stages of sentence verification. *Psychophysiology* 20: 400–409.

Ferguson, Heather J. and Anthony J. Sanford
  2008   Anomalies in real and counterfactual worlds: an eye-movement investigation. *Journal of Memory and Language* 58: 609–626.
Ferguson, Heather J., Anthony J. Sanford and Hartmut Leuthold
  2008   Eye-movements and ERPs reveal the time course of processing negation and remitting counterfactual worlds. *Brain Research* 1236: 113–125.
Fraenkel, Tamar and Yaacov Schul
  2008   The meaning of negated adjectives. *Intercultural Pragmatics* 5 (4): 517–540.
Gernsbacher, Morton Ann
  1990   *Language comprehension as structure building*. Hillsdale, NJ: Erlbaum.
Gernsbacher, Morton Ann, Boaz Keysar, Rachel W. Robertson and Necia K. Werner
  2001   The role of suppression and enhancement in understanding metaphors. *Journal of Memory and Language* 45: 433–450.
Giora, Rachel
  1997   Understanding figurative and literal language: The graded salience hypothesis. *Cognitive Linguistics* 7: 183–206.
Giora, Rachel
  2003   *On our Mind: Salience, Context and Figurative Language*. New York: Oxford University Press.
Giora, Rachel
  2006   Anything negatives can do affirmatives can do just as well, except for some metaphors. *Journal of Pragmatics* 38: 981–1014.
Giora, Rachel
  2007   "A good Arab is not a dead Arab – a racist incitement": On the accessibility of negated concepts. In *Explorations in Pragmatics: Linguistic, Cognitive and Intercultural Aspects*, I. Kecskés and L. R. Horn (eds.), 129–162. Berlin/New York: Mouton de Gruyter.
Giora, Rachel, Noga Balaban, Ofer Fein and Inbar Alkabets
  2005   Negation as positivity in disguise. In *Figurative Language Comprehension: Social and Cultural Influences*, Herbert L. Colston and Albert Katz (eds.), 233–258. Hillsdale, NJ: Erlbaum.
Giora, Rachel, Ofer Fein, Keren Aschkenazi and Inbar Alkabets-Zlozover
  2007   Negation in context: A functional approach to suppression. *Discourse Processes* 43: 153–172.
Giora, Rachel, Ofer Fein, Jonathan Ganzi, Natalie Alkeslassy Levi and Hadas Sabah
  2005   Negation as mitigation: the case of negative irony. *Discourse Processes* 39: 81–100.
Giora, Rachel, Vered Heruti, Nili Metuki and Ofer Fein
  2009   "When we say *no* we mean *no*": Interpreting negation in vision and language. *Journal of Pragmatics* 41 (11): 2222–2239.

Giora, Rachel, Dana Zimmerman and Ofer Fein
    2008   How can you compare! On negated comparisons as comparisons. *Intercultural Pragmatics* 5(4): 501–516.

Glucksberg, Sam
    1995   Commentary on nonliteral language: Processing and use. *Metaphor and Symbol* 10(1): 47–57.

Glucksberg, Sam and Boaz Keysar
    1990   Understanding metaphorical comparisons: Beyond similarity. *Psychological Review* 97(1): 3–18.

GUCK
    2001   http://gucky.livejournal.com/2001/02/28/ (2008-06-09).

Hasson, Uri and Sam Glucksberg
    2006   Does understanding negation entail affirmation? An examination of negated metaphors. *Journal of Pragmatics* 38: 1015–1032.

Horn, Laurence R.
    1989   *A Natural History of Negation*. Chicago: University of Chicago Press.

Horn, Laurence R.
    2001   *A Natural History of Negation* (Reprint with new introduction). Stanford: CSLI.

Horn, Laurence R.
    2002   Assertoric inertia and NPI licensing. In *Proceedings from the panels of the thirty-eight meeting of the Chicago Linguistic Society*, Vol. 38-2, M. Andronis, E. Debenport, A. Pycha and K. Yoshimura (eds.), 55-82. Chicago Linguistic Society.

Horn, Laurence R.
    2009   *Almost* et al.: Scalar adverbs revisited. In *Current Issues in Unity and Diversity of Languages* (Papers from CIL 18, Seoul, Korea). Seoul: Linguistic Society of Korea.

Irene
    2000   A respectable helpful person. http://palc.sd40.bc.ca/palc/StudentWriting eight.htm#Respectable

Israel, Michael
    2004   The pragmatics of polarity. In *The Handbook of Pragmatics*, L. Horn and G. Ward (eds.), 701–723. Oxford: Blackwell.

Jespersen, Otto
    1917   *Negation in English and other languages*. Copenhagen: Host.

Jespersen, Otto
    1924   *The philosophy of grammar*. London: Allen and Unwin.

Joan
    2008   20 Responses to "Bridget Moynahan Must Have Laughed, or Why I'm Almost Glad the Patriots Lost". (2008-03-28). http://www.collegiate-times.com/blogs/2008/02/05/bridget-moynahan-must-have-laughed-or-why-im-almost-glad-the-patriots-lost/.

Kaup, Barbara, Jana Lüdtke and A. Rolf Zwaan
    2006   Processing negated sentences with contradictory predicates: is a door that is not open mentally closed? *Journal of Pragmatics* 38: 1033–1050.
Keil, Frank, C.
    1979   *Semantic and conceptual development*. Cambridge, MA: Harvard University Press.
Kilpatrick, Carroll
    1973   Nixon Tells Editors, 'I'm Not a Crook'. http://www.washingtonpost.com/wp-srv/national/longterm/watergate/articles/111873-1.htm (1973-11-18)
Levine, William H. and Joel A. Hagaman
    2008   Negated concepts interfere with anaphor resolution. *Intercultural Pragmatics* 5(4): 471–500.
Levy, Gideon
    2008   He sold his soul. (2008-10-23). http://www.haaretz.com/hasen/spages/1030567.html
MacDonald, Maryellen C. and Marcel A. Just
    1989   Changes in activation levels with negation. *Journal of Experimental Psychology: Learning, Memory, and Cognition* 15: 633–642.
Mann, John
    2008   Review of Coming Up Violet.http://www.webcomicsnation.com/reviews.php?series=violet (2008-10-07).
Mayo, Ruth, Yaacov Schul and Eugene Burnstein
    2004   "I am not guilty" versus "I am innocent": the associative structure activated in processing negations. *Journal of Experimental Social Psychology* 40: 433–449.
Misgav, Uri
    2007   Not a quake and not a quiver (in Hebrew). (2007-04-23). http://www.haaretz.co.il/hasite/spages/851585.html.
Ortony, Andrew, Richard J. Vondruska, Mark A. Foss and Lawrence E. Jones
    1985   Salience, similes, and asymmetry of similarity. *Journal of Memory and Language* 24: 569–594.
Paradis, Carita and Caroline Willners
    2006   Antonymy and negation: the boundedness hypothesis. *Journal of Pragmatics* 38: 1051–1080.
Pearce, David and Wolfgang Rautenberg
    1987   Propositional logic based on the dynamics of disbelief. In *The logic of theory change*, A Fuhrmann and M. Morreau (eds.), 243–259. Berlin: Springer-Verlag.
Prado, Jérôme and Ira A. Noveck
    2006   How reaction time measures elucidate the matching bias and the way negations are processed. *Thinking and Reasoning* 12(3): 309–328.
Robusto
    2008   http://forums.ageofconan.com/showthread.php?t=67496. (2008-10-07).

Rubio Fernández, Paula
    2007   Suppression in metaphor interpretation: Differences between meaning selection and meaning construction. *Journal of Semantics*: 1–27.

Seanzky
    2008   No Shame, Whatsoever! http://www.seanzky.com/?p=72 (2008-09-27).

Shen, Yeshayahu
    1992   Metaphors and categories. *Poetics Today* 13: 771–794.

Shuval, Noa and Barbara Hemforth
    2008   Accessibility of negated constituents in reading and listening comprehension. *Intercultural Pragmatics* 5(4): 445–470.

Stringaris, Argyris K., Nicholas Medford, Rachel Giora, Vincent C. Giampietro, Michael J. Brammer and Anthony S. David
    2006   How metaphors influence semantic relatedness judgments: The role of the right frontal cortex. *NeuroImage* 33: 784–793.

Student
    2008   http://puihan1204.blogspot.com/2008/04/i-am-not-your-secretary.html (2008-04-10).

Summerfield, Karen Anne
    1998   http://www.storysite.org/story/changeintime~02.html (2008-09-28).

thisfred
    2004   Jabberwacky (thing). http://www.everything2.com/title/Jabberwacky (2004-04-04).

Victoria "cloroxcowgirl" B.
    2008   Don't Tell Anyone Else But… (2008-10-07). http://www.yelp.com/list_details?list_id=calP7srSYWbPSeV7hHKmcA.

Ward, Gregory
    1983   A pragmatic analysis of epitomization: Topicalization It's Not. *Papers in Linguistics* 17: 145–161.

York, Byron
    2008   'I Am Not President Bush.' http://article.nationalreview.com/?q=YjM0NzgwMDc3NDk3YjhkNzYyYjBiODQ0ZWVmNDJiMzE= (2008-10-16).

Zguris, Stephanie
    2008   (2008-09-28) https://www.msu.edu/~msupma/eboard.htm.

Negation in Classical Japanese

Yasuhiko Kato

1. Introduction

The aim of this paper is to elucidate some salient aspects of negation in Classical Japanese (henceforth, CJ) of the Heian period, especially the prose of the tenth and eleventh centuries,[1] and explore some of the core implications for current linguistic theory. Though the exposition is by no means exhaustive or systematic, we hope that it will bring forth some new materials for current research on linguistic theories in general and on negation in particular. Section 2 will present central facets of negation, including negative forms, sentence negation, negative imperatives, double negation, negative polarity items, and metalinguistic negation. Sections 3 and 4 examine in some detail the syntactic aspects of sentence negation and clause-internal focus, respectively. Section 5 will present a brief sketch of the left-periphery of CJ with special reference to the cases where paired negative forms appear. In the course of the discussion, some theoretical implications will be explored.

2. Facets of CJ Negation

2.1. Negative Forms

The Classical Japanese of the Heian period has at least four negative forms: *zu, nasi, mazi, zi*. As auxiliaries, these negatives exhibit inflection in the Irrealis form (*mizen-kei*), the Continuative form (*renyo-kei*), the Conclusive form (*shusi-kei*), the Attributive form (*rentai-kei*), and the Realis form (*izen-kei*). Of these forms, *zu* shows a suppletive paradigm consisting of *zu* and *na*, the latter of which fills in the attributive and irrealis forms of the former. *Na* and *nasi* have different origins, the latter being a reflex of *nafu*

---

[1] The term "Classical Japanese" (CJ) does not denote a particular period in Japanese history. In this paper, it refers to the language of the Heian period (10th to 11th centuries); CJ here thus corresponds to the period called "Chuko-go" of traditional Japanese grammars ("Kokugo-gaku"), which John Whitman (p.c.) calls "Early Middle Japanese" (EMJ). Other uses will be specified in context.

in the Eastern dialect of Old Japanese (hence OJ)(Yamada (1952: 150–158), Tokieda (1954: 128–149).[2]

While these negative forms share the meaning of negation, they may still be differentiated in their semantic contents. Thus, Tokieda (1954: 146, 148) proposes a modal-based distinction in that *mazi, zu,* and *zi* express different degrees of speaker's certainty, weakening in this order, with regard to possibly nonexistent facts or events under discussion. It might be the case that this common feature of non-existence has given rise to the truth-functional meaning of negation.

2.2. Sentence Negation: *(e) ... zu*

Sentence negation is expressed either by one of the negatives alone as in (1) or by a discontinuous or paired form of *e ... zu/nu* as in (2):

(1) Taketori Monogatari
  a. *tiyau-no uti-yori-mo   idasa-zu*
     curtain-inside-from-even let go-NEG
     '(They) did not even let [Kaguya-fime] go out from inside the curtain.'
  b. *yoru-fa yasuki i-mo ne-zu*
     at night soundly even sleep-NEG
     '(They) did not even sleep soundly at night.'

(2) Taketori Monogatari
  a. *kafi-wo-ba   e-tora-zu*
     shell-ACC-TOP *e*-obtain-NEG
     '(He) didn't/couldn't obtain the shell.'

   Kagero Nikki
  b. *e namida-fa todome-zu narinuru*
     e-tears-TOP stop-NEG come about-PERF
     '(It) came about that tears couldn't stop.'

As to these two types, at least the following issues will have to be addressed:

---

[2] For an extensive discussion of these negative forms, in comparison with Altaic languages, see Miller (1971: 245–285).

(3) (i) the semantic and functional differences between simple negation (1) and the *e/zu* type negation (2);
 (ii) the lexical properties of *e* in the *e/zu* pattern: its morphosyntactic, semantic and historical nature;
 (iii) the syntax of the sub-types of *e/zu* structure: their structures and derivations.

Leaving problem (3iii) for section 3, let us make brief comments on (3i,ii). In historical linguistic work on the Nara and the Heian periods, the preverbal *e* is regarded as an adverbial which developed from the semi-auxiliary *u* (Honda 1957: 1–2), with a meaning of potentiality or ability. While in OJ (including the *Man'yoshu*) *e* does not necessarily require the presence of the negative, in CJ it must co-occur with, and be licensed by, a negative suffix (but, according to Honda (1957: 17–18), its source *u* does not co-occur with the negative, at least in *The Tale of Genji*).[3] The lexical meaning of potentiality or ability attributed to *e* induces a semantic difference between the simple and the *e/zu* type negation, hence in typical cases the contrast as to whether the subject has an initial intention to do something but for some reason cannot, or has no initial intention at all (hence the simple and flat negation). For a detailed discussion of the semantic (and thus distributional) differences in question, see Honda (1957: 5–17).

It is expected from the semantic function of *e* that when it is absent, the resultant simple negation without *e* would be most naturally found in the description or simple statement of state of affairs; this is a thetic judgment in Kuroda's (1972) sense. Some typical examples:

(4) Taketori Monogatari
 a. *kono tigo-no   katati keura-naru koto yo-ni   naku,...*
  this infant-GEN figure beautiful-is fact world-in not.exist
  'There is no other infant who has a more beautiful figure than this one.'
 b. *nanino   sirusi-aru- beku-mo   mie-zu*
  anything-GEN signs-exist should-even be.noticed-NEG
  'Nothing that should be taken for a sign was noticed.'
 c. *ito-itaku afare-garase-tamafite, monomo kikosi-mesa-zu*
  so much grieve-polite   any word say-Polite-NEG
  '(He) grieved so much that (he) didn't say a single word.'

---

[3] See also Shibuya (1993) for a detailed survey on the history and various uses of the expression of *e*.

In (4a), *e* does not appear where the sentence means nonexistence, a typical state of affairs. (4b) and (4c) describe the scene in front of the speaker and the mental state of a person, respectively. In contrast, as is stated above (and amply observed below), the negative sentences with *e* tend to be deployed to describe the situations that involve various sorts of actions or intentions. A speculation is that the distinction in question may be a reflection of the mode of judgment in general, i.e. the thetic vs. categorial contrast (Kuroda 1972).

Let us note further a case where *e* alone (without *zu*) induces a negative meaning:

(5)　　Genji, Suma
　　a. *ka-bakari-no taimen-mo mata-fa e-simoya to omofu-koso, ...*
　　　 that-only-GEN meeting-also again-Top *e*-ever　that think-FOC
　　　 'As (I) think that such a meeting could ever (take place), ...'

　　Genji, Kagero
　　b. *e koso to notamafu-ni, ...*
　　　 e-such that told-Polite, ...
　　　 'As (she) said that (there) could (not be such), ...'

　　Genji, Suma
　　c. *saranaru koto-domo-fa e-namu.*
　　　 unworthy things to say　*e*-FOC
　　　 '(I) could (not write down) such unworthy things to say.'

Given these observations, one might conclude that *e* has a substantive negative meaning (in addition to the meaning of potential ability noted above). If this is the case, the *e-zu* pattern might be regarded as a case of negative concord as in Romance languages (Haegeman 1995, Zanuttini 1997, Watanabe 2004, among others).[4]

Alternatively, the cases in question may involve simply discourse deletion of the negative verb (V-*zu*)[5], which means that *zu* is present at the level of semantic interpretation as is the normal case of the *e-zu* pattern. Without further arguments, we will assume here the latter possibility. The syntactic status of *e* will be taken up in section 3.

---

[4] One difference between negative concord and the *e/zu* pattern is that, while in negative concord the missing verbs can be affirmative, all of the missing verbs in (5) are negative (John Whitman, p.c.).

[5] In (5a) and (5b), *e* belongs to *to* 'that' complement clause, whose V-*zu* is deleted.

2.3. Negative Imperatives: *na ... so*

CJ has another discontinuous form for negation, i.e., the *na...so* form for negative imperatives:

(6)   Taketori Monogatari
   a. *atari   yori-dani   na-ariki-so*
      around there-EMP NEG-walk-IMP
      'Don't walk around there.'

   Taketori Monogatari
   b. *mune-itaki-koto   na-si-tamafi-so*
      heart-breaking-thing NEG-do-Polite-IMP
      'Please don't do (such) a heart-breaking thing.'

   Genji, Yadoriki
   c. *fitori   tuki   na-mi-tamafi-so*
      by yourself moon NEG-look-Polite-IMP
      'Please don't gaze at the moon by yourself.'

Note that the locative adverbial in (6a) and the object phrase in (6b) appear to the left of *na*. It is not impossible, however, for these elements to appear between *na* and *so* (though being more restricted than in the case of *e ... zu*):

(7) Genji, Yugiri
    kono   figakoto   na   tuneni   notamafi-so
    this sort of wrong guess NEG always tell-IMP
    'Do not always make this sort of wrong guess.'

Note also that, just as in (7), the adverbial phrase *atari yori-dani* 'around there' in (6a) is interpreted as negated, though it appears to the left of (hence in a higher position than) the negative *na*. Some additional cases:

(8)   Genji, Agemaki
   a. *usirometaku   na   omofi-kikoe-tamafi-   so*
      with a guilty conscience NEG -think of-Polite-Polite-IMP
      'Please do not think with a guilty conscience (Don't worry about it).'

   Genji, Hotaru
   b. *mote-fanarete na-kikoe-tamafi-so*
      bluntly   NEG-reply-Polite-IMP
      'Do not reply bluntly.'

I will return to this sort of discrepancy of form and meaning, which is also found with regard to the *e/zu* structure, in section 3.5.

2.4. Double Negation

Instances of double negation are not rare in CJ. In fact, *The Tale of Genji* begins with a sentence with double negation as in (9), though the first negative is incorporated into a lexical word *yamugoto-naki* 'quite noble', which is a compound adjective of *yamugoto* 'things to be worried about' and *naki* 'not':

(9) Genji, Kiritsubo
 (…) *ito yamugoto-naki kifa-ni-fa ara-nu ga,…*
 quite to be worried-NEG birth-COP-TOP be-NEG though
 '(…) though (she) is not of quite a noble birth, …'

As Horn (1989, 1991) demonstrates with a vast range of evidence, double negation may have a semantic effect either of weakening or of strengthening. Both cases are attested in CJ. The examples in (10) and (11) represent weakening and strengthening effects, respectively:

(10) Genji, Tokonatu
 a. *kono fitobito-fa mina omofu kokoro naki nara-zi*
  these people-TOP all think heart not.exist be-NEG
  'It is not the case that all people did not think so in their hearts.'
 Genji, Sekiya
 b. *fitosire-zu, omofiyari-kikoe-nu-ni si-mo ara-zari-sikado*
  secretly worry-Polite-NEG-though EMP exist-NEG though
  'Though it was not the case that he did not think about it without letting anybody know it.'

(11) Genji, Kiritubo
 a. *yorosiki koto-dani kakaru wakare no kanasikara-nu-fa*
  ordinary case-even such farewell mournful-NEG-TOP
  naki-waza-naru-wo masite afare-ni ifu-kafi -nasi.
  not-exist fact be-PRT all the more mournful say-worth-NEG
  'Even in ordinary cases, it is not the case that such farewell (by death) does not make anyone feel sad. (For him, so young,) I'm so mournful that I have no words for consolation'

Tosa Nikki
b. *omofi-ide-nu   koto   naku,*   ...
   remember-NEG thing not.exist
   'There was nothing that I did not recall. (Everything is quite vivid in my mind)'

For factors, grammatical and/or discourse, that operate to differentiate these two cases, see Horn (1989, 1991).

2.5. Negative Polarity Items (NPIs)

As in present-day Japanese, NPIs in CJ include indeterminate -*mo* 'even' as in (12), adverbial -*mo* as in (13), and other adverbials as in (14):

(12) *Indeterminate -mo*

   Genji, Yomogifu
a. *kono fito-mo   mono-mo kikoe-yara-zu*
   this man also a word   say-NEG
   '(Because of tears) this man, too, couldn't say a single word.'

   Genji, Sifigamoto
b. *yononaka-no nifofi-mo nanto-mo oboe-zu namu.*
   this world   splendor   anything concern-NEG
   '(I) have no concern for status, nor anything in this world.'

   Genji, Wakana II
c. *sarani   nanigoto-mo obosi-wakare-zu*
   moreover anything   think-NEG
   'Moreover, (he) couldn't think anything.'

(13) *Adverbial -mo*

   Genji, Agemaki
a. *uti-mo   madoromi-tamafa-neba*
   a wink   sleep-Polite   -NEG
   '(She) did not sleep a wink.'

   Genji, Sifigamoto
b. *miko-no osumafi-wo mata-mo mi-zu   narinisi koto*
   Lord's   residence   again   see-NEG was   fact
   '(I) couldn't see the Lord's residence again.'

Taketori Monogatari
c. *motto-mo e-sira-zari-turu*
   at all   *e*-know-NEG-PERF
   '(I) didn't know (it) at all.'

(14) *Other Adverbials*

Genji, Agemaki
a. *mono-mo tuyu-bakari mawira-zu*
   anything at all   eat-Polite-NEG
   '(She) did not eat anything at all.'

Tosa Nikki
b. *mofara kaze yama-de*
   at all wind stop-NEG
   'The wind didn't stop at all.'

It seems that the class of NPIs in CJ has been largely preserved to present-day Japanese (cf. Kato 1985).[6] The full range of NPI expressions in CJ and their licensing conditions remain to be explored.

2.6. Metalinguistic Negation

It has been well established in the literature since Ota (1980) and Horn (1985, 1989) that in natural language there operates a non-truth-functional or metalinguistic use of negation. Though a coherent view of this type of negation has not yet been established, one may expect to find instances of metalinguistic negation in CJ, whose prose is highly sophisticated in nature. The present brief survey reveals, however, that metalinguistic use of negation is not prominent in CJ prose (which itself may require explanation).[7] One example that may have metalinguistic effects is:

---

[6] The word *motto-mo* in (13c) might not be used as an NPI in CJ, which suggests some confusion in text reconstruction (as noted by Tatuji Motohashi, p.c.). In CJ, however, the wh-*mo* series, such as *nani-mo, dae-mo,* are not attested. This might be a major difference between CJ and Mod J (observation due to John Whitman, p.c.).

[7] As Larry Horn (p.c.) suggests, one factor might be that "we don't have evidence of the kind of ironic exchanges in colloquial registers in which MN is most likely to occur."

(15) Genji, Hahakigi

misi yume-wo afu yo ariya to nageku
saw dream-ACC meet night exist-Q if I could have one lament
mani me-sae afadezo koromo feni-keru, nuru
while eyes-even close-NEG many days pass fall asleep
yo nake-reba
night not.exist-because

'While I have lamented, will I encounter a dream where I see (you), many days have passed, for I haven't had a night that I fell asleep.'

In (15), negation applies to the presuppositional part of the meaning of *yume* 'dream of you', i.e., to the presupposition that one has fallen asleep: I haven't dreamt of you, for I couldn't have fallen asleep at all these days.

In sections 2.1–2.6, we have observed so far the basic facets of negation attested in CJ prose. Though the range of the survey has been highly restricted, it should shed some light upon the phenomena of negative expression in a remote period of the tenth century.

3. Syntactic Aspects of Sentence Negation

We have seen in section 2.2 that (i) sentence negation in CJ is expressed either by *zu* or one of its allomorphs alone as in (1) above or by the discontinuous pair of elements *e... zu* as in (2), and that (ii) in the latter case, arguments of verbs (and adjuncts, as will be seen) may appear either to the left of (hence, in a higher position than) the first element *e* as in (2a) or between the two as in (2b). In this section, I will focus upon the discontinuous forms and present attested examples to show what kinds of constituents may appear to the left of, and between, the paired elements, and proceed to discuss their theoretical problems.

3.1. Some Basic Issues

To put it schematically, an argument or adjunct XP may appear either in what I will call the external position (16a) or the internal position (16b).[8] The alternation will be substantiated with observed data below.

---

[8] Of these two cases, the external pattern, where *e* is adjacent to V, seems to be more stable than the internal pattern, where XP intervenes between *e* and V. This is seen from the facts that (i) in Old Japanese of the eighth century only

(16) a. XP – e – V – zu   (external position)
     b. e – XP – V – zu   (internal position)

Note first that the V in (16) may not be a simple verbal stem, but rather a complex predicate that may consist of a number of verbal elements. Thus, in the external pattern,

(17)   Genji, Kiritsubo
   a. *sugasuga-to-mo e-mawir-   ase-   tatematuri-tamafa-nu*
      boldly-even   *e*-serve at court-Cause-Polite-Polite-   NEG
      nari-keri
      be-PAST
      'So (her mother) could not boldly do her service at court.'

   Genji, Tokonatsu
   b. *e-watari-mi-tatematuri-tamafa-zu*
      e-visit-Polite-Polite-Polite-polite-NEG
      '(The princess) would not visit (the palace).'

Hence, V in (16) will have to be represented as, say, V*, whose complexity is not upper bounded.

Given these two cases in (16), immediate problems to be considered include: (i) the syntactic properties of *e* and *zu*, and (ii) the derivational relations between the two patterns (cf. (3iii) above). Let us first consider the first issue, on which the derivational issue is crucially dependent.

The syntactic status of the negative *zu* is fairly straightforward: it is an auxiliary which appears in the head position immediately preceding inflectional endings (including the Tense marker). In contrast, the syntactic status of *e* is far from clear. At least three possibilities have been proposed in the literature. Namely, *e* could be (i) a functional head (Kato 2003a), (ii) a specifier of a functional projection (Kato 2003b; Whitman 2005 for Old Japanese), or (iii) a clitic-like adverbial (in adjoined position) (adapted from Tokieda 1954).[9] For the present purposes, however, two properties of *e* are relevant:

---

the external pattern is possible (due to Yuko Yanagida, p.c.) and (ii) *e* in the external pattern survived (as a prefix to V) into the Meiji era of the nineteenth century, especially in the style used for translating European literature (such as works by Ohgai Mori, among others).

[9] A parallel situation is found with respect to the Old Japanese prefix *i*, which has been analyzed as a clitic (Yanagida 2007) or a functional head (Ogawa 2009).

(18) (i) its edge property
(ii) its negative sensitivity

In (i) edge refers to the literal left boundary of any categories/projections, to which *e* is merged as an outer specifier or an adjunct. In (ii) negative sensitivity means the asymmetric dependency between *e* and the negative *zu*, i.e., the fact that *e* requires the presence of *zu*, like ordinary NPIs, but not vice versa. These two properties will be captured if we assume that:

(19) *e* is merged to a category/projection with the NEG-feature

where the NEG-feature is projected from lexical negative heads.

With the minimal specification of (19) that *e* eventually becomes the left-edge element, the derivational relation between the external and internal patterns is left with an array of explanatory possibilities. Before discussing the derivational issue, let us present attested data to show what kind of constituents can appear in the external and internal patterns.

3.2. External and Internal Patterns

3.2.1. Subjects

Subject phrases may appear in the external or the internal position. In the external position, they occur with case or focus particles. In particular, they can appear with only case particles and no focus markers, as in (20c). In the internal position, they never appear with case particles alone, as seen in (21).

(20)   Genji, Kiritsubo
  a. *fafagimi-mo tomini   e mono-mo notamafa-zu*
     mother-also right now *e* a word   say-NEG
     'Mother, too, couldn't say a word right now.'

  Tosa Nikki
  b. *fito   mina e-ara-de   warafu-yau nari*
     people all *e*-do so-NEG burst appearance is
     'All of the people couldn't do so, and burst into laughter.'

  Genji, Wakana II
  c. *matuno yo-ni   kudareru fito-no   e-akirame-fatu-maziku koso*
     degenerate age in silly people-SUBJ *e*-clarify-NEG-will
     'The silly people in this degenerate age could not clarify the matter.'

(21)   Murasaki-shikibu Nikki
   a. *usiro-no fosomiti-wo   e-fito-mo   tofora -zu*
      behind   narrow path-through *e* a person pass through-NEG
      'Nobody passes through the narrow path behind.'

   Kagero Nikki
   b. *e namida-fa todome-zu narinuru*   (=2b)
      e tears-Top stop-NEG   come about
      '(It) came about that tears couldn't stop.'

3.2.2. Object

Just as with subjects, objects in the external position may occur with case and/or focus particles. They can appear with only case particles as in (22c), which is not possible in the internal position.

(22)   Taketori Monogatari
   a. *kono tama   tafayasuku e-tora-zi-wo*
      this treasure easily   *e*-obtain-NEG-ACC
      '(He) didn't obtain this treasure easily.'

   Taketori Monogatari
   b. *kafi-wo-ba   e-tora-zu*   (=2a)
      shell-ACC-TOP *e*-obtain-NEG
      '(He) didn't/couldn't obtain the shell.'

   Taketori Monogatari
   c. *ano kuni-no   fito-wo   e-tatakafa-nu-nari*
      that country's people-ACC *e*-fight-NEG
      '(They) didn't/couldn't fight against the soldiers of that country'

(23)   Genji, Kiritubo
   a. *fafagimi-mo tomini   e mono-mo   notamafa-zu*   (=20a)
      mother   right away *e* a word (ACC) speak-NEG
      'Mother couldn't speak a word right away.'

   Genji, Yugao
   b. *e-sasi-irafe-mo   kikoe-zu*
      e-insert-reply-also send-NEG
      '(He) could not return the poem in reply.'

3.2.3. VP Complements

The VP complements with *to* 'that' appear in the external position. As far as our survey is concerned, no example is attested for the internal position.[10]

(24)　　Genji, Ukifune
   a. *nikusi-to-fa e-obosi-fate-nu　　nameri*
      hateful　　*e*-think of-finish-NEG　seem
      '(She) does not think of (him) decisively as hateful.'

   Genji, Yuugiri
   b. *ikani subeki koto-to-mo　e omofi-e-zu*
      how to do　　　　　　　　*e*-think-potential-NEG
      '(They) don't know how to do (about it) at all.'

3.2.4. Datives and Nominal Adjuncts

As with subjects and objects, *ni*-phrases can appear in the external position without focus particles as in (25b). In (26a), though they occur without a focus particle, this may be due to the fact that *kokoro-ni makase-*'at one's disposal' forms a phrasal idiom.

(25)　　Genji, Kasifagi
   a. *win-ni-mo　　mada e-mausi-tamafa-zari-keri*
      lord-DAT-FCS　yet　　*e*-speak to-Polite-NEG-PAST
      '(He) could not have spoken to the lord, yet.'

   Genji, Fatune
   b. *kono koro-no fito-ni　e-simo masara-zari-kemu kasi*
      younger generation　　*e*-exceed-NEG-seem
      'It is not the case that (old men) did not exceed the younger generation.'

---

[10] In (24b), the sequence *e-omofi-e-zu* involves two occurrences of *e*, which indicates that pre-verbal *e* is merged into the position independently of post-verbal auxiliary *e*. Additional examples:

(i) *monomo　e-kiko-e-zu*　　　　　　　　　　　　(Genji, Utsusemi)
    a word　*e*-say-potential-NEG
    '(I) couldn't say a word.'

(ii) *kurai-wo　　　e-yuzuri-kiko-e-nu　　　koto*　(Genji, Hujino-uraba)
     Imperial Throne *e*-abdicate-polite-potential that
     'that (the Emperor) could not abdicate the Throne (to Genji)'

(26) Genji, Suetumuhana
    a. *fitorimi-wo e kokoro-ni makase-nu   fodo koso, ...*
       oneself   *e* for one's disposal do-NEG when
       'If she is at the age when she cannot decide nor act for herself, ...'

    Genji, Fujibakama
    b. *e sono suzino fitokazu-ni-fa monosi-tamafa-de*
       e that sort of person count as-Polite-NEG
       '(He) does not count (me) as that sort of person.'

3.2.5. VP Adverbials

A wide range of VP adverbials may appear in either position:

(27) Taketori Monogatari
    a. *kokoro-no mama-ni-mo e-seme-zu*
       heart-GEN will-at-even *e*-blame-NEG
       '(I) couldn't blame (her) with my heart's will.'

    Taketori Monogatari
    b. *kono tama   tafayasuku e-tora-zi-wo*
       this treasure easily   *e*-obtain-NEG
       '(He) didn't/couldn't obtain this treasure easily.'

    Makurano sosi, 260
    c. *futomo   e-ori-zu*
       immediately *e*-come down-NEG
       '(I) couldn't come down immediately.'

    Genji, Yadoriki
    d. *kufasiku-simo e-sira-zu-ya*
       in detail-Emph *e*-know-NEG-Q
       'Wouldn't (I) know in detail?'

(28) Makura no sosi, 176
    a. *sarani e futomo miziroga-neba*
       at all *e* easily   move-NEG-since
       '... since (I) couldn't move easily/in a hurry.'

    Genji, Sawarabi
    b. *e ito-kau-made-fa ofase-nu waza-zo*
       e very-as much as help-NEG way-indeed
       '(One) may not help others so much.'

Genji, Utsusemi

c. *rei-nara-nu   fito faberite, e tikau-mo yori-fabera-zu*
   unusual person stayed there e closer   go-polite-NEG
   'Because an unusual person stayed there, (I) couldn't go closer to (her).'

3.2.6. *Interim Summary*

The observations so far presented in 3.2.1–3.2.5 reveal at least that:

(29) (i) XP may appear either in the external or the internal position of the paired forms, where XP is a (clausal) argument or an adjunct.

   (ii) In the external position, XP may appear with case particles and/or focus particles; in the internal position, XP must appear with focus particles.

The above statement (29ii) is substantiated by the fact that structural case particles *no* and *wo* (but not *ni*) do not appear in the internal position. The absence of case particles here might result from some process of readjustment to give proper sequences of case and focus particles, such as the deletion of case particles *ga* and *o* before the focus particle *wa* in present-day Japanese.

Let us turn to the issues on the derivational relation between the external and the internal forms (section 3.3–3.5). The proper treatment of focus-related matters will be discussed in section 4.

3.3. Edge Positions and Movement

Let us first assume two hypotheses concerning the clausal architecture:

(30) VP-internal subject hypothesis:
   Arguments of a verb, including subject and object, are base-generated in VP. (Kitagawa 1986, Kuroda 1988, Koopman and Sportiche 1991)

(31) Split-IP hypothesis:
   Functional heads, including Tense and Neg, have their own functional projections.   (Pollock 1989)

Let us assume, along the line of (31), that the Neg element occupies a higher position than VP, and more specifically that Neg selects vP, a functional projection above VP, i.e. (32) below. As to the paired elements of e and zu, we have proposed (33):

(32) Neg head selects vP

(33) (i) zu is a negative head, and
    (ii) e is merged to a category/projection with the NEG-feature   (=19)

Assuming that the neg-feature in (33ii) is projected from the neg-head zu, e can only be merged (as a specifier or an adjunct) with a projection containing zu, to give structure (34):

(34) $[_{ZP}$ e- $[_{ZP +neg}$ $[_{vP}$ ... $[_{VP}$ V $]]$ zu $]$ $]$

The spec/head status of e and zu as proposed in Kato (2003b) is subsumed under (34). Whitman (2005) argued for the spec-head analysis for Old Japanese based on the comparative data of French *ne...pas* and Korean *an*.[11] Note in passing that proposal (33ii) means that e is licensed by the neg-feature in terms of the operation Merge, or to put it differently, by the sister relation instead of standard c-command.

Given the VP-internal subject hypothesis (30) and structure (34), it immediately follows that the external frame of XP-e-V-zu, where XP is an argument or a VP adverbial, can only be derived in terms of a leftward movement of XP over e as in (35):

(35) [XP [e- [[$_{VP}$ (XP) V] zu]]]

The exact nature of the triggering forces and the landing sites of the XP movement are related to what kind of semantic or information-based properties the moved XPs have. We will see some of their effects in section 5.

To repeat, the external pattern of negative structures, together with theoretical assumptions of the VP-internal subject hypothesis and the Split-IP

---

[11] To decide the syntactic status of e, Whitman (2005) emphasises the inflectional parallelism between *ne...pas* in French and *e...zu* in CJ. These two constructions, however, may not be identical with respect to other morphosyntactic properties. For one thing, in CJ two patterns of negative sentences co-exist, i.e., with and without e. From a historical point of view, *ne...pas* represents a definite step in the so-called Jespersen's cycle (cf. Jespersen 1917, Horn 1989, and van der Auwera's chapter in the present volume) in the historical development of *ne* > *ne... pas* > *pas*, but *e...zu* does not in that its development is: $zu > e... zu > zu$.

Negation in Classical Japanese 273

hypothesis, which are independently motivated, entail that there operates in CJ an instance of movement (internal Merge) that moves arguments and/or adjuncts from within VP to higher positions. (See also Kato 2003a for the possibility that the movement in question involves a step in the PF component.)

3.4. On Island Effects

While the movement analysis of the basic cases of *e-zu* structure is well motivated on theoretical and empirical grounds, we should point out that there are a few cases which seem to cast doubt upon the movement analysis outlined above. These involve apparent violations of so-called island constraints (Ross 1967), which I will refer to for expository purposes. Observe that in (36)–(38), each of the (b) examples should show illicit movement as indicated:

(36) Taketori Monogatari
    a. *Kaguya-fime-wo   e-tatakafi-tome-zu   nari-nuru koto*
       Kaguya-hime-ACC *e*-fight-and-protect-NEG become-PERF
       '(We) couldn't (fight with the enemy and couldn't) stop Kaguyahime being taken away'

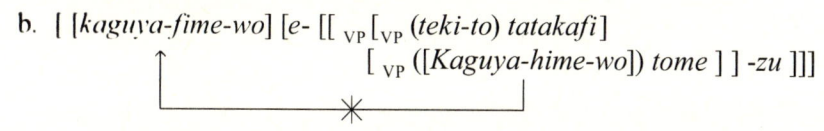

    b. [[*kaguya-fime-wo*] [*e*- [[$_{VP}$[$_{VP}$ (*teki-to*) *tatakafi*]
                           [$_{VP}$ ([*Kaguya-hime-wo*]) *tome*]] -*zu*]]]

    c. [[*kaguya-fime-wo*] [*e*- [[$_{VP}$ ([*Kaguya-hime-wo*]) *tatakafi-tome*]]
                                                    -*zu*]]]

(37) Tosa Nikki
    a. *kono kaditori-fa fi-mo   e-fakara-nu   katawi nari-keri*
       this boatman   day-even *e*-calculate-NEG uneducated-be
       'Lit. This boatman was an uneducated man who couldn't calculate what day it was.'

    b. [*kono kaditori-fa fi-mo* [$_{NP}$ [$_{IP}$ e [[$_{VP}$___*hakaru*] nu]] *katawi*] *nari-keri*]

    c. [[$_{NP}$ *kono kaditori-fa fi-mo* [$_{IP}$ e [[$_{VP}$___*hakaru*] nu]] *katawi*] *nari-keri*]

(38)   Genji, Wakamurasaki

    a. *koko-nifa tune-nimo e-mawira-nu-ga   obotuka -nakere-ba, ...*
        here-to   always   *e*-visit-NEG -NOM worry
        'I'm worried about the fact that (I) wouldn't visit here always, ...'

    b. [*koko-nifa* [$_{NP}$ [$_{IP}$ e [$_{VP}$ __ *mawira*] nu]]] -*ga* ...

    c. [$_{NP}$ [*koko-nifa* [$_{IP}$ e [$_{VP}$ __ *mawira*] nu]]] -*ga* ...

In (36b), the object of the second conjunct VP appears to the left of the entire VP, which induces a violation of the Coordinate Structure Constraint (Ross 1967). In this case, however, the whole argument hinges upon the proper treatment of the complex expression *tatakai-tome* 'fight-stop.' If the verbal complex in question is not yet a compound word in the syntax and each element appears in separate VPs, then we have an instance of island violation as is indicated in (36b). If, on the other hand, the complex is formed in the lexicon, so that it is inserted into syntax as such, or is formed in syntax before movement (presumably by a sort of reanalysis (Kato 2002)), so that no conjoined VPs are involved (at least at the stage of movement) as in (36c), the movement in question does not induce a CSC violation.

Likewise, if we assume that the object is extracted from within a relative clause as in (37b) and a VP adverbial moves out of the subject phrase as in (38b), we have violations of the Complex NP Constraint and the Sentential Subject Constraint, respectively. But if the moved phrases still remain within relative clause as in (37c) or within the sentential subject phrase as in (38c), then we have no island violations in these cases.[12]

Finally, in (39), if we assume movement, the extracted phrase might correspond to multiple gaps, which undermines the movement analysis. However, if the gaps here should be identified as *pro*, so that no movement is involved, we have no island problem.[13]

---

[12] These possibilities were pointed out to me by John Whitman (p.c.).

[13] This point is also from John Whitman (p.c.).

(39)   Tosa Nikki
   a. *hiyakusan-wo,   aru mono yonoma-tote, funayakata-ni*
      spiced sake-ACC a man   night during board of a boat
      sasifasameri-kereba, kaze-ni   fuki-nara-sase-te, umi-ni irete,
      put between   wind-by   be blown   sea soak
      *e-noma-*zu   *nari-nu*
      *e-*drink-NEG become

   'A man put spiced sake between the boards of the boat during the night, and it was blown by the wind and soaked in the sea, so that it became that we could not drink it.'

   b. [spiced sake [put ___ between, ___ be blown, ___ be soaked, we couldn't drink ___]

In sum, the arguments are all dependent upon the structures posited, especially the landing sites of moved phrases, whose exact nature could not be established insofar as our attested data are concerned. I will continue to assume, however, that XP movement operates in these cases, with structures of (c) options in (36)–(38), and *pro* in (39).

3.5. On Reconstruction Effects

More positive evidence for the XP movement comes from the fact that the moved XP phrases exhibit so-called reconstruction effects, i.e., that the moved XPs behave as if they are in situ. More specifically, while an external XP appears to the left of, hence in a higher position than, the negative, it is still interpreted as if it is in the scope of negation. This scope-internal, hence reconstruction, property is most clearly seen from the fact that (i) adverbials in the external position may be interpreted as negated (3.5.1) and (ii) negative polarity items (NPIs) can also appear in the external position (3.5.2).

3.5.1. Adverbial Interpretation

As seen in the examples in (27) and (28) above, adverbials are readily interpreted as negated in both external and internal positions. Some of them are reproduced below:

(40) a. *kono tama   tafayasuku e-tora-zi-wo* (=27b)
       this treasure easily   *e*-obtain-NEG
       '(He) didn't/couldn't obtain this treasure easily.'

   b. *kufasiku-simo   e-sira-zu-ya* (=27d)
      in detail-EMPH *e*-know-NEG-Q
      'Wouldn't (I) know in detail?'

(41) a. *sarani e futomo miziroga-neba* (=28a)
       at all *e* easily move-NEG-since
       '... since (I) couldn't move easily.'

   b. *e ito-kau-made-fa ofase-nu waza-zo* (=28b)
      e so much-TOP   help-NEG
      '(One) may not help others so much.'

As the glosses indicate, the scope relations between the negative and adverbials are invariably [¬ [Adv]], i.e., negation takes wide scope over adverbials. Incidentally, the same applies to the negative imperative form *na...so*:

(42)   Taketori Monogatari

   a. *atari   yoridani na-ariki-so* (=6a)
      around there   NEG-walk-IMP
      'Don't walk around there (but you may walk in other places).'

   b. *kofa-dakani na-notamafi-so*
      loudly   NEG-speak-IMP
      'Don't speak loudly.'

3.5.2. Negative Polarity Items (NPIs)

Like other nominal and adverbial XPs, NPIs can appear in both external and internal positions:

(43)   Genji, Utsusemi

   a. *itofosiute, monomo e-kikoe-zu*
      so sorry   a word *e*-say-NEG
      '(I'm) so sorry (for her) that I couldn't say a word.'

       Genji, Asagao

   b. *monomo e-kikoe-yara-zu*
      a word *e*-say-NEG
      '(I) couldn't say a word.'

Genji, Otome
c. *tare-tomo e-omofi-tado-rare-zu*
   anybody *e*-identify-NEG
   '(I) couldn't identify anybody.'

Genji, Yadoriki
d. *sanomino   e-mote-kaku-sare-nu-niya*
   for any length of time *e*-keep secret-NEG
   '(We) would not keep (it) secret for any length of time.'

(44) a. *usiro-no fosomiti-wo   e fitomo   tofora-zu*   (=21a)
        behind   narrow path-through *e* a person pass through-NEG
        'Nobody passes through the narrow path behind.'

     b. *fafagimi-mo tomini   e mono-mo notamafa-zu*   (=20a)
        mother-also right now *e* a word say-NEG
        'Mother, too, couldn't say a word right now.'

3.5.3. Implications

As is observed, adverbials and NPIs behave uniformly with respect to negation in both external and internal positions, i.e., as if they are in the scope of negation. Note that the standard definition of scope of negation is as in (45); and in the external position, the XP occupies the position to the left of *e* as in (46):

(45) The scope of negation is its c-command domain.
                                          (Klima 1964, Reinhart 1976)

(46) [XP ... [e- [$_{ZP+neg}$ [$_{vP}$ ... [$_{VP}$ (XP) V]] zu]]]

Our observation shows that, while XPs appear outside of the scope of negation, as is defined by (45), it behaves as if they are in the scope of negation, specifically, as if it occupies the position of (XP) in (46). Here we have a standard case of reconstruction, whose effect indicates the presence of the XP movement. We will leave open, however, the exact nature of reconstruction, i.e., whether the moved phrase is literally undone at LF (Saito 1989) or the movement is a PF operation that has no semantic or licensing effects (Kato 2003a). For problems on reconstruction with respect to NPI licensing in general, see Kato (1994).

4. Clause-Internal Focus

Let us turn to the previous observation stated in section 3.2.6, which reduces to (47) below. I will propose here that it is a consequence of (48). (For the clause internal focus position in Modern Japanese, see Yanagida 1995.)

(47) XP may appear in the external position without focus particles, but not in the internal position.

(48) The internal position is a focus-related position.

In general, the focus is marked by syntactic position, overt focus marker, and/or some phonological means. The observation (47) shows that the internal position is the one that requires some overt marking, i.e., some focus particle in this case.

A refinement of (48) comes from a restriction on focus elements observed in (49) and (50) below:

(49) a. *usiro-no fosomiti-wo*   e *fito-mo tofora*-zu   (=21a)
    behind   narrow path-through   *e* a person pass through-NEG
    'Nobody passes through the narrow path behind.'

  b. *fafagimi-mo tomini*   e *mono-mo notamafa*-zu   (=20a)
    mother-also   right now   *e* a word   say-NEG
    'Mother, too, couldn't say a word right now.'

(50) a. *fitorimi-wo* e *kokoro-ni makase*-nu   *fodo koso*, ...   (=26a)
    oneself   *e* for one's   disposal do-NEG when
    'If she is at the age when she cannot decide nor act for herself, ...'

  b. e *namida-fa todome-zu narimuru*   (=21b)
    e tears-TOP   stop-NEG   come about
    '(It) came about that tears couldn't stop.'

As is observed, exactly one element appears between *e* and verbal stems: a subject in (49a) and (50b), a direct object in (49b), and an indirect object in (50a). Other related elements such as a locative in (49a), a subject in (49b), and an object in (50a) appear in the external position. It seems that, if more than one argument (or adjunct) appears with the verb, these constituents should move to the left of *e*- (i.e., out of VP), leaving behind at most one phrase. This leads to the generalization in (51):

(51) At most one phrase may appear in the internal position, i.e., between *e*- and the verbal stem.

Though the source of the ban on multiple foci here is by no means clear,[14] it shows that the XP movement to the external position is motivated by focus-related factors.

5. Left Periphery

5.1. Some Attested Orders

As a preliminary to explicate the internal structure of the so-called left periphery (Rizzi 1997, Cinque 2006, Endo 2007, among others) in CJ, I will first present some observations concerning attested data (exclusively from *The Tale of Genji*). Some typical patterns will be presented with regard to the constituent orders above negative projection (where > stands for 'precede').[15] As to the particle *wo*, I will leave it unglossed here and return to consider its function below.

(52) *Topic > S-adverbial > Focus* (Kiritubo)
Genzi-no kimi-*fa* (TOP), ufeno tuneni mesi-matufaseba(S-ADV), kokoro-yasuku satozumi-*mo* (FOC) *e*-sitamafa-*zu*
'Since the Emperor always called him, Genzi could not stay at his home comfortably.'

(53) *Topic > S-adverbial > wo* (Miwotukusi)
Ofokisaki (TOP), nafo onnayami omoku-ofasimasu-uti-nimo(S-ADV) tufini konofito-*wo e*-kieta-*zu* narinamu kototo kokoro nayami-obosi-keredo, ...
'The late empress, in addition to her serious illness, was not happy with the idea that she could not reduce the person to submission.'

(54) *wo > S-adverbials* (Siigamoto)
misute-gataki-*wo*, ikeru kagiri-fa akekure (S-ADV), *e*-sakera-*zu* mitate-maturu-wo
'(I would) take care of the child, insofar as I live.'

---

[14] John Whitman (p.c.) suggests that the ban on multiple foci in this case indicates that *e* has a property of "an inverted approximative adverb like English 'hardly' (cf. Horn 2002)." Though *e* and *hardly* differ from each other with respect to their negative sensitivity, I leave it open for further studies to pursue the above common features.

[15] For a more comprehensive list, see Kato 2003b.

(55) *Topic > wo > Focus* (Wakana 1)
fito-bito (Top), arafa-*wo* futomo (Foc) *e*-mituke-*nu* naru-besi
'It seems that ladies do not immediately become aware of being seen from outside.'

(56) *Focus > wo* (Sakaki)
kamudatime-nado-*mo* (FOC), yono tutumasisa-*wo e* simo fabakari-tamafa-*de*, ...
'Even the nobles did not pay attention to the constrained air around them.'

(57) *wo > Focus* (Wakana 1)
tada kono on-arisama-*wo*, uti-sofite-*mo* (FOC) *e*-mitate-matura-*nu* obotuka-nasa-ni, ...
'I'm just worried that I couldn't take close care of her daily life.'

(58) *Focus > Acc (Ø)* (Wakana 1)
miko-*mo* (FOC), wefi-naki (ACC) *e*-todome-tamafa-*zu*
'The prince couldn't stop crying, either.'

(59) *Acc (Ø) > Focus* (Makibasira)
on-kaferi (ACC) kokoni-*fa* (FOC), *e*-kikoe-*zi* to ...
'It's quite impossible for me to send a reply.'

(60) *wo+Focus > Focus* (Hananoen)
Genzi-no kimino on-*wo-ba* (ACC+FOC), kauzi-*mo* (FOC) *e*-yomi-yara-*zu*, ...
'Even the lecturer couldn't read aloud the poem by Genzi (due to his admiration).'

(61) *Focus > e > Focus* (Yugiri)
ito-ofokameredo, fito-*fa* (FOC) *e* mafo-*nimo* (FOC) mi-*zu*
'Though many things seem to be written, the ladies couldn't really see them.'

(62) *Topic > WH+Focus* (Suetsumu hana)
kimi-*fa* (TOP), *tare*-to-*mo* (WH+FOC) *e*-mi-waki-tamafa-*de*, ...
'The lord couldn't identify the person.'

(63) *WH+Focus* (Kasiwagi)
nani-goto-*mo* (WH+FOC) *e*-wakimafe-fabera-*zu* tote, ...
'(I feel that I) couldn't make anything clear.'

5.2. Implications

A brief look into these patterns will reveal that the left periphery in CJ has at least the following properties that any theory must explain:

(64) (i) fixed hierarchical positions for discourse-related (informational) interpretation
 (ii) multiple occurrence of constituents of the same kind
 (iii) free occurrence of the preverbal negative *e* and adverbial elements
 (iv) function of the particle *wo* as a focus marker

Let us first note the particle *wo* in (64 iv), which is an accusative Case marker in Modern Japanese. That it functions as a focus, however, does not necessarily mean that it has a dual function; rather we adopt here Yanagida's (2003, 2005) hypothesis that *wo* is in fact a focus marker in Old Japanese, and (non-focused) accusative phrases in SOV order bear the unmarked form without any Case marker. It seems that *wo* still retains its focus property in CJ. This is squarely consonant with our attested data.

As to the properties (64 i–iii) of the hierarchical positions, our observations have revealed that the hierarchy of orders is as in (65), which shows in itself that the upper part of clause structure in CJ is fairly rich in its contents.

(65) Topic > S-adverbial > Focus > *e* > XP-*wo*/ Ø
 > S-adverbial > (Wh-) Focus > [$_{zP}$ *e* ... -V-*zu*]

Coupled with the results obtained in sections 3 and 4, it follows from (65) that in CJ there are at least two positions where focus phrases can appear. That is,

(66) (a) the internal position (to the right of *e* within zP)
 (b) the left periphery (above zP)

where zP is a projection of a negative element *zu*.

If this is the case, natural questions to be addressed are (i) why CJ provides multiple positions for focus phrases, and (ii) whether or not the multiplicity in (65) is unique to CJ. Our speculation is that the different positions correlate with different discourse-related functions of focus. If so, it must hold cross-linguistically.

Turning to (65) itself, it should be noted that the hierarchy is, as it were, the maximal array of the left periphery. That is to say that not all combinations of constituents are actually attested in the text, nor do all constituents have the same status with regard to their roles in structure-building in this domain. It seems that these constituents fall into (at least) two categories: (i) those that determine the basic frame of this domain in terms of their specific discourse-related functions, and (ii) those that are attached to appropriate positions in this domain either by Internal or External Merge. Topic, XP-mo, and (Wh-)Focus belong to category (i), and S-adverbials and *e* to category (ii).

Viewed in this light, the skeleton of the left periphery is of the form (67), with possible operations of merge (or interpretation), which attach or associate S-adverbials and *e* into designated positions as in (65) above: [16]

(67) Topic > Focus > XP-*wo*/ Ø > (Wh-) Focus > IP

6. Concluding Remarks

In this paper, I have presented a preliminary survey of negation in Classical Japanese, with special reference to some of its salient features. Given the basic observations, I have focused upon the syntactic properties of sentential negation, suggested a movement analysis within a general framework of minimalist syntax, and provided a preliminary sketch of clause-internal focus and the left periphery in CJ. Needless to say, the findings of the present survey, which is far from comprehensive or systematic, should be supplemented with and/or revised by future research in this domain of inquiry.

Acknowledgements

Factual observations and theoretical considerations presented in this paper are partly based upon my previous works on CJ negation, including Kato (2002, 2003a,b). In the course of preparing the present version, I have been much indebted to Ken Hinomizu, Laurence R. Horn, Tatuji Motohashi, Sadayoshi Ogawa, Yuko Yanagida, and John Whitman, for their helpful comments and discussions. All remaining errors and inadequacies are my own. This work is supported in part by research fund at Sophia Linguistic Institute for International Communication, Sophia University, Tokyo.

---

[16] For the left periphery – specifically, the topic-focus articulation – in Old Japanese, see Watanabe (2005).

Abbreviations

Symbols in glosses include: ACC (accusative), COP (copula), DAT (dative), EMP (emphatic), FOC (focus), GEN (genitive), IMP (imperative), NEG (negative), NOM (nominative), SUBJ (subject), TOP (topic), PAST (past tense), PERF (perfective), POLITE (politeness), Q (question).

Texts

Genji Monogatai [The Tale of Genji].
    1976   Critical edition by Joji Ishida and Yoshiko Shimizu. Tokyo: Shinchosha.
Genji Monogatari [The Tale of Genji].
    1993   Critical edition by Shigeshi Yanai et al. Tokyo: Iwanami Shoten.
Taketori Monogatari [The Tale of Bamboo Cutter].
    1970   Critical edition by Atsuyoshi Sakakura. Tokyo: Iwanami Shoten.
Tosa Nikki, Kagero Nikki, Murasaki-shikibu Nikki, Sarashina Nikki. Shin Nihon Koten Bungaku Taikei [A New Collection of Japanese Classical Literature].
    1989   Critical edition by Masaharu Hasegawa et al. Tokyo: Iwanami Shoten.

References

Baek, Judy Yoo-Kyung
    1998   Negation and Object Shift in Early Child Korean. *MIT Working Papers in Linguistics* 25: 73–86.
Chomsky, Noam
    1995   *The Minimalist Program*. Cambridge, MA: MIT Press.
Chomsky, Noam
    2001   Derivation by phase. In *Ken Hale: A Life in Language*, Michael Kenstowicz (ed.), 1–52 Cambridge, MA: MIT Press.
Chomsky, Noam
    2004   Beyond explanatory adequacy. In *Structures and Beyond*. A. Belletti (ed.), 104–131 Oxford: Oxford University Press.
Cinque, Guglielmo
    2006   *Restructuring and Functional Heads*. Oxford: Oxford University Press.
Endo, Yoshio
    2007   *Locality and Information Structure: A Cartographic Approach to Japanese*. Amsterdam/Philadelphia: John Benjamins.
Haegeman, Liliane
    1995   *The Syntax of Negation*. Cambridge: Cambridge University Press.
Honda, Yoshihiko
    1957   Genji Monogatari Fukushi 'e' Kou. [On the Adverb 'e' in the Tale of Genji]. *The Journal of Kumamoto Women's University* 9 (1): 1–22.

Horn, Laurence R.
　1985　　Metalinguistic negation and pragmatic ambiguity. *Language* 61: 121–174.
Horn, Laurence R.
　1989　　*A Natural History of Negation*. Chicago: The University of Chicago Press. Reissued in 2001, Stanford: CSLI Publlications.
Horn, Laurence R.
　1991　　Duplex negatio affirmat: The economy of double negation. In *Papers from the 27th Regional Meeting of the Chicago Linguistic Society. Part Two: The Parasession on Negation*. L. Dobrin et al. (eds.) Chicago: Chicago Linguistic Society, 80–106.
Horn, Laurence R.
　2002　　Assertoric Inertia and NPI Licensing. *CLS* 38(2), The Panels: 55–82.
Jespersen, O.
　1917　　*Negation in English and Other languages*. Copenhagen: A. F. Host.
Kato, Yasuhiko
　1985　　*Negative Sentences in Japanese*. Sophia Linguistica 19, Monograph. Tokyo: Sophia University.
Kato, Yasuhiko
　1994　　Negative polarity and movement. *Formal Approaches to Japanese Linguistics* 1 (MIT Working Papers in Linguistics 24): 101–120.
Kato, Yasuhiko
　2000　　Interpretive asymmetries of negation. In *Negation and Polarity: Syntactic and Semantic Perspectives*. Laurence R. Horn and Yasuhiko Kato (eds.), 62–87. Oxford: Oxford University Press.
Kato, Yasuhiko
　2002　　Negation in Classical Japanese: A Preliminary Survey. *Sophia Linguistica* 49: 99–119.
Kato, Yasuhiko
　2003a　　Negation in Classical Japanese: A minimalist perspective. In *Empirical and Theoretical Investigations into Language: A Festschrift for Masasu Kajita*. S. Chiba et al. (eds.), 314–325. Tokyo: Kaitakusha.
Kato, Yasuhiko
　2003b　　Negation in Classical Japanese: Projection and Movement *Sophia Linguistica* 50: 91–102.
Kitagawa, Y.
　1986　　Subjects in Japanese and English. PhD diss., University of Massachusetts, Amherst.
Klima, E.
　1964　　Negation in English. In *The Structure of Language: Readings in the Philosophy of Language*. J. A. Fodor and J. J. Katz (eds.), 246–323. Englewood Cliffs: Prentice-Hall.
Koopman, H. and D. Sportiche
　1991　　The Positions of Subjects. *Lingua* 85: 211–258.

Kuroda, S.-Y.
  1972   The categorial and the thetic judgment: Evidence from Japanese syntax. *Foundations of Language* 9: 153–185.
Kuroda, S.-Y.
  1988   Whether we agree or not: A comparative syntax of English and Japanese. *Lingvisticoe Investigationes* XII, 1: 1–47.
Martin, Samuel E.
  1987   *The Japanese Language Through Time.* New Haven: Yale University Press.
Miller, R. Andrew
  1971   *Japanese and the Other Altaic Languages.* Chicago: The University of Chicago Press.
Motohashi, Tatsushi
  1989   Case theory and the history of the Japanese language. PhD diss., University of Arizona.
Motohashi, Tatsushi
  2001   A handbook of historical Japanese grammar. Ms., Sophia University.
Nishio, Toraya
  1972   Uchikeshi no Jodoshi [Auxiliaries of negation]. In *Hinshi-betsu Nihon Bunpo Koza, Jodoshi I* [Lectures on Japanese grammar on category types, auxiliaries I], 105–130. Tokyo: Meiji Shoin.
Ogawa, Sadayoshi
  2009   Jodai Nihongo no doshi-setto-ji to setu-kozo (Verbal prefix and clausal structure in Old Japanese). Ms. Tokyo Metropolitan University.
Ota, Akira
  1980   *Hitei-no Imi* (The Meaning of Negation). Tokyo: Taishu-kan.
Payne, John R.
  1985   Negation. In *Language Typology and Syntactic Description. Vol. 1: Clausal Structure*, T. Shopen (ed.), 197–242. Cambridge: Cambridge University Press.
Pollock, J. Y.
  1989   Verb movement, UG and the structure of IP. *Linguistic Inquiry* 20: 365–424.
Rizzi, L.
  1997   The fine structure of the Left Periphery. In *Elements of Grammar.* L. Haegeman (ed.), 281–338. Dordrecht: Kluwer.
Reinhart, Tanya
  1976   The Syntactic Domain of Anaphora. PhD diss., MIT.
Ross, John Robert
  1967   Constraints on Variables in Syntax. PhD diss., MIT.
Saito, Mamoru
  1989   Scrambling as semantically Vacuous A'-movement. In *Alternative Conceptions of Phrase Structure.* M. R. Baltin and A. Kroch (eds.), 182–200. Chicago: University of Chicago Press.

Shibuya, Katumi
　1993　*Nihongo Kano-hyogen-no Shoso to Hatten.* [Aspects and Developments of Potential Expressions in Japanese.] Memoirs of the Faculty of Letters, Osaka University vol. XXXIII, Part 1.

Tokieda, Motoki
　1954　*Nihon Bunpoo – Bungo-hen* (Japanese Grammar – Classical Prose). Tokyo: Iwanami Shoten.

Uesaka, Nobuo
　1999　*Taketori Monogatari Zen-Hyoshaku: Honbun Hyoshaku-hen* [The Complete Commentary on Taketori Monogatari: Commentary on the Main Texts]. Tokyo: Yubun Shoin.

Watanabe, Akira
　2004　The genesis of negative concord: Syntax and morphology of Negative Doubling. *Linguistic Inquiry* 35(4): 559–612.

Watanabe, Akira
　2005　Topic-focus articulation in Old Japanese. Ms., University of Tokyo.

Watanabe, Minoru
　1981　*Heiancho Bunshoshi* [The History of Prose in the Heian Period]. Tokyo: University of Tokyo Press. Reprint 2000, Tokyo: Chikuma Shobo.

Whitman, John
　2005　Preverbal negative elements in Korean and Japanese. In *The Oxford Handbook of Comparative Syntax.* G. Cinque and R. S. Kayne (eds.), 880–902. New York: Oxford University Press.

Yamada, Takao
　1952　*Heiancho Bunposhi* [A Historical Grammar of the Heian Period]. Tokyo: Hobun-kan.

Yanagida, Yuko
　1995　Focus projection and Wh-head movement. PhD diss., Cornell University.

Yanagida, Yuko
　2003　Obligatory movement and head parameter: Evidence from Early Classical Japanese. *Sophia Linguistica* 50: 103–116.

Yanagida, Yuko
　2005　*The Syntax of FOCUS and WH-Questions in Japanese: A Cross-Linguistic Perspective.* Tokyo: Hituzi-Shobo.

Yoshimura, Akiko
　1999　*Hitei Kyokusei Gensho* [Negative Polarity Phenomena]. Tokyo: Eihosha.

Zanuttini, Raffaella
　1997　*Negation and Clausal Structure: A Comparative Study of Romance Languages.* Oxford: Oxford University Press.

Negation and polarity in the new millennium:
A bibliography

Laurence R. Horn

Alphabetization principles employed here override *van, van der, von, de,* and similar nomenclatural prefixes. The abbreviations *BLS, CLS, NELS, SALT,* and *WCCFL* represent volumes of papers or proceedings from the Berkeley Linguistics Society, the Chicago Linguistic Society, the North East Linguistic Society, Semantics and Linguistic Theory conferences, and West Coast Conference on Formal Linguistics respectively. NLLT is Natural Language and Linguistic Society. Thanks to Johan van der Auwera, Jack Hoeksema, and Yasuhiko Kato for comments and additions, and to Pierre Larrivée, Frank Richter, Henriëtte de Swart, and Hedde Zeijlstra for their own earlier compilations.

Abbott, Barbara
    2006   Where have some of the presuppositions gone? In Birner and Ward (eds.), 1–20.
Abels, Klaus
    2002   Expletive (?) negation. In *Proceedings of Formal approaches to Slavic Linguistics 10,* J. Toman (ed.), 1–21. Bloomington, IN: Michigan Slavic Publications.
Abels, Klaus
    2003   Who gives a damn about minimizers in questions? *SALT XIII,* 1–18.
Abels, Klaus
    2005   "Expletive negation" in Russian: A conspiracy theory. *Journal of Slavic Linguistics* 13: 5–74.
Alonso-Ovalle, Luis and Elena Guerzoni
    2002   Double negatives, negative concord and metalinguistic negation. *CLS* 38 (1): *The Main Session,* 15–31.
Amaral, Patricia
    2007   The Meaning of Approximative Adverbs: Evidence from European Portuguese. PhD dissertation, The Ohio State University.
An, Duk-Ho
    2007   On the distributions of NPIs in Korean. *Natural Language Semantics* 15: 317–350.
Anderssen, Jan
    2007   Generalized domain widening überhaupt. *WCCFL* 25: 58–66.

Anderwald, Liselotte
    2002   *Negation in Non-Standard British English: Gaps, Regularizations and Asymmetries.* London: Routledge.
Anderwald, Lieselotte
    2003   Non-standard English and typological principles: The case of negation. In *Determinants of Grammatical Variation in English*, G. Rohdenburg and B. Mondorf (eds.), 507–529. Berlin/New York: Mouton de Gruyter.
Anderwald, Lieselotte
    2004   Local markedness as a heuristic tool in dialectology: The case of *amn't*. In *Dialectology Meets Typology*, B. Kortmann (ed.), 47–67. Berlin/New York: Mouton de Gruyter.
Anderwald, Lieselotte
    2005   Negative concord in British dialects. In Iyeiri (ed.), 113–137.
Andorno, Cecilia
    2008   Entre énoncé et interaction: le role des particules d'affirmation et negation dans les lectes d'apprenants. *Acquisition et Interaction en Langue Étrangère* 26: 173–190.
Andronis, Mary et al. (eds.)
    2002   *CLS* 38 (2): *The Panels.* (Negation and Polarity Items, 3–158.) Chicago: Chicago Linguistic Society.
Antoine, Fabrice
    2000   From understatement to overstatement in French. *French Studies Bulletin* 75: 13–17.
Armstrong, Nigel
    2002   Variable deletion of French *ne:* A cross-stylistic perspective. *Language Sciences* 24: 153–173.
Armstrong, Nigel and Alan Smith
    2002   The influence of linguistic and social factors on the recent decline of French *ne*. *Journal of French Language Studies* 12: 23–41.
Ashby, William
    2001   Un nouveau regard sur las chute du *ne* en français parlé tourangeau. *Journal of French Language Studies* 11: 1–22.
Atkinson, Jo, Ruth Campbell, Jane Marshall, Alice Thacker and Bencie Woll
    2004   Understanding 'not': neuropsychological dissociations between hand and head markers of negation in BSL. *Neuropsychologia* 42: 214–229.
Atlas, Jay D.
    2005   *Logic, Meaning and Conversation.* Oxford: Oxford University Press.
Atlas, Jay D.
    2007   On a pragmatic explanation of negative polarity licensing. In N. Burton-Roberts (ed.), 10–23.
van der Auwera, Johan
    2001   On the typology of negative modals. In Hoeksema et al. (eds.), 23–48.

van der Auwera, Johan
    2006   Why languages prefer prohibitives. *Wai Guo Yu* [Journal of Foreign Languages] 161: 1–25.
van der Auwera, Johan
    2009   The Jespersen cycles. In van Gelderen (ed.), 35–71.
van der Auwera, Johan and Annemie Neuckermans
    2004   Jespersen's cycle and the interaction of predicate and quantifier negation in Flemish. In *Typology Meets Dialectology: Dialect Grammar from a Cross-Linguistic Perspective*, B. Kortmann (ed.), 453–478. Berlin/New York: Mouton de Gruyter.
van der Auwera, Johan, Ludo Lejeune and Valentin Goussev
    2005   The prohibitive. In Haspelmath et al. (eds.), 290–293.
van der Auwera, Johan, Ludovic De Cuypere and Annemie Neuckermans
    2006   Negative indefinites: A typological and diachronic perspective on a Brabantic construction. In *Types of Variation: Diachronic, Dialectal and Typological Interfaces*, T. Nevalainen et al. (eds.), 305–319. Amsterdam: John Benjamins.
Babyonyshev, Maria and Dina Brun
    2001   Specificity matters: A new look at the new genitive of negation in Russian. *Proceedings of Formal Approach to Slavic Linguistics* 10.
Babyonyshev, Maria, Jennifer Ganger, David Pesetsky and Kenneth Wexler
    2001   The maturation of grammatical principles: Evidence from Russian unaccusatives. *Linguistic Inquiry* 32: 1–44.
Barbarczy, Anna
    2006   Negation and word order in Hungarian child language. *Lingua* 116: 377–392.
Barbiers, Sjef
    2002   Microvariation in negation in varieties of Dutch. In Barbiers et al. (eds.), 13–40.
Barbiers, Sjef, Leonie Cornips and Susanne van der Kleij (eds.)
    2002   *Syntactic Microvariation*. Amsterdam: Meertens Institute Electronic Publications in Linguistics. http://www.meertens.knaw.nl/books/synmic.
Barbiers, Sjef, Margreet van der Ham, Olaf Koeneman and Maria Lekakou (eds.)
    2008   *Microvariations in Syntactic Doubling*. Bingley, U.K.: Emerald.
Baumeister, Roy, Ellen Bratslavlavsky, Catrin Finkenauer and Kathleen Vohs
    2001   Bad is stronger than good. *Review of General Psychology* 5: 323–370.
Bayer, Josef
    2006   *Nothing/nichts* as negative polarity survivors? In *Between 40 and 60 Puzzles for Krifka: a web festschrift for Manfred Krifka*, H.-M. Gärtner et al. (eds.). Berlin: ZAS. Downloadable from author's web site, http://ling.uni-konstanz.de/pages/home/bayer/publikationen_e.html.
Bayer, Josef
    2009   Nominal negative quantifiers as adjuncts. *Journal of Comparative Germanic Linguistics* 12: 5–30.

Beal, Joan and Karen Corrigan
    2005   *No, nay, never:* Negation in Tyneside English. In Y. Iyeiri (ed.), 139–157.
Beaver, David
    2004   Five only pieces. *Theoretical Linguistics* 30: 45–64.
Beaver, David and Brady Clark
    2002   Monotonicity and focus sensitivity. *Proceedings of SALT XII.*
Beaver, David and Brady Clark
    2008   *Sense and Sensitivity: How Focus Determines Meaning.* Oxford: Oxford University Press.
Becker, Angelika
    2005   The semantic knowledge base for the acquisition of negation and the acquisition of finiteness. In *The Structure of Learner Varieties*, H. Hendriks (ed.), 263–314. Berlin/New York: Mouton de Gruyter.
Belletti, Adriana
    2001   Specualtions on the possible source of expletive negation in Italian comparative clauses. In *Current Issues in Italian Syntax: Essays Offered to Lorenzo Renzi*, G. Cinque and G. Salvi (eds.), 19–38. Amsterdam: North-Holland.
Bendjaballah, Sabrina
    2000   The 'negative preterite' in Kabyh Berber. *Folia Linguistica* 34: 185–223.
Benmakhlouf, Ali
    2001   G. Frege sur la negation comme opposition sans force. *Revue de Métaphysique et de Morale* 2: 7–19.
Benmamoun, Elabbas
    2006   Licensing configurations: The puzzle of head negative polarity items. *Linguistic Inquiry* 37: 141–149.
Benoist, Jocelyn
    2001   La théorie phénoménologique de la negation, entre acte et sens. *Revue de Métaphysique et de Morale* 2: 21–35.
Bernini, Giuliano
    2000   Negative items and negation strategies in non-native Italian. *Studies in Second Language Acquisition* 22: 399–440.
Béziau, Jean-Yves
    2003   New light on the square of opposition and its nameless corner. *Logical Investigations* 10: 218–232.
Bezuidenhout, Anne, Robin Morris and Cinta Widmann
    2009   The DE-blocking hypothesis: The role of grammar in scalar reasoning. In Sauerland and Yatsuhiro (eds.), 124–144.
Biberauer, Theresa
    2007   Afrikaans negation and polarity: a comparative investigation of the two *nie*s. In Zeijlstra and Soehn (eds.), 1–7.

Biberauer, Theresa
2008   Doubling vs omission: Insights from Afrikaans negation. In Barbiers et al. (eds.), 103–140.

Biberauer, Theresa
2009   Jespersen off course? The case of contemporary Afrikaans negation. In van Gelderen (ed.).

Birner, Betty and Gregory Ward (eds.)
2006   *Drawing the Boundaries of Meaning: Neo-Gricean Studies in Pragmatics and Semantics in Honor of Laurence R. Horn.* Amsterdam: John Benjamins.

Blaszczak, Joanna
2001   *Investigation into the Interaction Between Indefinites and Negation.* Berlin: Akademie Verlag.

Blaszczak, Joanna
2002   What do bagels and Polish *kolwiek*-pronouns have in common? In Andronis et al. (eds.), 3–18.

Bond, Oliver
2006   A questionnaire for negation: capturing variation in African languages. http://www.soas.ac.uk/linguistics/research/research-projects/negtyp/39436.pdf.

Borroff, Marianne
2006   Degree phrase inversion in the scope of negation. *Linguistic Inquiry* 37: 514–521.

Borsley, Robert and Bob Morris Jones
2005   *Welsh Negation and Grammatical Theory.* Cardiff: University of Wales Press.

Bošković, Željko
2008   Licensing negative constituents and negative concord. *NELS* 38: 125–138.

Brann, Eva
2001   *The Ways of Naysaying: no, not, nothing and nonbeing.* Lanham, MD: Rowman & Littlefield.

Breitbarth, Anne
2009   A hybrid approach to Jespersen's cycle in West Germanic. *Journal of Comparative Germanic Linguistics* 12.

Branco, António and Berthold Crysmann
2001   Negative concord and the distribution of quantifiers. In d'Hulst et al. (eds.), 39–62.

Broadbent, Judith
2009   The *amn't* gap: The view from West Yorkshire. *Journal of Linguistics* 45: 251–284.

Bresnan, Joan, Ashwini Deo and Devyani Sharma
2007   Typology in variation: a probabilistic approach to *be* and *n't* in the *Survey of English Dialects*. *English Language and Linguistics* 11: 301–346.

Brown, Sue and Adam Przepiorkowski (eds.)
    2007   *Negation in Slavic.* Bloomington, IN: Slavica Publishers.

Buchanan, Michiko Todoroko
    2007   Two types of negative polarity items: Evidence from VP ellipsis in Japanese. In Zeijlstra and Soehn (eds.), 16–21.

Bulatovic, Andjelka Gina
    2002   Looking for *any*: Serbo-Croatian *bilo,* Bulgarian *kakvoto ida e* and the nonveridicality hypothesis. In Andronis et al. (eds.), 19–28.

Büring, Daniel
    2005   Negative inversion. *NELS* 35.

Burridge, Kate and Margaret Florey
    2002   *Yeah no, he's a good kid:* A discourse analysis analysis of *yeah no* in Australian English. *Australian Journal of Linguistics* 22: 149–171.

Burton-Roberts, Noel
    2007   Varieties of semantics and encoding: Negation, narrowing/loosening and numericals. In Burton-Roberts (ed.), 90–114.

Burton-Roberts, Noel (ed.)
    2007   *Pragmatics.* Houndmills: Palgrave Macmillan.

Byloo, Pieter
    2009   Modality and Negation: A Corpus-Based Study. PhD dissertation, University of Antwerp.

Cameron-Faulkner, Thea, Elena Lieven and Anna Theakston
    2007   What part of no do children not understand? A usage-based account of multiword negation. *Journal of Child Language* 33: 251–282.

Carston, Robyn
    2002   Negation. *Thoughts and Utterances,* Chapter 4. Oxford: Blackwell.

Catalani, Luigi
    2001   *Die Negation im Mittelfranzösischen.* Frankfurt: Peter Lang.

Cheshire, Jenny
    2009   Negation. *Variation in an English Dialect: A Sociolinguistic Study,* Chapter 5. Cambridge: Cambridge University Press.

Chierchia, Gennaro
    2004   Scalar implicatures, polarity phenomena and the syntax/ pragmatics interface. In A. Belletti (ed.), *Structures and Beyond,* 39–103. Oxford: Oxford University Press.

Chierchia, Gennaro
    2006   Broaden your views: Implicatures of domain widening and the 'logicality' of language. *Linguistic Inquiry* 37: 535–590.

Choi, Jinyoung
    2007   Free Choice and Negative Polarity: A Compositional Analysis of Korean Polarity Sensitive Items. PhD dissertation, University of Pennsylvania.

Chung, Inkie
  2007 Suppletive negation in Korean and Distributed Morphology. *Lingua* 117: 95–148.
Cirillo, Buchanan, Michiko Todoroko
  2007 The neg stranding hypothesis. In Zeijlstra and Soehn (eds.), 22–28.
van Compernolle, Rémi
  2008 Morphosyntactic and phonological constraints on negative particle variation in French-language chat discourse. *Language Variation and Change* 20: 317–339.
Corblin, Francis, Viviane Déprez, Henriëtte de Swart and Lucia Tovena
  2004 Negative concord. In Corblin and de Swart (eds.).
Corblin, Francis and Henriëtte de Swart (eds.)
    Handbook of French Semantics. Stanford: CSLI Publications.
Corblin, Francis and Lucia Tovena
  2001 On the multiple expression of negation in Romance. In d'Hulst et al. (eds.), 87–115.
Corblin, Francis and Lucia Tovena
  2003 L'expression de la négation dans les langues romanes. In *Les langues romanes: problèmes de la phrase simple*, D. Godard (ed.), 279–341. Paris: CNRS Editions.
Cornips, Leonie and Paula Fikkert (eds.)
  2003 *Linguistics in the Netherlands 2003.* Amsterdam: John Benjamins.
Crain, Stephen, Amanda Gardner andrea Gualmini and Beth Rabbin
  2002 Children's command of negation. In *Proceedings of the Third Tokyo Conference on Psycholinguistics*, Y. Otsu (ed.), 71–95. Tokyo: Hituzi.
de Cuypere, Ludovic, Johan van der Auwera and Klaas Willems
  2007 Double negation and iconicity. In *Insistent Images*, E. Tabakowska et al. (eds.), 301–320. Amsterdam: John Benjamins.
Cyffer, Norbert, Erwin Ebermann and Georg Ziegelmeyer (eds.)
  2009 *Negation Patterns in West African Languages and Beyond.* Amsterdam: John Benjamins.
David-Ménard, Monique
  2001 La négation comme sortie de l'ontologie. *Revue de Métaphysique et de Morale* 2: 59–67.
Davis, Henry
  2005 On the syntax and semantics of negation in Salish. *International Journal of American Linguistics* 71: 1–55.
Dayal, Veneeta
  2004 The universal force of free choice *any*. *Linguistic Variation Yearbook* 4: 5–40.
De Decker, Paul, Erik Larsson and Andrea Martin
  2005 Polarity judgments: An empirical view. Poster for NYU Polarity Workshop, posted at http://www.nyu.edu/gsas/dept/lingu/events/polarity/posters.html.

Déprez, Viviane
    2000   Parallel (a)symmetries and the internal structure of negative expressions. *NLLT* 18: 253–342.

Déprez, Viviane and France Martineau
    2004   Micro-parametric variation and negative concord. In *Contemporary Approaches to Romance Linguistics: Selected Papers from the 33rd Linguistic Symposium on Romance Languages*, J. Auger et al. (eds.), 139–158. Amsterdam: John Benjamins.

Devos, Maud and Johan van der Auwera
    to appear   Jespersen cycles in Bantu: Double and triple negation. In *Atti del Convegno della Società Italiana di Glottologia (Palermo 2008)*.

Dewaele, Jean-Marc and Vera Regan
    2002   Maîtriser la norm sociolinguistique en interlangue française: le cas de l'omission variable de *ne*. *Journal of French Language Studies* 12: 123–148.

den Dikken, Marcel
    2002   Direct and parasitic polarity item licensing. *Journal of Comparative Germanic Linguistics* 45: 33–66.

den Dikken, Marcel
    2006   Parasitism, secondary triggering and depth of embedding. In Zanuttini et al. (eds.), 175–198.

den Dikken, Marcel and Anastasia Giannakidou
    2002   From hell to polarity: 'Aggressively non-D-linked' wh-phrases as polarity items. *Linguistic Inquiry* 33: 31–61.

Dimroth, Christine
    2008   Age effects on the process of L2 acquisition? Evidence from the acquisition of negation and finiteness L2 German. *Language Learning* 58: 117–150.

Dimroth, Christine
    to appear   Stepping stones and stumbling blocks: why negation accelerates and additive particles delay the acquisition of finiteness in German. In *Functional Elements: Variation in Learner Systems*, C. Dimroth and P. Jordens (eds.), Berlin/New York: Mouton de Gruyter

Doetjes, Jenny
    2005   The cameleontic nature of French *ni*: Negative coordination in a negative concord language. *Proceedings of Sinn und Bedeutung* 9: 72–86.

Donohue, Mark
    2006   Negative grammatical functions in Skou. *Language* 82: 383–398.

Dowty, David
    2006   Resumptive negation as assertion revision. Unpublished ms., The Ohio State University.

Drago, Antonino
2001   Vasiliev's paraconsistent logic interpreted by means of the dual role played by the double negation law. *Journal of Applied Non-Classical Logic* 11: 281–94.

Drago, Antonino
2003   Traduzione, doppia negazione ed ermeneutica. *Studium* 99: 769–780.

Drago, Antonino
2005   A. N. Kolmogoroff and the relevance of the double negation law in science. In *Essays on the Foundations of Mathematics and Logic*, G. Sica (ed.), 57–81. Milan: Polimetrica.

Drenhaus, Heiner
2002   On the acquisition of German nicht 'not' as sentential negation: Evidence from negative polarity items. *BUCLD 26: Proceedings of the 26th Annual Boston University Conference on Language Development* Somerville: Cascadilla Press, 166–174.

Drenhaus, Heiner, Joanna Blaszczak and Juliane Schütte
2007   Some psycholinguistic comments on NPI licensing. *Proceedings of Sinn und Bedeutung* 11: 180–193.

Drenhaus, Heiner, Peter beim Graben, Stefan Frisch and Douglas Saddy
2006   Diagnosis and repair of negative polarity constructions in the light of symbolic resonance analysis. *Brain and Language* 96: 255–268.

Drenhaus, Heiner, Douglas Saddy and Stefan Frisch
2005   Processing negative polarity items: When negation comes through the backdoor. In *Linguistic Evidence: Empirical, Theoretical and Computational Perspectives*, S. Kepser, S. and M. Reis (eds.), 145–165. Berlin/New York: Mouton de Gruyter.

Drozd, Kenneth
2001   Metalinguistic sentence negation in child English. In Hoeksema et al. (eds.), 49–78.

Drury, John and Karsten Steinhauer
2009   Brain potentials for logical semantics/pragmatics. In Sauerland and Yatsuhiro (eds.), 186–215.

Dryer, Matthew
2005   Negative morphemes. In *The World Atlas of Language Structures*, M. Haspelmath et al. (eds.), 454–457. Oxford: Oxford University Press.

Dryer, Matthew
to appear   Verb-object-negative order in Central Africa. In *Negation Patterns in West Africa*, N. Cyffer et al. (eds.). Amsterdam: John Benjamins.

Duffley, Patrick and Pierre Larrivée
in press   Anyone for non-scalarity? *English Language and Linguistics*.

Dundes, Alan
2002   Much ado about 'sweet bugger all': Getting to the bottom of a puzzle in British folk speech. *Folklore* 113: 35–49.

Dunn, J. Michael and Chunlai Zhou
    2005   Negation in the context of gaggle theory. *Studia Logica* 80: 235–264.

É. Kiss, Katalin
    2002   Negative quantifiers and specificity. In Kenesei and Siptár (eds.), 39–61.

Eckardt, Regine
    2003   Eine Runde im Jespersen-Zyklus: Negation, emphatische Negation und negativ-polare Elemente im Altfranzösischen. Online publication in KOPS, http://kops.ub.uni-konstanz.de/volltexte/2003/991

Eckardt, Regine
    2005   Too poor to mention: Subminimal events and negative polarity items. In C. Maienborn and A. Wöllstein (eds.), *Event arguments: Foundations and Applications,* 301–330. Tübingen: Niemeyer.

Eckardt, Regine
    2006   From step to negation: The development of French complex negation patterns. *Meaning Change in Grammaticalization: An Inquiry into Semantic Reanalysis,* Chapter 5. Oxford: Oxford University Press.

ver Eecke, Wilfried
    2004   Ontology of denial. In *Rereading Freud: Psychoanalysis through Philosophy*, J. Mills (ed.), 103–125. Albany: State University of New York Press.

ver Eecke, Wilfried
    2006   *Denial, Negation and the Forces of the Negative: Freud, Hegel, Lacan, Spitz and Sophocles.* Albany: State University of New York Press.

Eilam, Aviad
    2007   The crosslinguistic realization of -*ever*: Evidence from Modern Hebrew. *CLS* 43 (2): 39–53.

Ernst, Thomas
    2005   On speaker-oriented adverbs as positive polarity items. Poster for NYU Polarity Workshop, posted at http://www.nyu.edu/gsas/dept/lingu/events/polarity/posters.html.

Erschler, Michiko Todoroko
    2007   The structure of events and the genitive of negation with measure adverbials in Russian. In Zeijlstra and Soehn (eds.), 29–35.

Espinal, M. Teres
    2000   On the semantic status of n-words in Catalan and Spanish. *Lingua* 110: 557–580.

Espinal, M. Teresa
    2007   Licensing expletive negation and negative concord in Catalan and Spanish. In Floricic (ed.), 49–74.

Etchemendy, Nancy
    2000   *The Power of Un.* New York: Scholastic.

Everett, Anthony and Thomas Hofweber (eds.)
 2000 *Empty Names, Fiction and the Puzzles of Non-Existence*. Stanford: CSLI Publications.

Falaus, Anamaria
 2007a Double negation and negative concord: the Romanian puzzle. In *Selected Papers from the 36th Linguistics Symposium on Romance Languages*, J. Camacho and V. Déprez (eds.), 135–48. Amsterdam: John Benjamins.

Falaus, Anamaria
 2007b Le paradoxe de la double negation dans une langue à concordance negative stricte. In Floricic (ed.), 75–91.

Falkenberg, Gabriel
 2001 Lexical sensitivity in negative polarity verbs. In Hoeksema et al. (eds.), 79–98.

Farkas, Donka
 2006 Free choice in Romanian. In Birner and Ward (eds.), 71–94.

Feigenbaum, Susanne
 2002 A contrastive analysis of French and Hebrew prepositions: The case of *sans, bli-belo* and *lelo*. In *Prepositions in their Syntactic, Semantic and Pragmatic Context*, S. Feigenbaum and D. Kurzon (eds.), 171–191. Amsterdam: John Benjamins.

Feigenbaum, Susanne
 2003 L'antonyme en extension: le cas de *sans*. In *La syntaxe raisonnée: mélanges de linguistique générale et française offerts à Annie Boone*, P. Hadermann et al. (eds.), 185–194. Brussels: De Boeck Duculot.

Ferguson, Heather, Anthony Sanford and Hartmut Leuthold
 2008 Eye-movements and ERPs reveal the time course of processing negation and remitting counterfactual worlds. *Brain Research* 1236: 113–125.

von Fintel, Kai and Sabine Iatridou
 2007 Anatomy of a modal construction. *Linguistic Inquiry* 38: 445–483.

Fitzmaurice, Susan M.
 2002 The textual resolution of structural ambiguity in eighteenth-century english: a corpus linguistic study of patterns of negation. In *Using Corpora to Explore Linguistic Variation*, R. Reppen et al. (eds.), 227–247. Amsterdam: John Benjamins.

Floricic, Franck
 2005 La négation dans les langues romanes. *Lalies 25: Actes des Sessions de literature et linguistique*, 163–194. Paris: Presses de l'École Normale Supérieure.

Floricic, Franck (ed.)
 2007 *La négation dans les langues romanes*. Amsterdam: John Benjamins.

Floricic, Franck and Françoise Mignon
    2007   Négation et réduplication intensive en français et en italien. In Floricic (ed.), 117–136.

Floricic, Franck and Lucia Molinu
    2008   L'Italie et ses dialectes. *Lalies 28: Actes des sessions de littérature et linguistique*, 5–107. Paris: Presses de l'École Normale Supérieure.

Fonseca-Greber, Bonnie
    2007   The emergence of emphatic *ne* in conversational Swiss French. *Journal of French Language Studies* 17: 249–275.

Fraenkel, Tamar and Yaacov Schul
    2008   The meaning of negated adjectives. *Intercultural Pragmatics* 5: 517–540.

Frampton, John
    2001   The *amn't* gap, ineffability and anomalous *aren't*: Against morphosyntactic competition. *CLS* 37(2): 1–13.

Freeman, Jason
    2004   Syntactical analysis of the "So don't I" construction. *Cranberry Linguistics* 2 (University of Connecticut Working Papers in Linguistics 12): 25–38.

Franco, Jon and Alazne Landa
    2006   Preverbal n-words and anti-agreement effects. In *Selected Proceedings of the 9th Hispanic Linguistics Symposium*, N. Sagarra and and A. J. Toribio (eds.), 34–42. Somerville, MA: Cascadilla.

Furtado da Cunha, Maria Angélica
    2007   Grammaticalization of the strategies of negation in Brazilian Portuguese. *Journal of Pragmatics* 39: 1638–1653.

Furukawa, Yukio
    2007   Unembedded 'negative' quantifiers. In *New Frontiers in Artificial Intelligence*, T. Washio et al. (eds.), 232–245. Berlin: Springer.

Gadet, Françoise
    2000   Des corpus pour *ne...pas*. In *Corpus: méthodologie et applications linguistiques*, M. Bilger (ed.), 156–167. Paris: Champion.

Gajewski, Jon
    2005a   *Only John* vs. *Nobody but John:* A solution. Poster for NYU Polarity Workshop, posted at http://www.nyu.edu/gsas/dept/lingu/events/polarity/posters.html.

Gajewski, Jon
    2005b   Neg Raising: Polarity and Presuppositions. PhD dissertation, MIT.

Gajewski, Jon
    2007   Neg-raising and polarity. *Linguistics and Philosophy* 30: 289–328.

van Gass, Kate
    2007   Multiple n-words in Afrikaans. *SPIL Plus* 35: 167–186.

van Gelderen, Elly
    2004   *Grammaticalization as Economy.* Amsterdam: John Benjamins.
van Gelderen, Elly
    2008   Negative cycles. *Linguistic Typology* 12: 195–243.
van Gelderen, Elly
    2009   Cyclical change, an introduction. In van Gelderen (ed.).
van Gelderen, Elly (ed.)
    2009   *Cyclical Change.* Amsterdam: John Benjamins.
Gennari, Silvia and Maryellen MacDonald
    2006   Acquisition of negation and quantification: Insights from adult production and comprehension. *Language Acquisition* 13: 125–168.
Geurts, Bart and Rick Nouwen
    2007   *At least* et al.: The semantics of scalar modifiers. *Language* 83: 533–559.
Geurts, Bart and Frans van der Silk
    2005   Monotonicity and processing load. *Journal of Semantics* 22: 97–117.
Giannakidou, Anastasia
    2000a   Negative…concord? *Natural Language and Linguistic Theory* 18: 457–523.
Giannakidou, Anastasia
    2000b   Negative concord and the scope of universals. *Transactions of the Philological Society* 98: 87–120.
Giannakidou, Anastasia
    2001a   The meaning of free choice. *Linguistics and Philosophy* 24: 659–735.
Giannakidou, Anastasia
    2001b   Varieties of polarity items and the (non)veridicality hypothesis. In Hoeksema et al. (eds.), 99–129.
Giannakidou, Anastasia
    2002a   Licensing and sensitivity in polarity items: From downward entailment to (non)veridicality. In Andronis et al. (eds.), 29–53.
Giannakidou, Anastasia
    2002b   UNTIL, aspect and negation: A novel argument for two *untils*. *SALT XII*: 84–103.
Giannakidou, Anastasia
    2004   Review of Horn (2001). *Journal of Linguistics* 40: 426–433.
Giannakidou, Anastasia
    2006a   *Only,* emotive factive verbs and the dual nature of polarity dependency. *Language* 82: 575–603.
Giannakidou, Anastasia
    2006b   N-words and negative concord. In *The Syntax Companion, Vol. 3*, M. Everaert et al. (eds.), 327–391. London: Blackwell.
Giannakidou, Anastasia
    2006c   Polarity, questions and the scalar properties of *even*. In Birner and Ward (eds.), 95–116.

Giannakidou, Anastasia
   2007a   The landscape of *even*. *Natural Language and Linguistic Theory* 25: 39–81.

Giannakidou, Anastasia
   to appear Positive polarity items and negative polarity items: variation, licensing and compositionality. In *Semantics: An International Handbook of Natural Language Meaning*, C. Maienborn et al. (eds.). Berlin/New York: Mouton de Gruyter.

Giannakidou, Anastasia and Lisa Cheng
   2006   (In)definiteness, polarity and the role of wh-morphology in free choice. *Journal of Semantics* 23: 135–183.

Giora, Rachel
   2006   Anything negatives can do affirmatives can do just as well, except for some metaphors. *Journal of Pragmatics* 38: 981–1014.

Giora, Rachel
   2007   "A good Arab is *not* a dead Arab – a racist incitement": on the accessibility of negated concepts. In *Explorations in Pragmatics*, I. Kecskes and L. Horn (eds.). Berlin/New York: Mouton de Gruyter.

Giora, Rachel, Noga Balaban, Ofer Fein and Inbar Alkabets
   2005   Negation as positivity in disguise. In *Figurative Language Comprehension: Social and Cultural Influences*, H. Colston and A. Katz (eds.), 233–258. Hillsdale, NJ: Erlbaum.

Giora, Rachel, Vered Heruti, Nili Metuki and Ofer Fein
   2009   "When we say *no* we mean *no*": Interpreting negation in vision and language. *Journal of Pragmatics* 41: 2222–2239.

Giora, Rachel, Ofer Fein, Keren Aschkenazi and Inbar Alkabets-Zlozover
   2007   Negation in context: A functional approach to suppression. *Discourse Processes* 43: 153–172.

Giora, Rachel, Ofer Fein, Jonathan Ganzi, Natalie Alkeslassy Levi and Hadas Sabah
   2005   Negation as mitigation: the case of negative irony. *Discourse Processes* 39: 81–100.

Giora, Rachel, Dana Zimmerman and Ofer Fein
   2008   How can you compare! On negated comparisons as comparisons. *Intercultural Pragmatics* 5: 501–516.

Giuliano, Patricia
   2003   Negation and relational predicates in French and English as second languages. In *Information Structure and the Dynamics of Language Acquisition*, C. Dimroth and M. Starren (eds.), 119–158. Amsterdam: John Benjamins.

Giuliano, Patricia
   2004   *La négation linguistique dans l'acquisition d'une langue étrangère: un débat conclu?* Bern: Peter Lang.

Godard, Danièle
    2004   French negative dependency. In Corblin and de Swart (eds.), 351–389.
Godard, Danièle and Jean-Marie Marandin
    2007   Aspects pragmatiques de la negation renforcée en italien. In Floricic (ed.), 137–160.
Gualmini, Andrea
    2007   Negation and polarity: The view from child language. In Zeijlstra and Soehn (eds.), 43–49.
Gualmini, Andrea
    2009   Experimental pragmatics and parsimony: The case of scopally ambiguous sentences containing negation. In Sauerland and Yatsuhiro (eds.), 145–161.
Gualmini, Andrea and Stephen Crain
    2005   The structure of children's linguistics knowledge. *Linguistic Inquiry* 36: 463–474.
Gualmini, Andrea and Bernhard Schwarz
    2007   Negation and downward entailingness: Consquences for learnability theory. In Zeijlstra and Soehn (eds.), 50–56.
Guerzoni, Elena
    2002   *Even*-NPIs in questions. *NELS* 32: 153–170.
Guerzoni, Elena
    2006   Intervention effects on NPIs and feature movement: Towards a unified account of intervention. *Natural Language Semantics* 14: 359–398.
Guerzoni, Elena and Yael Sharvit
    2007   A question of strength: On NPIs in interrogative clauses. *Linguistics and Philosophy* 30: 361–391.
Guidetti, Michèle
    2005   Yes or no? How young French children combine gestures and speech to agree and refuse. *Journal of Child Language* 32: 911–924.
Gunlogson, Christine
    2003   *True to Form: Rising and Falling Declaratives as Questions in English*. New York: Routledge.
Gutiérrez-Rexach, Javier and Scott Schwenter
    2002   Propositional NPIs and the scalar nature of polarity. In *From Words to Discourse: Trends in Spanish Semantics and Pragmatics*, J. Gutiérrez-Rexach (ed.), 237–262. Oxford: Elsevier.
Haegeman, Liliane
    2000   Negative preposing, negative inversion and the split CP. In Horn and Kato (eds.), 21–61.
Haegeman, Liliane
    2001   Approaches to OV: West Flemish negation as evidence for double movement. In *Progress in Grammar: Articles at the 20th Anniversary of the Comparison of the Grammatical Models Group in Tilburg*,

M. van Oostendorp and E. Anagnostopoulou (eds.). Online at http://www.meertens.nl/books/progressingrammar.

Haegeman, Liliane
2002a   Some notes on DP-internal negative doubling. In Barbiers et al. (eds.).

Haegeman, Liliane
2002b   West Flemish negation and the derivation of SOV order in West Germanic. *Nordic Journal of Linguistics* 25: 154–189.

Hagstrom, Paul
2002   Implications of child error for the syntax of negation in Korean. *Journal of East Asian Linguistics* 11: 211–242.

Han, Chung-hye
2001   Force, negation and imperatives. *The Linguistic Review* 18: 289–325.

Han, Chung-hye and Chungmin Lee
2007   On negative imperatives in Korean. *Linguistic Inquiry* 38: 373–395.

Han, Chung-hye and Chungmin Lee
2008   A morphological constraint on negation in imperatives in Korean. In *Japanese/Korean Linguistics* 13, M. Endo Hudson et al. (eds.), 130–140. Stanford: CSLI Publications.

Han, Chung-hye, Jeffrey Lidz and Julien Musolino
2007   Verb-raising and grammar competition in Korean: Evidence from negation and quantifier scope. *Linguistic Inquiry* 38: 1–47.

Han, Chung-hye and Maribel Romero
2001   Negation, focus and alternative questions. *WCCFL* 20: 262–275.

Han, Chung-hye and Maribel Romero
2004   Disjunctions, focus and scope. *Linguistic Inquiry* 35: 179–217.

Han, Chung-hye, Dennis Storoshenko and Yasuko Sakurai
2004   Scope of negation and clause structure in Japanese. *BLS* 30.

Han, Chung-hye, Dennis Storoshenko and Yasuko Sakurai
2008   An experimental investigation into the syntax of negation in Japanese. *Language Research* [Seoul] 44: 1–31.

Hansen, Anita and Isabelle Malderez
2004   Le *ne* de négation parisienne: une étude en temps réel. *Langage et Sociétés* 107: 5–30.

Hansen, Maj-Britt Mosegaard
2009   The grammaticalization of negative reinforcers in Old and Middle French: a discourse-functional approach. In *Current Trends in Diachronic Semantics and Pragmatics*, M.-B. Mosegaard Hansen and J. Visconti (eds.), Bingley: Emerald.

Hartung, Simone
2007   Forms of negation in polar questions. In Zeijlstra and Soehn (eds.), 57–63.

Haspelmath, Martin
2005   Negative indefinite pronouns and predicate negation. In Haspelmath et al. (eds.), 466–469.

Haspelmath, Martin, Matthew Dryer, David Gil and Bernard Comrie (eds.)
    2005   *The World Atlas of Language Structures*. Oxford: Oxford University Press. (Online version, 2008: http://wals.info/.)

Hasson, Uri and Sam Glucksberg
    2006   Does understanding negation entail affirmation? An examination of negated metaphors. *Journal of Pragmatics* 38: 1015–1032.

Hatano, Etsuko
    2002   Early acquisition of negation in pre-verbal to two word utterance infancy. In Kato (ed.), 145–160.

Hawkins, Roger
    2001   The second language acquisition of negation and verb movement. *Second Language Syntax: A Generative Persepctive*, Chapter 3. London: Blackwell.

van Hecke, Tine
    2007   La negation de la modalité déontique. In Floricic (ed.), 161–176.

Hendricks, Petra
    2004   *Either, both and neither* in coordinate structures. In *The Composition of Meaning: From Lexeme to Discourse*, A. ter Meulen and W. Abraham (eds.), 115–138. Amsterdam: John Benjamins.

Herburger, Elena
    2000   *What Counts: Focus and Quantification*. Cambridge, MA: MIT Press.

Herburger, Elena
    2001   The negative concord puzzle revisited. *Natural Language Semantics* 9: 289–333.

Herburger, Elena
    2003   A note on Spanish *ni siquiera, even and* the analysis of NPIs. *Probus* 15: 237–256.

Herburger, Elena and Simon Mauck
    2007   A new look at Ladusaw's puzzle. In Zeijlstra and Soehn (eds.), 64–70.

Herdan, Simona and Yael Sharvit
    2006   Definite and non-definite superlatives and NPI licensing. Poster for NYU Polarity Workshop, posted at http://www.nyu.edu/gsas/dept/lingu/events/polarity/posters.html.

Heritage, John
    2002   The limits of questioning: negative interrogatives and hostile question content. *Journal of Pragmatics* 34: 1427–1446.

Hintikka, Jaakko
    2002   Negation in logic and natural language. *Linguistics and Philosophy* 25: 585–600.

Hoeksema, Jack
    2000   Negative polarity items: Triggering, scope and c-command. In Horn and Kato (eds.), 115–146.

Hoeksema, Jack
   2001   Rapid change among expletive polarity items. In L. Brinton and D. Lundström (ed.), *Selected Papers from the 14th International Conference on Historical Linguistics,* 175–186. Amsterdam: John Benjamins.

Hoeksema, Jack
   2002a   Minimaliseerders in het standaard-Nederlands. *TABU* 32: 105–174.

Hoeksema, Jack
   2002b   Polarity-sensitive scalar particles in Early Modern and Present-Day Dutch. *Belgian Journal of Linguistics* 16: 53–64.

Hoeksema, Jack
   2003   Partitivity, degrees and polarity. *Verbum* XXV: 81–96.

Hoeksema, Jack
   2005   In days, weeks, months, years, ages: a class of temporal negative polarity items. In *Rejected Papers: Feestbundel voor Ron van Zonneveld*, D. Gilbers and P. Hendriks (eds.), Groningen: Rijksuniversiteit.

Hoeksema, Jack
   2007a   Parasitic licensing of negative polarity items. *Journal of Comparative Germanic Linguistics* 10: 163–182.

Hoeksema, Jack
   2007b   Dutch *enig:* From nonveridicality to downward entailment. In Zeijlstra and Soehn (eds.), 8–15.

Hoeksema, Jack
   2008   There is no number effect in the licensing of negative polarity items: a reply to Guerzoni and Sharvit. *Linguistics and Philosophy* 31: 397–407.

Hoeksema, Jack
   2009   Jespersen recycled. In van Gelderen (ed.), 15–34.

Hoeksema, Jack and Donna Jo Napoli
   2008   Just for the hell of it: A comparison of two taboo-term constructions. *Journal of Linguistics* 44: 347–378.

Hoeksema, Jack and Hotze Rullmann
   2001   Scalarity and polarity: A study of scalar adverbs as polarity items. In Hoeksema et al., 129–172.

Hoeksema, Jack, Hotze Rullmann, Victor Sanchez-Valencia and Ton van der Wouden (eds.)
   2001   *Perspectives on Negation and Polarity Items.* Amsterdam: John Benjamins.

Holmberg anders
   2003   Yes/no questions and the relation between tense and polarity in English and Finnish. *Linguistic Variation Yearbook* 3: 43–68.

Holmberg anders
   2007   Null subjects and polarity focus. *Studia Linguistica* 61: 212–236.

Horn, Laurence R.
    2000a   Pick a theory (not just *any* theory): Indiscriminatives and the free-choice indefinite. In Horn and Kato (eds.), 147–192.

Horn, Laurence R.
    2000b   *any* and *(-)ever:* Free choice and free relatives. *IATL 15 (Proceedings of the 15th Annual Conference of the Israeli Association for Theoretical Linguistics),* 71–111.

Horn, Laurence R.
    2000c   From IF to IFF: Conditional perfection as pragmatic strengthening. *Journal of Pragmatics* 32: 289–326.

Horn, Laurence R.
    2001a   *A Natural History of Negation* [orig. published University of Chicago Press, 1989], reissued with new introductory material. Stanford: Center for the Study of Language and Information.

Horn, Laurence R.
    2001b   Flaubert triggers, squatitive negation and other quirks of grammar. In Hoeksema, et al. (eds.), 173–202.

Horn, Laurence R.
    2002a   Assertoric inertia and NPI licensing. In Andronis et al. (eds.), 55–82.

Horn, Laurence R.
    2002b   Uncovering the un-word: A study in lexical pragmatics. *Sophia Linguistica* 49: 1–64.

Horn, Laurence R.
    2002c   The logic of logical double negation. In Kato (ed.), 79–112.

Horn, Laurence R.
    2005a   Airport '86 revisited: Toward a unified indefinite *any*. In *Reference and Generality: The Partee Effect*, G. Carlson and F. J. Pelletier (eds.), 179–205. Cambridge: MIT Press.

Horn, Laurence R.
    2005b   An un-paper for the unsyntactician. In *Polymorphous Linguistics: Jim McCawley's Legacy*, S. Mufwene et al. (eds.), 329–365. Cambridge: MIT Press.

Horn, Laurence R.
    2005c   Diagnosing a diagnostic: Revisiting the Quantifier Constraint on exceptives. Poster for NYU Polarity Workshop, posted at http://www.nyu.edu/gsas/dept/lingu/events/polarity/posters.html.

Horn, Laurence R.
    2006a   The border wars: A neo-Gricean perspective. In K. Turner and K. von Heusinger (eds.), *Where Semantics Meets Pragmatics,* 21–48. London: Elsevier.

Horn, Laurence R.
    2006b   Speaker and hearer in neo-Gricean pragmatics. *Journal of Foreign Languages* 164: 2–25.

Horn, Laurence R.
    2006c   Contradiction. *Stanford Encyclopedia of Philosophy*. Posted at http://plato.stanford.edu/archives/fall2006/entries/contradiction.

Horn, Laurence R.
    2007   Neo-Gricean pragmatics: a Manichaean manifesto. In N. Burton-Roberts (ed.), 158–183.

Horn, Laurence R.
    2009a   *Almost* et al.: Scalar adverbs revisited. In *Current Issues in Unity and Diversity of Languages* (Papers from CIL 18, Seoul, Korea). Seoul: Linguistic Society of Korea.

Horn, Laurence R.
    2009b   Hypernegation, hyponegation and *parole* violations. Paper presented at BLS 35. To appear in proceedings of the conference.

Horn, Laurence R.
    2009c   WJ-40: Implicature, truth and meaning. *International Review of Pragmatics* 1: 3–34.

Horn, Laurence R.
    2009d   Negation. Entry in L. Cummings (ed.), *The Pragmatics Encyclopedia*. New York: Routledge.

Horn, Laurence R.
    to appear   Lexical pragmatics and the geometry of opposition. In *Histoire d'*O: New Perspectives on the Square of Opposition*, J.-Y. Béziau and G. Payette (eds.), Downloadable at http://www.yale.edu/linguist/faculty/horn_pub.html.

Horn, Laurence R. and Yasuhiko Kato
    2000   Negation and polarity at the millennium. Introduction to Horn and Kato (eds.), 1–20.

Horn, Laurence R. and Yasuhiko Kato (eds.)
    2000   *Negation and Polarity: Syntactic and Semantic Perspectives*. Oxford: Oxford University Press.

Howe, Darin
    2005   Negation in African American Vernacular English. In Iyeiri (ed.), 173–203.

Hoyt, Frederick
    2006   Long distance negative concord and restructuring in Palestinian Arabic. In *Concord Phenomena and the Syntax-Semantics Interface*, P. Dekker and H. Zeijlstra (eds.). 27–32. ESSLLI Malaga publication.

Hsiao, Katherine Pei-Yi
    2007   Polarity and 'bipolar' constructions: Epistemic biases in questions. In Zeijlstra and Soehn (eds.), 71–77.

Huddleston, Rodney and Geoffrey Pullum
    2002   Negation. *Cambridge Grammar of the English Language*, Chapter 9, 785–849. Cambridge: Cambridge University Press.

Hudson, Richard
2000 *I amn't. Language 76: 297–323.
d'Hulst, Yves, Johan Rooryck and Jan Schroten (eds.)
2001 Romance Languages and Linguistic Theory 1999. Amsterdam: John Benjamins.
Huot, Hélène
2007 La préfixation negative en français moderne. In Floricic (ed.), 177–203.
Iatridou, Sabine and Ivy Sichel
2008 Negative DPs and scope diminishment: Some basic patterns. NELS 38: 411–424.
Ingham, Richard
2000 Negation and OV order in Late Middle English. Journal of Linguistics 36: 13–38.
Ingham, Richard
2006a On two negative concord dialects in early English. Language Variation and Change 18: 241–266.
Ingham, Richard
2006b Negative concord and the loss of the negative particle ne in Late Middle English. Studia Anglica Posnaniensia: An International Review of English Studies 42: 77–97.
Ingham, Richard
2007 A structural constraint on multiple negation in Late Middle and Early Modern English. In To Make His Englissh Sweete upon His Tonge, M. Krygier and L. Sikorska (eds.), 55–67. Frankfurt: Peter Lang.
Inkova, Olga
2006 La negation explétive: un regard d'ailleurs. Cahiers Ferdinand de Saussure 59: 107–129.
Ippolito, Michela
2006 Remarks on only. Proceedings of SALT XVI.
Ippolito, Michela
2008 On the meaning of only. Journal of Semantics 25: 45–91.
Isac, Daniela
2004 Focus on negative concord. In Selected Papers from Romance Languages and Linguistic Theory 2002, R. Bok-Bennema et al. (eds.), 119–140. Amsterdam: John Benjamins.
Israel, Michael
2001 Minimizers, maximizers and the rhetoric of scalar reasoning. Journal of Semantics 18: 297–331.
Israel, Michael
2004 The pragmatics of polarity. In The Handbook of Pragmatics, L. Horn and G. Ward (eds.), 701–723. Oxford: Blackwell.

Israel, Michael
   2006   Saying less and meaning less. In Birner and Ward (eds.), 137–156.
Israel, Michael
   to appear *Pragmatics, Polarity and the Logic of Scales.* Cambridge: Cambridge University Press.
Iyeiri, Yoko
   2001   *Negative Constructions in Middle English.* Fukuoka: Kyushu University Press.
Iyeiri, Yoko
   2005   "I not say" once again: A study of the early history of the "*not* + finite verb" type in English. In Iyeiri (ed.), 59–81.
Iyeiri, Yoko (ed.)
   2005   *Aspects of English Negation.* Amsterdam: John Benjamins.
Jacobson, Pauline
   2006   I can't seem to figure this out. In Birner and Ward (eds.), 157–175.
Jäger, Agnes
   2005   Negation in Old High German. *Zeitschrift für Sprachwissenschaft* 24: 227–262.
Jäger, Agnes
   2007   On the diachrony of polarity types of indefinites. In Zeijlstra and Soehn (eds.), 78–84.
Jäger, Agnes
   2008   *History of German Negation.* Amsterdam: John Benjamins.
Jagueneau, Liliane
   2007   Négation simple et négation discontinue en occitan limousin. In Floricic (ed.), 99–116.
Jang, Youngjun and Yung-Hye Kwon
   2002   Indefinite nouns plus two types of conjoiners. In Andronis et al. (eds.), 83–95.
Jaspers, Dany
   2005   *Operators in the Lexicon: On the Negative Logic of Natural Language.* (Universiteit Leiden dissertation.) Utrecht: LOT.
Jayez, Jacques and Lucia Tovena
   2005   Free-choiceness and non-individuation. *Linguistics and Philosophy* 28: 1–71.
Jefferson, Gail
   2002   Is *no* an acknowledgment token? Comparing American and British uses of (+)/(–) tokens. *Journal of Pragmatics* 34: 1345–1383.
Joe, Jieun and Chungmin Lee
   2002   A 'removal' type of negative predicates. In *Japanese/Korean Linguistics, Vol. 10*, N. Akatsuka and S. Strauss (eds.), 559–572. Stanford: CSLI Publications.

Johannessen, Janne Bondi
2003   Negative polarity verbs in Norwegian. *Working Papers in Scandinavian Syntax 71*, 33–73. Department of Scandinavian Languages, Lund University

Jones, Steve
2002   *Antonymy: A corpus-based perspective.* London: Routledge.

Joseph, Kate and Julian Pine
2002   Does error-free use of French negation constitute evidence for very early parameter setting? *Journal of Child Language* 29: 71–89.

Kaiser, Elsi
2006   Negation and the left periphery in Finnish. *Lingua* 116: 314–350.

Kamide, Norihiro
2004   Quantized linear logic, involutive quantales and strong negation. *Studia Logica* 77: 355–384.

Kamide, Norihiro
2007   Gentzen-type methods for bilattice negation. *Studia Logica* 80: 265–289.

Kallel, Amel
2007   The loss of negative concord: Internal factors. *Language Variation and Change* 19: 27–49.

Kappus, Martin
2000   Topics in German Negation. PhD dissertation, University of Stony Brook.

Kato, Yasuhiko
2000   Interpretive asymmetries of negation. In Horn and Kato (eds.), 62–87.

Kato, Yasuhiko
2002   Negation in English and Japanese: Some (a)symmetries and their theoretical implications. In Kato (ed.), 1–21.

Kato, Yasuhiko (ed.)
2002   *Proceedings of the Sophia Symposium on Negation.* Tokyo: Sophia University.

Kato, Yasuhiko
2003   Negation in Classical Japanese: Projection and movement. *Sophia Linguistica* 50: 91–102.

Kaufmann, Anita
2002   Negation and prosody in British English: a study based on the London-Lund corpus. *Journal of Pragmatics* 34: 1473–1494.

Kaup, Barbara
2001   Negation and its impact on the accessibility of text information. *Memory & Cognition* 7: 960–967.

Kaup, Barbara
2009   How are pragmatic differences between positive and negative sentences captured in the processes and representation in language comprehension? In Sauerland and Yatsuhiro (eds.), 162–185.

Kaup, Barbara, Jana Lüdtke and Rolf Zwaan
  2006 Processing negated sentences with contradictory predicates: is a door that is not open mentally closed? *Journal of Pragmatics* 38: 1033 1050.
Kaup, Barbara, Richard Yaxley, Carol Madden, Rolf Zwaan and Jana Lüdtke
  2007 Experiential simulations of negated text information. *Quarterly Journal of Experimental Psychology* 60: 976–990.
Kaup, Barbara and Rolf Zwaan
  2003 Effects of negation and situational presence on the accessibility of text information. *Journal of Experimental Psychology: Learning, Memory and Cognition* 29: 439–446.
Kearns, John
  2006 Conditional assertion, denial and suppositon as illocutionary acts. *Linguistics and Philosophy* 29: 455–485.
Keenan, Edward
  2002 Some properties of natural language quantifiers: Generalized quantifier theory. *Linguistics and Philosophy* 25: 253–326.
van Kemenade, Ans
  2000 Jespersen's cycle revisited: Formal properties of grammaticalization. In S. Pintzuk et al. (eds.), *Diachronic Syntax: Models and Mechanisms*, 51–74. Oxford: Oxford University Press.
Kemmerer, David and Saundra Wright
  2002 Selective impairment of knowledge underlying *un*-prefixation: Further evidence for the autonomy of grammatical semantics. *Journal of Neurolinguistics* 15: 403–432,
Kenesei, István and Péter Siptár (eds.)
  2002 *Approaches to Hungarian 8: Papers from the Budapest Conference.* Budapest: Akadémiai Kiadó.
Kennedy, Christopher
  2001a On the monotonicity of polar adjectives. In Hoeksema et al. (eds.), 201–232.
Kennedy, Christopher
  2001b Polar opposition and the ontology of 'degrees'. *Linguistics and Philosophy* 24: 33–70.
Kennedy, Christopher and Louise McNally
  2005 Scale structure, degree modification and the semantics of gradable predicates. *Language* 81: 345–381.
Kim, Ae-Ryung
  2002 Two positions for Korean negation. In *Japanese/Korean Linguistics 10*, A. Akatsuka and S. Strauss (eds.), 587–600. Stanford: CSLI Publications.
Kim, Jong-Bok
  2000a *The Grammar of Negation: A Constraint-Based Account.* Stanford: CSLI Publications.

Kim, Jong-Bok
  2000b   On the prefixhood and scope of short form negation. In S. Kuno et al. (eds.), *Harvard Studies in Korean Linguistics 8*, 403–418. Cambridge, MA: Harvard University Department of Linguistics.
Kim, Jong-Bok and Ivan Sag
  2002   Negation without head-movement. *Natural Language and Linguistic Theory* 20: 339–412.
Kim, Shin-Sook and Peter Sells
  2007   Generalizing the immediate scope constraint on NPI-licensing. In H. Zeijlstra and J.-P. Soehn (eds.), 85–91.
Kiparsky, Paul and Cleo Condoravdi
  2006   Tracking Jespersen's Cycle. In *Proceedings of the 2nd International Conference of Modern Greek Dialects and Linguistic Theory*, M. Janse et al. (eds.), 172–197. Patras: University of Patras Press.
Kishimoto, Hideki
  2007   Negative scope and head raising in Japanese. *Lingua* 117: 247–288.
Kishimoto, Hideki
  2008   On the variability of negative scope in Japanese. *Journal of Linguistics* 44: 379–435.
Kjellmer, Göran
  2001   *No Work Will Spoil a Child*: On Ambiguous Negation, Corpus Work and Linguistic Argument. *International Journal of Corpus Linguistics* 5: 121–132.
Klein, Henny
  2001   Polarity sensitivity and collocational restrictions on adverbs of degree. In Hoeksema et al. (eds.), 223–236.
Klooster, Wim
  2003   Negative raising revisited. In J. Koster and H. van Riemsdijk (eds.), *Germania et Alia, A Linguistic Webschrift for Hans den Besten*. http://odur.let.rug.nl/~koster/DenBesten/contents.htm.
Krifka, Manfred
  2007   Negated antonyms: Creating and filling the gap. In *Presupposition and Implicature in Compositional Semantics,* U. Sauerland and P. Stateva (eds.), 163–177. Houndsmills: Palgrave Macmillan.
Lahiri, Utpal
  2001   *Even*-incorporated NPIs in Hindi definites and correlatives. In Hoeksema et al. (eds.), 237–264.
Laing, Margaret
  2002   Corpus-provoked questions about negation in Early Middle English. *Language Sciences* 24: 297–321.
Larrivée, Pierre
  2001   *L'interpretation des sequences negatives: portée et foyer des négations en français*. Brussels: Duculot.

Larrivée, Pierre
    2002a   La polysemie de certains indéfinis. In *Actes du IXe Colloque de l'Association internationale de psychomécanique du langage*, R. Lowe (ed.), 485–496. Québec: Presses de l'Université Laval.

Larrivée, Pierre
    2002b   *À propos de l'organisation du sens:* qui que ce soit *et la polarité negative*. Brussels: Duculot.

Larrivée, Pierre
    2004   *L'association negative: Depuis la syntaxe jusqu'à l'interprétation*. Geneva: Droz.

Larrivée, Pierre
    2005   'Quelqu'un n'est pas venu.' *Journal of French Language Studies* 15: 279–296.

Larrivée, Pierre
    2006   Indéfinis et termes polarisés. In *Indéfinis et prédications en français*, F. Corblin et al. (eds.), 205–216. Paris: Presses Universitaires de Paris-Sorbonne.

Larrivée, Pierre
    2007a   *Du tout au rien: Libre-choix et polarité négative*. Paris: Champion.

Larrivée, Pierre
    2007b   La scalarité des indéfinis à sélection arbitraire. *Travaux de linguistique* 54: 94–107.

Lechner, Winfried
    2000   Bivalent coordination in German. *Snippets* 1: 11–12.

Lee, Chungmin
    2002   Negative polarity in Korean and Japanese. In *Japanese/Korean Linguistics, Vol. 10*, In N. Akatsuka and S. Strauss (eds.), 481–494. Stanford: CSLI Publications.

Lee, Chungmin
    2006   Contrastive topic/focus and polarity in discourse. In *Where Semantics Meets Pragmatics*, K. von Heusinger and K. Turner (eds.), 381–420. London: Elsevier.

van Lente, Erica
    2003   The acquisition of negative concord: The case of *pas...non plus*. In Cornips and Fikkert (eds.), 105–115.

Levine, William and Joel Hagaman
    2008   Negated concepts interfere with anaphor resolution. *Intercultural Pragmatics* 5: 471–500.

Levinson, Dmitry
    2007a   Negative polarity and semantic negativity. Stanford University dissertation proposal. Downloadable at http://www.stanford.edu/~dmitryle/.

Levinson, Dmitry
    2007b   Licensing of negative polarity particles. In Zeijlstra and Soehn (eds.), 92–98.

Levinson, Stephen
2000   *Presumptive Meanings: The Theory of Generalized Conversational Implicature.* Cambridge: MIT Press.

Lewandowska-Tomaszczky, Barbara
2005   The nature of negation: literal or not literal. In *The Literal and Nonliteral in Language and Thought*, S. Coulson and B. Lewandowska-Tomaszczyk (eds.), 87–102. Frankfurt: Peter Lang.

Liberman, Mark
2004   Caring less with stress. Language Log post, July 8, 2004. http://158.130.17.5/~myl/languagelog/archives/001182.html.

Lichte, Timm
2005   Corpus-based acquisition of complex negative polarity items. In J. Gervain (ed.), *Proceedings of the Tenth ESSLLI Student Session*, 157–168.

Lichte, Timm and Laura Kallmeyer
2006   Licensing German negative polarity items in LTAG. *Proceedings of the 8th International Workshop on Tree Adjoining Grammar and Related Formalisms*, 81–90.

Lichte, Timm and Manfred Sailer
2004   Extracting negative polarity items from a partially parsed corpus. In *Proceedings of the Third Workshop on Treebanks and Linguistic Theories (TLT 2004)*, S. Kübler et al. (eds.), 89–101. Seminar für Sprachwissenschaft, University of Tübingen.

Lichte, Timm and Jan-Philipp Soehn
2007   The retrieval and classification of negative polarity items using statistical profiles. In *Roots: Linguistics in Search of its Evidential Base*, S. Featherston and W. Sternefeld (eds.), 249–266. Berlin / New York: Mouton de Gruyter.

Löbner, Sebastian
2000   Polarity in natural language: Predication, quantification and negation in particular and characterizing sentences. *Linguistics and Philosophy* 23: 213–308.

Löbner, Sebastian
to appear   Dual opposition in lexical meaning. To appear in *Semantics: An International Handbook of Natural Language Meaning*, C. Maienborn et al. (eds.), Berlin / New York: Mouton de Gruyter

Lucas, Christopher
2007   Jespersen's cycle in Arabic and Berber. *Transactions of the Philological Society* 105: 398–431.

Martin, John
2004   *Themes in Neoplatonic and Aristotelian Logic: Order, Negation and Abstraction.* Aldershot, Hampshire: Ashgate.

Martineau, France and Raymond Mougeon
2003   Sociolinguistic research on the origin of *ne* deletion in European and Quebec French. *Language* 79: 118–152.

Martineau, France and Marie-Thérèse Vinet
 2005 Microvariation in French negation markers: a historical perspective. In *Grammatical Variation and Parametric Variation*, M. Batllori et al. (eds.), 194–205. Oxford: Oxford University Press.

Martins, Ana Maria
 2000 Polarity items in Romance: Underspecification and lexical change. In *Diachronic Syntax: Models and Mechanisms*, S. Pintzuk et al. (eds.), 191–219. Oxford: Oxford University Press.

Martins, Ana Maria
 2006 Emphatic affirmation and polarity: Contrasting European Portuguese with Brazilian Portuguese, Spanish, Catalan and Galician. In *Romance Languages and Linguistic Theory*, J. Doetjes and P. González (eds.), 197–223. Amsterdam: John Benjamins.

Mayo, Ruth, Yaacov Schul and Eugene Burnstein
 2004 "I am not guilty" versus "I am innocent": the associative structure activated in processing negations. *Journal of Experimental Social Psychology* 40: 433–449.

Masayuki, Ohkado
 2005 On grammaticalization of negative adverbs, with special reference to Jespersen's Cycle recast. In Iyeiri (ed.), 59–81.

Mathieu, Éric
 2001 On the nature of French n-words. *UCL Working Papers in Linguistics* 13: 319–352.

Matos, Gabriella
 2001 Negative concord and the minimalist approach. In d'Hulst et al. (eds.), 245–280.

Mazzon, Gabriella
 2004 *A History of English Negation*. Harlow: Longman.

Medina Granda, Rosa
 2007 Occitano antiguo *ge(n)s:* su ausencia en ciertos contextos negativos. In Floricic (ed.), 1–27.

Merchant, Jason
 2000 Antecedent-contained deletion in negative polarity items. *Syntax* 3: 144–150.

Miestamo, Matti
 2000 Towards a typology of standard negation. *Nordic Journal of Linguistics* 23: 65–88.

Miestamo, Matti
 2005a Symmetric and asymmetric standard negation. In *The World Atlas of Language Structures*, Martin Haspelmath et al. (eds.), 458–461. Oxford: Oxford University Press.

Miestamo, Matti
 2005b Subtypes of asymmetric standard negation. In *The World Atlas of Language Structures*, Martin Haspelmath et al. (eds.), 462–465. Oxford: Oxford University Press.

Miestamo, Matti
 2005c *Standard Negation: The Negation of Declarative Verbal Main Clauses in a Typological Perspective*. Berlin/New York: Mouton de Gruyter.
Miestamo, Matti
 2006 Negation. In *Handbook of Pragmatics: The 2006 Installment*, J.-O. Östman and J. Verschueren (eds.), 1–25. Amsterdam: John Benjamins
Miestamo, Matti
 2007 Negation – an overview of typological research. *Language and Linguistics Compass* 1: 552–570.
Miestamo, Matti and Johan van der Auwera
 2007 Negative declaratives and negative imperatives: similarities and differences. In *Linguistics Festival, May 2006, Bremen*, A. Ammann (ed.), 59–77. Bochum: Brockmeyer.
Miller, Philip
 2003 Negative complements in direct perception reports. *CLS* 38 (1): 287–303.
Mitchell, Erika
 2006 The morpho-syntax of negation and the positions of NegP in the Finno-Ugric languages. *Lingua* 116: 228–244.
Mittwoch, Anita
 2001 Perfective sentences under negation and durative adverbials: A double-jointed construction. In Hoeksema et al. (eds.), 283–330.
Molnárfi, László
 2002 Die Negationsklammer im Afrikaans: Mehrfachnegation aus formaler und funktionaler Sicht. In *Issues in Formal German(ic) Typology*, W. Abraham and C. J.-W. Zwart (eds.), 223–261. Amsterdam: John Benjamins.
Muehleisen, Victoria and Maho Isono
 2009 Antonymous adjectives in Japanese discourse. *Journal of Pragmatics* 41: 2185–2203.
Murphy, M. Lynne, Carita Paradis, Caroline Willners and Steven Jones
 2009 Discourse functions of antonymy: A cross-linguistic investigation of Swedish and English. *Journal of Pragmatics* 41: 2159–2184.
Musolino, Julien, Stephen Crain and Rosalind Thornton
 2000 Navigating negative quantificational space. *Linguistics* 38: 1–32.
Nakamura, Fujio
 2005 A history of the negative interrogative *do* in seventeenth- to nineteenth-century diaries and correspondance. In Iyeiri (ed.), 93–111.
Nakanishi, Kimiko
 2006a *Even, only and* negative polarity in Japanese. *Proceedings of SALT XVI*.
Nakanishi, Kimiko
 2006b The semantics of *even* and negative polarity items in Japanese. *WCCFL* 25: 288–296.

Nelson, Gerald
 2004 Negation of lexical *have* in conversational English. *World Englishes* 23: 299–308.

Neuckermans, Annemie
 2008 *Negatie in de Vlaamse dialecten volgens de gegevens van de Syntactische Atlas van de Nederlandse dialecten (SAND).* PhD dissertation, University of Ghent.

Nevalainen, Terttu
 2006 Negative concord as an English "Vernacular Universal": Social history and linguistic typology. *Journal of English Linguistics* 34: 257–278.

Newmeyer, Frederick J.
 2004 Negation and modularity. In Birner and Ward (eds.), 241–261.

Nishiguchi, Sumiyo
 2005 Negative *also:* A bipolar item in Japanese. Poster for NYU Polarity Workshop, posted at http://www.nyu.edu/gsas/dept/lingu/events/polarity/posters.html.

Nishimura, Hideo
 2005 Decline of multiple negation revisited. In Iyeiri (ed.), 83–92.

Noveck, Ira
 2009 Meaning and inference linked to negation: An experimental pragmatics approach. In Sauerland and Yatsuhiro (eds.), 113–123.

Noveck, Ira, Raphaële Guelminger, Nicolas Georgieff and Nelly Labruyere
 2007 What autism can reveal about *every ... not* sentences. *Journal of Semantics* 24: 73–90

Oakley, Todd
 2005 Negation and blending: a cognitive-rhetorical approach. In *The Literal and Nonliteral in Language and Thought*, S. Coulson and B. Lewandowska-Tomaszczyk (eds.), 103–128. Frankfurt: Peter Lang.

Oda, Toshiko
 2002 Exclamatives and negative islands. In Andronis et al. (eds.), 97–109.

Ogawa, Sadayoshi
 2002 Contrastive negation in medieval and modern French. In Kato (ed.), 23–50.

Orlandini, Anna and Paolo Poccetti
 2007 Il y a *nec* et *nec:* trois valeurs de la negation en latin et dans les langues de l'Italie ancienne. In Floricic (ed.), 29–47.

Østbø, Christine Bjerkan
 2007 Two negative markers in Scandinavian. In Zeijlstra and Soehn (eds.), 99–105.

Ouali, Hamid
 2003 Sentential negation in Berber: a comparative study. In *Linguistic Description. Typology and Representation of African Languages*, J. Mugany (ed.), 243–256. Trenton: Africa World Press.

Ouali, Hamid
 2005 Negation and negative polarity items in Berber. *BLS* 30: 330–340.
Pakendorf, Brigitte and Ewa Schalley
 2007 From possibility to prohibition: a rare grammaticalization pathway. *Linguistic Typology* 11: 515–540.
Palacios Martinez, Ignacio
 2003 Multiple negation in Modern English: A preliminary corpus-based study. *Neuphilologische Mitteilungen* 104: 477–498.
Palma, Silvia
 2000 La négation dans les proverbes. *Langages* 139: 59–68.
Pappas, Dino Angelo
 2004 A sociolinguistic and historical investigation of the "So don't I" construction. *Cranberry Linguistics* 2 (University of Connecticut Working Papers in Linguistics 12), 53–62.
Paradis, Carita and Caroline Willners
 2006 Antonymy and negation: the boundedness hypothesis. *Journal of Pragmatics* 38: 1051–1080.
Paradis, Carita and Caroline Willners
 2009 Negation and approximation as configuration construal in space. In C. Paradis et al. (eds.), *Conceptual Spaces and the Construal of Spatial Meaning: Empirical Evidence from Human Communication.* Oxford: Oxford University Press.
Parsons, Terence
 2006 The traditional Square of Opposition. *Stanford Encyclopedia of Philosophy.* Revised posting at http://plato.stanford.ePdu/archives/win2006/entries/square.
Partee, Barbara H.
 2008 Negation, intensionality and aspect: Interaction with NP semantics. In *Theoretical and Crosslinguistic Approaches to the Semantics of Aspect*, S. Rothstein (ed.), 291–317. Amsterdam: John Benjamins.
Partee, Barbara H. and Vladimir Borschev
 2004 The semantics of Russian Genitive of Negation: The nature and role of Perspectival Structure. *SALT XIV,* 212–234.
Partee, Barbara H. and Vladimir Borschev
 2007 Existential sentences, BE and the Genitive of Negation in Russian. In *Existence: Semantics and Syntax*, I. Comorovski and K. von Heusinger (eds.), 147–190. Dordrecht: Springer.
Payne, John and Erika Chisarik
 2000 Negation and focus in Hungarian: An Optimality Theory account. *Transactions of the Philological Society* 98: 185–230.
Penka, Doris
 2005 A crosslinguistic perspective on negative terms. Poster for NYU Polarity Workshop, posted at http://www.nyu.edu/gsas/dept/lingu/events/polarity/posters.html.

Penka, Doris
2006 A cross-linguistic perspective on n-words. *Proceedings of BIDE05.*
Penka, Doris
2007 Negative Indefinites. University of Tübingen dissertation.
Penka, Doris and Hedde Zeijlstra
2005 Negative indefinites in Dutch and German. Ms., University of Tübingen, posted at http://ling.auf.net/lingbuzz/000192.
Pereltsvaig, Asya
2000 Monotonicity-based vs. veridicality-based approaches to negative polarity: evidence from Russian. *Formal Approaches to Slavic Linguistics* 8: 328–346. Ann Arbor: Michigan Slavic Publishers.
Pereltsvaig, Asya
2007 Negative polarity items in Russian and the "bagel problem." In Brown and Przepiorkowski (eds.).
Pfau, Roland
2002 Applying morphosyntactic and phonological readjustment rules in natural language negation. In *Modality and Structure in Signed and Spoken Languages*, R. P. Meier et al. (eds.), 263–295. Cambridge: Cambridge University Press.
Pitts, Alyson
2009 Metamessages of Denial: The Pragmatics of English Negation. PhD dissertation, University of Cambridge.
Pons Bordería, Salvador and Scott Schwenter
2005 Polar meaning and 'expletive' negation in approximative adverbs: Spanish *por poco (no)*. *Journal of Historical Pragmatics* 6: 262–282.
Portner, Paul and Raffaela Zanuttini
2000 The force of negation in Wh exclamatives and interrogatives. In *Negation and Polarity*, L. Horn and Y. Kato, 193–231. Oxford: Oxford University Press.
Postal, Paul
2004a The structure of one type of American English vulgar minimizer. In *Skeptical Linguistic Essays,* 159–172. New York: Oxford University Press. Downloadable at http://www.nyu.edu/gsas/dept/lingu/people/faculty/postal/papers/skeptical/.
Postal, Paul
2004b A remark on English double negatives. In *Syntaxe, Lexique, et Lexique-Grammaire*, E. Laporte et al. (eds.), 497–508. Amsterdam: John Benjamins.
Postma, Gertjan
2001 Negative polarity and the syntax of taboo. In Hoeksema et al. (eds.), 283–330.
Postma, Gertjan
2002 De enkelvoudige clitische negatie in het Middelnederlands en de Jespersen cyclus. *Nederlandse Taalkunde* 7: 44–87.

Potts, Christopher
2000   When even *no*'s neg is splitsville. Jorge's Hankamer's Web Fest, http://ling.ucsc.edu/Jorge/.
Potts, Christopher
2002   The syntax and semantics of *as*-parentheticals. *NLLT* 20: 623–689.
Prado, Jérôme and Ira Noveck
2006   How reaction time measures elucidate the matching bias and the way negations are processed. *Thinking & Reasoning* 12: 309–328.
Priest, Graham
2002   *Beyond the Limits of Thought*. Oxford: Oxford University Press.
Progovac, Ljiljana
2000a   Coordination, c-command and "logophoric" n-words. In Horn and Kato (eds.), 88–114.
Progovac, Ljiljana
2007   Negative and positive feature checking and the distribution of polarity items. In Brown and Przepiorkowski (eds.), 88–114.
Puskás, Genoveva
2002   On negative licensing contexts and the role of n-words. In Kenesei and Siptár (eds.), 81–107.
Puskás, Genoveva
2006   Negation in Finno-Ugric: an introduction. *Lingua* 116: 203–227.
Pustejovsky, James
2000   Events and the semantics of opposition. In *Events as Grammatical Objects*, C. Tenny and J. Pustejovsky (eds.), 445–482. Stanford: CSLI Publications.
van Raemdonck, Dan
2003   De la syntaxe incidentielle à l'interprétation pragmatique: Le cas de la négation. In *Parcours énonciatifs et parcours interprétatifs: Théories et applications*, A. Ouattara (ed.), 57–68. Gap: Ophrys.
Ramat, Paolo
2006   Italian negatives from a typological/areal point of view. In *Scritti in onore di Emanuele Banfi in occasione del suo 60° compleanno*, N. Grandi and G. Iannàccaro (eds.), 355–370. Cesena: Caissa Italia.
Ramchand, Gillian
2003   Two types of negation in Bengali. In *Clause Structure in South Asian Languages*, A. Mahajan and V. Srivastav (eds.), 39–66. Dordrecht: Kluwer.
Rapp, Irene and Arnim von Stechow
2000   *Fast* 'almost' and the visibility parameter for functional adverbs. *Journal of Semantics* 16: 149–204.
Richter, Frank and Manfred Sailer
2004   Polish negation and lexical resource semantics. In *Electronic Notes in Theoretical Computer Science* 53, L. Moss and R. Oehrle (eds.), 309–321. London: Elsevier.

Richter, Frank and Jan-Philipp Soehn
  2006   "Braucht niemanden zu scheren": A survey of NPI licensing in German. In *Proceedings of the 13th International Conference on Head-Driven Phrase Structure Grammar*, S. Müller (ed.), 421–440. Stanford: CSLI Publications.
Roberge, Paul
  2000   Etymological opacity, hybridization and the Afrikaans brace negation. *American Journal of Germanic Linguistics & Literatures* 12: 101–176.
Roberts, Ian and Anna Roussou
  2003   *Syntactic Change: A Minimalist Approach.* Cambridge: Cambridge University Press.
Romero, Maribel and Chung-hye Han
  2004   On negative yes/no questions. *Linguistics and Philosophy* 27: 609–658.
van Rooy, Robert
  2003   Negative polarity items in questions: Strength as relevance. *Journal of Semantics* 20: 239–273.
van Rooij, Robert and Katrin Schulz
  2007   *Only:* meaning and implicature. In *Questions in Dynamic Semantics*, M. Aloni et al. (ed.), 193–223. London: Elsevier.
Rooryck, Johan
  2008   A compositional analysis of French negation. Unpublished ms., Leiden University.
Rothschild, Daniel
  2006   Non-monotonic NPI-licensing, definite descriptions and grammaticalized implicatures. *Proceedings of SALT XVI*, 228–240.
Rowlett, Paul
  2001   French *ne* in non-verbal contexts. In d'Hulst et al. (eds.), 335–354.
Rowlett, Paul (ed.)
  2000   *Papers from the Salford negation conference.* Special issue, *Transactions of the Philological Society*, 98(1).
Rullmann, Hotze
  2003a A note on the history of *either.* In Andronis et al. (eds.), 111–125.
Rullmann, Hotze
  2003b Additive particles and polarity. *Journal of Semantics* 20: 329–401.
Saddy, Douglas, Heiner Drenhaus and Stefan Frisch
  2004   Processing polarity items: Contrastive licensing costs. *Brain and Language* 90: 495–502.
Sadock, Jerrold
  2006   Motors and switches: an exercise in syntax and pragmatics. In Birner and Ward (eds.), 317–325.

Sæbø, Kjell Johan
    2001   The semantics of Scandinavian free choice items. *Linguistics and Philosophy* 24: 737–787.

Sæbø, Kjell Johan
    2004   Natural language corpus semantics: The free choice controversy. *Nordic Journal of Linguistics* 17: 197–218.

Sailer, Manfred
    2006   *Don't believe* in lexical resource semantics. In *Empirical Issues in Syntax and Semantics* 6, O. Bonami and P. Cabredo Hofherr (eds.), 375–403.

Sailer, Manfred
    2007a   Complement anaphora and negative polarity items. In *Proceedings of Sinn und Bedeutung* 11, E. Puig-Waldmüller (ed.), 494–508.

Sailer, Manfred
    2007b   NPI licensing, intervention and discourse representation structures in HPSG. In *Proceedings of the 14th International Conference on Head-Driven Phrase Structure Grammar*, S. Müller (ed.), 214–234. Stanford: CSLI Publications.

Sailer, Manfred
    2007c   Dynamic intervention – a DRT-based characterization of interveners in NPI licensing. In Zeijlstra and Soehn (eds.), 106–112.

Sailer, Manfred and Frank Richter
    2002   Collocations and the representation of polarity. In *Proceedings of the Seventh Symposium on Logic and Language*, G. Alberti et al. (eds.), 129–138. Pécs, Hungary.

Sauerland, Uli.
    2004   Scalar implicatures in complex sentences. *Linguistics and Philosophy* 27: 367–391.

Sauerland, Uli and Kazuko Yatsuhiro (eds.)
    2009   *Semantics and Pragmatics: From Experiment to Theory*. Houndsmills: Palgrave Macmillan.

Schaffer, Barbara
    2002   CAN'T: The negation of modal notions in ASL. *Sign Language Studies* 3: 34–53.

Schapansky, Nathalie
    2000   *Negation, Referentiality and Boundedness in Gwenedeg Breton*. Munich: Lincom Europa.

Schapansky, Nathalie
    2002   The syntax of negation in French: Contrariety versus contradiction. *Lingua* 112: 793–826.

Schwarz, Bernhard
    2005   Scalar additive particles in negative contexts. *Natural Language Semantics* 13: 125–168.

Schwarz, Bernard and Rajesh Bhatt
    2006   Light negation and polarity. In Zanuttini et al. (eds.), 175–197.
Schwarzschild, Roger and Karina Wilkinson
    2002   Quantifiers in comparatives: a semantics of degree based on intervals. *Natural Language Semantics* 10: 1–41.
Schwenter, Scott
    2002a   Discourse context and polysemy: Spanish *casi*. In *Romance Philology and Variation: Selected Papers from the 30th Linguistic Symposium on Romance Languages*, C. Wiltshire and J. Camps (eds.), 161–175. Amsterdam: John Benjamins.
Schwenter, Scott
    2002b   Pragmatic variation between negatives: Evidence from Romance. *Papers from NWAV (New Ways of Analyzing Variation)* 30, 249–263. *University of Pennsylvania Working Papers in Linguistics* 8.3.
Schwenter, Scott
    2002c   Discourse markers and the PA/SN distinction. *Journal of Linguistics* 38: 43–70.
Schwenter, Scott
    2005   The pragmatics of negation in Brazilian Portuguese. *Lingua* 115: 1427–1456.
Schwenter, Scott
    2006   Fine-tuning Jespersen's Cycle. In Birner and Ward (eds.), 327–344.
Sells, Peter
    2000   Negation in Swedish: Where it's not at. In *Online Proceedings of the LFG-00 Conference*, M. Butt and T. H. King (eds.), Stanford: CSLI Publications.
Sells, Peter
    2001a   Negative polarity licensing and interpretation. In *Harvard Studies in Korean Linguistics* 9, S. Kuno et al. (eds.), 3–22. Cambridge: Harvard University.
Sells, Peter
    2001b   Three aspects of negation in Korean. *Journal of Linguistic Studies* (Cheju Linguistics Circle) 6: 1–15.
Sells, Peter
    2004   Negative Imperatives in Korean. *Proceedings of the 10th Harvard International Symposium on Korean Linguistics*. Cambridge: Harvard University.
Sells, Peter
    2006   Interaction of negative polarity items in Korean. *Proceedings of the 11th Harvard International Symposium on Korean Linguistics*, 724–737. Cambridge: Harvard University.
Seuren, Pieter
    2000   Presupposition, negation and trivalence. *Journal of Linguistics* 31: 261–297.

Seuren, Pieter
2001   Presupposition and negation. *A View of Language*, Ch. 15. Oxford: Oxford University Press.

Seuren, Pieter
2006   The natural logic of language and cognition. *Pragmatics* 16: 103–138.

Shuval, Noa and Barbara Hemforth
2008   Accessibility of negated constituents in reading and listening comprehension. *Intercultural Pragmatics* 5: 445–470.

Silberstein, Dagmar
2001   Facteurs interlingues et specifiques dans l'acquisition non-guidée de la négation en anglais L2. *Acquisition et Interaction en Langue Etrangère* 14: 25–58.

Sobin, Nicholas
2003   Negative inversion as nonmovement. *Syntax* 6: 183–212.

Soulez, Antonia
2001   De la négation à la dénégation chez Wittgenstein. *Revue de Métaphysique et de Morale* 2: 37–58.

Sprouse, Jon
2005   The accent projection principle: Why the hell not? *Penn Working Papers in Linguistics* 12(1): 349–359.

Storjohann, Petra
2009   Plesionymy: A case of synonymy or contrast? *Journal of Pragmatics* 41: 2140–2158.

Stromswold, Karin and Kai Zimmermann
2000   Acquisition of *nein* and *nicht* and the VP-internal subject stage in German. *Language Acquisition* 8: 101–127.

Surányi, Balázs
2002   Negation and the negativity of n-words in Hungarian. In Kenesei and Siptár (eds.), 39–61.

Surányi, Balázs
2006a Predicates, negative quantifiers and focus: specificity and quantificationality of n-words. In *Event Structure and the Left Periphery*, K. É. Kiss (ed.), 255–286. Dordrecht: Springer.

Surányi, Balázs
2006b Quantification and focus in negative concord. *Lingua* 116: 272–313.

de Swart, Henriëtte
2000   Scope ambiguities with negative quantifiers. In *Reference and Anaphoric Relations*, K. von Heusinger and U. Egli (eds.), 109–132. Dordrecht: Kluwer.

de Swart, Henriëtte
2001   Négation et coordination: la conjonction *ni*. In *Adverbial Modification*, R. Bok-Bennema et al. (eds.), 109–124. Amsterdam: Rodopi.

de Swart, Henriëtte
    2006   Marking and interpretation of negation: a bi-directional OT approach. In Zanuttini et al. (eds.), 199–218.

de Swart, Henriëtte
    2009   Negation in early L2: a window on language genesis. In *Language Evolution: The View from Restricted Linguistic Systems*, R. Botha and H. de Swart (eds.), 59–100. Utrecht: LOT Publications.

de Swart, Henriëtte
    in press   *Expression and Interpretation of Negation.* Dordrecht: Springer.

de Swart, Henriëtte and Ivan Sag
    2002   Negation and negative concord in Romance. *Linguistics and Philosophy* 25: 373–417.

Szabolcsi, Anna
    2002   Hungarian disjunctions and positive polarity. In Kenesei and Siptár (eds.), 217–241.

Szabolcsi, Anna
    2004   Positive polarity – negative polarity. *Natural Language and Linguistic Theory* 22: 409–452.

Szabolcsi, Anna, Lewis Bott and Brian McElree
    2008   The effect of negative polarity items on inference verification. *Journal of Semantics* 25: 411–450.

Szabolcsi, Anna and Bill Haddican
    2004   Conjunction meets negation: a study in cross-linguistic variation. *Journal of Semantics* 21: 219–250.

Tagliamonte, Sali and Jennifer Smith
    2002   *Either it isn't or it's not*: NEG/AUX contraction in British dialects. *English World-Wide* 23: 251–281.

Takizawa, Naohiro
    2005   A corpus-based study of the *haven't* NP pattern in American English. In Iyeiri (ed.), 159–171.

Taleghani, Azita
    2008   *Aspect and Negation in Persian.* Amsterdam: John Benjamins.

Tam, Clara and Stephanie Stokes
    2001   Form and function of negation in early developmental Cantonese. *Journal of Child Language* 28: 371–391.

Tammet, Daniel
    2006   *Born on a Blue Day: A Memoir.* New York: Free Press.

Thomson, Hanne-Ruth
    2006   Negation patterns in Bengali. *Bulletin of the School of Oriental and African Studies* 69: 243–265.

Tieken-Boon van Ostade, Ingrid
    2000   *The Two Versions of Malory's Morte d'Arthur: Multiple Negation and the Editing of the Text.* Cambridge: D. S. Brewer.

Tomić, Olga Mišeska
- 2001   The Macedonian negation operator and cliticization. *NLLT* 19: 647–682.

Tonhauser, Judith
- 2001   An approach to polarity sensitivity and negative concord by lexical underspecification. In *Proceedings of the 7th International HPSG Conference*, D. Flickinger and A. Kathol (eds.), 285–304. Stanford: CSLI Publications.

Tovena, Lucia
- 2001a   Neg-raising: Negation as failure. In Hoeksema et al. (eds.), 331–356.

Tovena, Lucia
- 2001b   The phenomenon of polarity sensitivity: Questions and answers. *Lingua e Stile* 36: 131–167.

Tovena, Lucia
- 2003   Distributional restrictions on negative determiners. In *Meaning Through Language Contrast*, K. Jaszczolt and K. Turner (eds.), 3–28. Amsterdam: John Benjamins

Tovena, Lucia
- 2008   Negative quantification and existential sentences. In *Existence: Semantics and Syntax*, I. Comorovski and K. von Heusinger (eds.), 191–222. Dordrecht: Springer.

Tovena, Lucia, Viviane Déprez and Jacques Jayez
- 2005   Polarity sensitive items. In Corblin and de Swart (eds.) 391–415.

Tseng, Meylysa, Jung-Hee Kim and Benjamin Bergen
- 2006   Can we simulate negation? The simulation effects of negation in English intransitive sentences. *BLS* 32.

Tsurska, Olena
- 2009   The negative cycle in Early and Modern Russian. In van Gelderen (ed.).

Tubau, Susagna
- 2007   N-words and PIs in non-standard British English. In Zeijlstra and Soehn (eds.), 113–119.

Tubau, Susagna
- 2008   Negative Concord in English and Romance: Syntax-Morphology Interface Conditions on the Expression of Negation. PhD dissertation, University of Amsterdam.

Ukaji, Masatomo
- 2002   Some aspects of negation in the history of English. In Kato (ed.), 169–192.

Unsworth, Sharon, Andrea Gualmini and Christina Helder
- to appear   Children's interpretation of indefinites in sentences containing negation: A re-assessment of the cross-linguistic picture. *Language Acquisition* 15 (4).

Vakarelov, Dimiter
    2005   Nelson's negation on the base of weaker versions of intuitionistic negation. *Studia Logica* 80: 393–430.

Vasishth, Shravan
    2002   Word order, negation and negative polarity in Hindi. *Research on Language and Computation* 3.

Vasishth, Shravan, Richard Lewis, Sven Brüssow and Heiner Drenhaus
    2008   Processing polarity: How the ungrammatical intrudes on the grammatical. *Cognitive Science* 32: 685–712.

Véronique, Daniel
    2005   Syntactic and semantic issues in the acquisition of negation in French. In *Focus on French as a Foreign Language*, J.-M. Dewaele (ed.), 114–134. Clevedon: Multilingual Matters.

Veselinova, Lyuba
    2008   Negation in Slavic languages. Summary posting on Linguist List, 19.1733. http://www.ling.su.se/staff/ljuba/LL_summary.pdf.

Vinet, Marie-Thérèse
    2000   La polarité pos/neg, *-tu (pas)* et les questions oui/non. *Revue québécoise de linguistique* 2: 169–176.

Vlachou, Evangelia
    2003   Weird polarity indefinites in French. In Cornips and Fikkert (eds.), 89–100.

Vlachou, Evangelia
    2006   Le puzzle des indéfinis en *qu-*. In *Indéfinis et prédication*, F. Corblin et al. (eds.), 235–249. Paris: Presses de l'Université de Paris-Sorbonne.

Vlachou, Evangelia
    2007   *Free Choice In and Out of Context: Semantics and Distribution of French, Greek and English Free Choice Items*. Utrecht: LOT.

Vogeleer, Svetlana
    2001   French negative sentences with *avant* ('before')-phrases and *jusqu'à* ('until')-phrases. In d'Hulst et al. (eds.), 355–382.

Wagner, Michael
    2006   Association by movement: Evidence from NPI-licensing. *Natural Language Semantics* 14: 297–324.

Walker, James A.
    2005   The *ain't* constraint: *Not*-contraction in early African American English. *Language Variation and Change* 17: 1–17.

Wallage, Phillip
    2005   Negation in Early English: Parametric Variation and Grammatical Competition. PhD dissertation, University of York.

Wallage, Phillip
    2008   Jespersen's cycle in Middle English: Parametric variation and grammatical competition. *Lingua* 118: 643–674.

Warner, Anthony
  2005   Why DO dove: Evidence for register variation in Early Modern English negatives. *Language Variation and Change* 17: 257–280.
Watanabe, Akira
  2001   Decomposing the neg-criterion. In d'Hulst et al. (eds.), 383–406.
Watanabe, Akira
  2002   Feature-checking and neg-factorization in negative concord. In Kato (ed.), 51–77.
Watanabe, Akira
  2004   The genesis of negative concord: Syntax and morphology of negative doubling. *Linguistic Inquiry* 35: 559–612.
Weiß, Helmut
  2002a   Three types of negation: A case study in Bavarian. In Barbiers et al. (eds.), 305–332.
Weiß, Helmut
  2002b   A quantifier approach to negation in natural languages, or why negative concord is necessary. *Nordic Journal of Linguistics* 25: 125–153.
Werle, Adam
  2002   A typology of negative indefinites. In Andronis et al. (eds.), 127–141.
Whitman, John
  2005   Preverbal elements in Korean and Japanese. In *Oxford Handbook of Comparative Syntax*, G. Cinque and R. Kayne (eds.), 880–902. Oxford: Oxford University Press.
Wible, David and Eva Chen
  2000   Linguistic limits on metalinguistic negation: Evidence from Mandarin and English. *Language and Linguistics* (Academia Sinica, Taiwan) 1: 233–255.
Willis, David
  2006   Negation in Middle Welsh. *Studia Celtica* 40: 63–88.
Willis, David
  to appear A minimalist approach to Jespersen's cycle in Welsh. In *Syntactic Variation and Change*, D. Jonas (ed.), Oxford: Oxford University Press.
van der Wouden, Ton
  2001   Three modal verbs. In *Zur Verbmorphologie germanischer Sprachen*, S. Watts et al. (eds.), 189–210. Tübingen: Niemeyer.
van der Wouden, Ton
  2005   Negatief-polair *meer*. *TABU* 34: 47–70.
Wurmbrand, Susi
  2008   *Nor:* Neither disjunction nor paradox. *Linguistic Inquiry* 39: 511–522.
Xiao, Richard and Anthony McEnery
  2008   Negation in Chinese: a corpus-based study. *Journal of Chinese Linguistics* 36: 274–330.

Yaeger-Dror, Malcah
    2002   Introduction to special issue on negation and disagreement. *Journal of Pragmatics* 34: 1333–1343.
Yaeger-Dror, Malcah, Lauren Hall-Lew and Sharon Deckert
    2002   *It's not* or *isn't it?* Using large corpora to determine the influences on contraction strategies. *Language Variation and Change* 14: 79–118.
Yamada, Masamichi
    2000   Negation in Japanese Narratives: A Functional Analysis. PhD dissertation, Georgetown University, Washington, DC.
Yamada, Masamichi
    2003   *The Pragmatics of Negation*. Tokyo: Hituzi Syobo.
Yamanashi, Masa-aki
    2000   Negative inference, space construal and grammaticalization. In Horn and Kato (eds.), 243–254.
Yamanashi, Masa-aki
    2002   Space and negation: A cognitive analysis of indirect negation and natural logic. In Kato (ed.), 133–144.
Yamashita, Hideaki
             Prosody and the syntax of *shita*-NPIs in Tokyo Japanese. In Zeijlstra and Soehn (eds.), 120–126.
Yang, Jun Hui and Susan Fischer
    2002   Expressing negation in Chinese sign language. *Sign Language and Linguistics* 5: 167–202.
Yoon, Suwon
    2008   Expletive negation in Japanese and Korean. In *Japanese/Korean Linguistics, Vol. 18*, M. den Dikken and W. McClure (eds.). Stanford: CSLI Publications.
Yoshimura, Akiko
    2002   A cognitive-pragmatic approach to metalinguistic negation. In Kato (ed.), 113–132.
Zanuttini, Raffaella
    2001   Sentential negation. In *The Handbook of Contemporary Syntactic Theory*, M. Baltin and C. Collins (eds.), 511–535. Oxford: Blackwell.
Zanuttini, Raffaella
    to appear La negazione in italiano antico. In *Grande Grammatico dell'Italiano Antico,* L. Renzi and G. Salvi (eds.).
Zanuttini, Raffaella, Héctor Campos, Elena Herburger and Paul Portner (eds.)
    2006   *Cross-Linguistic Research in Syntax and Semantics: Negation, Tense and Clausal Architecture (GURT 2004)*. Washington: Georgetown University Press.
Zanuttini, Raffaela and Paul Portner
    2000   The characterization of exclamative clauses in Paduan. *Language* 76: 123–132.

Zeijlstra, Hedde
    2002   What the Dutch Jespersen Cycle may reveal about negative concord. In Andronis et al. (eds.), 143–156.

Zeijlstra, Hedde
    2004   *Sentential Negation and Negative Concord.* PhD dissertation, University of Amsterdam. Utrecht: LOT.

Zeijlstra, Hedde
    2006   The ban on true negative imperatives. *Empirical Issues in Syntax and Semantics* 6: 405–424.

Zeijlstra, Hedde
    2007   Negation in natural language: On the form and meaning of negative elements. *Language and Linguistics Compass* 1: 498–518.

Zeijlstra, Hedde
    2008   Emphatic multiple negative expressions in Dutch – a product of loss of negative concord. In Barbiers et al. (eds.).

Zeijlstra, Hedde
    2009   On French negation. Paper presented at BLS 35. To appear in proceedings of the conference.

Zeijlstra, Hedde
    to app. a   *Not* in the first place. *Natural Language and Linguistic Theory.*

Zeijlstra, Hedde
    to app. b   Negative concord is syntactic agreement. *Linguistics.*

Zeijlstra, Hedde
    to app. c   On the lexical status of negative indefinites in Dutch. *Journal of Comparative Germanic Linguistics.*

Zeijlstra, Hedde and Jean-Philipp Soehn (eds.)
    Proceedings of the Workshop on Negation and Polarity. Eberhard Karls Universität Tübingen. Posted at http://www.sfb441.uni-tuebingen.de/negpol/negpol07.pdf.

Zepter, Alex
    2003   How to be universal when you are existential: Negative polarity items in the comparative: Entailment along a scale. *Journal of Semantics* 20: 193–237.

Ziegeler, Debra
    2000   What *almost* can reveal about counterfactual inferences. *Journal of Pragmatics* 32: 1743–1776.

Ziegeler, Debra
    2006   Proximative aspect. *Interfaces with English Aspect,* Ch. 4. Amsterdam: John Benjamins.

Contributors

Östen Dahl
Department of Linguistics
Stockholm University
106 91 Stockholm
Sweden
oesten@ling.su.se

Christine Dimroth
Language Acquisition Group
Max-Planck-Institut für
Psycholinguistik
PO Box 310
6500 AH Nijmegen
The Netherlands
christine.dimroth@mpi.nl

Ofer Fein
School of Behavioral Sciences
Academic College of Tel Aviv-Yaffo
Tel Aviv-Yaffo, 68114
Israel
oferf@mta.ac.il

Rachel Giora
Department of Linguistics
Tel Aviv University
Tel Aviv, 69978
Israel
giorar@post.tau.ac.il

Anja Neukom-Hermann
Englisches Seminar
University of Zurich
Plattenstrasse 47
8032 Zurich
Switzerland
anja.neukom@es.uzh.ch

Jack Hoeksema
Department of Dutch Language and
Culture, Faculty of Letters
University of Groningen
PO Box 716
9700 AS Groningen
The Netherlands
j.hoeksema@rug.nl

Laurence R. Horn
Department of Linguistics
Yale University
PO Box 208366
New Haven, CT 06520-8366
USA
laurence.horn@yale.edu

Yasuhiko Kato
Department of Linguistics
Sophia University
7-1 Kioicho, Chiyoda-ku
Tokyo 102-8554
Japan
yhkato@sophia.ac.jp

Nili Metuki
Psychology Department

Bar Ilan University
Ramat Gan, 52900
Israel

nili.metuki@gmail.com

Pnina Stern
School of Cultural Studies, The Program of Cognitive Studies of Language Use

Tel-Aviv University
Tel Aviv, 69978
Israel

pninas@barak.net.il

Gunnel Tottie
Englisches Seminar

University of Zurich
646 Vincente Avenue
Berkeley, CA 94707
USA

gtottie@mac.com

Johan van der Auwera
Center for Grammar, Cognition, and Typology (CGCT)

University of Antwerp
Prinsstraat, 13
2000 Antwerpen
Belgium

johan.vanderauwera@ua.ac.be

Index of subjects

all, 3, 5–6, 76–77, 100, 149–180, 195, 198, 201, 212, 214, 230
ambiguity, 34, 54, 126, 131, 150–151, 159, 161, 164, 179
anaphoric negation, 40, 48–49, 51, 54–55
anti-additivity, 195
any, 30, 187–189, 192–193, 195, 198, 210
asymmetry (negation/affirmation), 1, 3–4, 12, 14, 17–18, 27, 75, 86–89, 101, 138, 199, 227–228, 231–232, 234, 251, 267

be, 123, 163, 200, 216–217

Classical Japanese, 7, 257, 282
classification of negation constructions, 9–12, 14–15, 19, 30, 43, 218
collective reading, 5, 153, 178
constituent negation, 4, 31, 40, 51, 57, 63, 95, 98, 232, 279,
contradictory negation, 26, 49, 114–115, 117–119
contrary negation, 98, 114–115, 119, 133
corpus linguistics, 149
corpus-based research, 5–6, 149–153, 178–179, 189, 196, 199, 205, 211, 236, 241, 246, 250–251

degree adverb, 192, 197, 212–214
denial, 4, 6, 31–32, 43–48, 54–55, 59, 67, 76, 114, 168, 170, 176
distributive reading, 1, 5–6, 50, 56–57, 114, 120, 153, 157, 160, 168, 188–191, 196–197, 209, 218, 241, 247
Division of Pragmatic Labor, 116–117, 140
double negative particles, 7, 12–13, 19–20, 24, 26, 113, 116–118, 120, 127, 132, 137, 212–213, 257

ever, 192–193, 195–196, 198–199, 234
existential negation, 3, 28, 30, 32, 75, 77, 88, 93–95, 179, 193, 198,

finiteness, 3–4, 11–12, 26, 40, 52, 57, 58, 61, 64–66, 68–71, 87–88
focus, 3–4, 6, 10–12, 27, 29, 31, 34, 40, 42, 45, 51, 59–60, 65, 71, 101, 121, 232, 234, 257, 265, 267–269, 271, 278–282
formulaic expressions, 5, 159, 162–164, 168, 173–174, 177–180

grammaticalization, 3, 19–20, 31–33, 73, 76, 88–92, 125, 190–191, 205

have, 150, 193–194, 206, 216–217
hypernegation, hyponegation, 6, 111–112, 120, 134, 137–140

islands, 29, 211, 273–274

Jespersen cycle, 3, 73–76, 80–81, 83, 85, 88–89, 94, 96–99, 101

left periphery, 257, 279, 281–282
litotes, 119, 209, 211–212

Merge, 272–273, 282
metalinguistic negation, 4, 7, 46, 54–56, 168, 170, 213–214, 257, 264
metaphoric/literal, 4, 77, 96, 117, 191, 212, 225–226, 231–251, 267, 277
morphological (affixal) negation, 2–3, 6, 11–12, 14–19, 24, 56–58, 61, 111, 119, 149–150, 208
multiple negation, 1, 6, 111–141, 150

need, 192, 198, 200–202

negation, implicit, 6, 63, 83, 88, 138, 213, 251
negative absorption, 97
negative adverb, 1–2, 30, 33, 75, 80, 88, 95, 100–101, 122, 127, 132, 138, 192, 197, 200–201, 206, 210, 212–214, 218, 261, 263, 275–276, 279, 281
negative concord, 1–2, 4, 40, 98–99, 111, 120, 129–130, 140, 150, 260
negative particles, 3, 12, 14–15, 19–21, 25–26, 32, 81–82, 90, 98, 100, 140, 267, 281
negative polarity, 1–3, 7, 30–31, 40, 71, 77, 84–85, 96, 120, 123, 138, 150, 187–218, 234, 257, 263, 275–276
negative polarity item, 1, 3, 7, 31, 40, 71, 85, 96, 123, 138, 150, 187–218, 234, 257, 263, 275–276
negative pronoun, 10, 30, 33, 74–75, 88, 95–99, 100–101, 245
negative verbs, 3, 12, 17, 20–21, 25, 28, 260
nonexistence, 3–4, 43–44, 47, 55, 206, 235, 260
non-verbal negation, 27–28, 75, 88, 93, 101

pleonastic negation, 3, 6, 111, 121–122, 124–130, 135, 138–139, 141
positive polarity item, 5, 187–218
prohibitive(s), 3, 6, 10, 27, 43, 75, 88–93, 95, 101

quantifier scope, 58, 150–151, 178, 188, 194, 200

reconstruction, 264, 275, 277
rejection, 4, 43–47, 55, 59, 226, 231
resonance, 231, 245–249
resumptive negation, 80, 128, 129
retention, 228, 230, 232

salience, 253, 255
scope of negation, 2, 4, 30, 40, 51, 53, 62–63, 68, 151, 157, 178, 187–188, 194, 226, 230–233, 275, 277
sentence negation, 3, 7, 51, 53, 55–56, 58, 60, 66, 114, 120, 126, 257–258, 265
standard negation, 3, 10–11, 13, 15–16, 19, 21, 23, 26, 29, 33, 73–75, 88, 94, 101, 130
suppression, 4, 226–228, 230, 250
symmetric vs. asymmetric negation, 12, 14

topic, 3–4, 7, 10, 28, 31, 34, 39, 44, 46, 56, 62–63, 65, 75, 85, 150, 176, 179, 228, 233, 235, 240, 251, 279–282
truth-functional negation, 4, 6, 42–43, 45–46, 48, 54–55, 125, 132, 166, 193, 258

XP movement, 272, 275, 277, 279

Index of languages

Afrikaans, 20, 80–81, 91
Altaic, 258
American English, 150
American Indian, 160
Amharic, 122
Arabic, 97

Bantu, 84, 91, 176
Basque, 132
British English, 150, 152
Bulgarian, 94–95

Cantonese, 41, 70, 91
Carib, 15, 22
Catalan, 132, 133
Celtic, 2, 19
Chinese, 138
Chukchi, 15
Coptic, 95
Creole, 63, 90
Croatian, 91
Czech, 15, 17, 28

Danish, 203
Dravidian, 18
Dutch, 6, 41, 57–59, 68, 71, 77–78, 80–85, 90, 99–100, 121, 129, 132, 139, 190, 196, 198–201, 203, 206, 208–215, 217

Early Modern English, 121
Egyptian, 76, 95
English, 1–6, 10–11, 13–14, 22, 24, 29–30, 33, 41, 44, 46–47, 50–54, 57–63, 65–68, 74, 76–78, 86–87, 90, 95–96, 98–101, 111–113, 116, 118–123, 125–126, 128, 132, 134–136, 139, 149–150, 152, 155, 168, 177, 180, 188, 196–198, 200–201, 203, 205–206, 209–212, 214, 216, 236, 241–243, 246–247, 249–251, 279

Old ~, 98, 100, 121, 131, 177, 180
Estonian, 21
Evenki, 21

Finnish, 21, 41, 65
Flemish, 83, 132
French, 5–6, 13, 19–20, 29–31, 33, 41, 46, 49–50, 59–62, 66–68, 70–71, 74–80, 82, 85, 96, 112, 119, 122, 126, 128–129, 131–132, 140, 190, 203–204, 213, 272
Old ~, 79

German, 5–6, 29, 41, 49–53, 56–63, 66, 68, 69–71, 77, 121–122, 137, 139, 174, 190–191, 196, 198–201, 203, 206, 209–210, 212–213, 215, 236, 241–242, 244, 246, 248–251
Swiss ~, 69, 139
Germanic, 2, 6, 19–20, 29, 66, 80, 98
Greek, 2, 27, 29, 121–122, 132, 189
Gur, 91

Hebrew, 41, 230, 236, 241, 250
Hindi, 25, 32
Hungarian, 41, 63, 64

Igbo, 17
Indonesian, 19, 27
Iranian, 13, 33
Italian, 30, 41, 59, 62, 65, 80, 84, 97, 112, 120, 132–133

Japanese, 3, 7, 22, 41, 46–47, 53, 66, 69, 71, 133, 257, 263–266, 271–272, 278, 281–282

Korean, 22, 41, 46, 67, 69, 133, 272

Latin, 2, 33, 75, 77, 91, 113, 122

Index of languages

Mandarin, 41, 91, 138
Mandinka, 26
Mayan, 19
Middle English, 84–85, 100, 121, 177

Oceanic, 30, 91
Old English, 98, 100, 121, 131, 177, 180
Old French, 79

Penutian, 77
Polish, 15, 28, 41, 70
Portuguese, Brazilian, 80–82, 85, 129, 132–133

Romance, 2, 6, 19, 66, 80, 120–123, 260
Russian, 27–31, 41, 70, 94, 112, 121, 123, 236, 241, 243–244, 246–251

Salish, 20
Sanskrit, 122
sign language, 10, 172

Slavic, 13, 17, 31, 33, 120, 122
Spanish, 30, 65, 80, 91, 112, 132, 138–140
  Mexican ~, 92
Squamish, 20
Swahili, 18
Swedish, 18, 30, 41, 59–60, 65, 68, 80

Tagalog, 100
Tamil, 17, 41, 46, 69, 71
Tonga, 20
Tungusic, 21
Turkish, 3, 14, 28, 41, 54, 64, 70

Welsh, 29
West African languages, 19

Yakut, 92
Yaqui, 26
Yiddish, 123
Yoruba, 29

Index of persons

Abels, Klaus, 121
Ackema, Peter, 208
Aksu-Koç, Ayhan, 41, 54, 64
Ali, Latif H., 97
Alkabets, Inbar, 227, 232
Alkabets-Zlozover, Inbar, 228
Alonso-Ovalle, Luis, 121
Amaral, Patricia, 138, 141
Amritavalli, R., 41, 69
Anderson, Diane E., 41
Anderson, John M., 11
Anderson, Judi Lynn, 92
Anderwald, Lieselotte, 2, 150
Andorno, Cecilia, 62, 64
Antes, Gertraud, 67
Antinucci, Francesco, 41, 45, 71
Ariel, Mira, 235
Aschkenazi, Keren, 228
Austin, Frances, 112
Auwera, Johan van der, 2–3, 27, 73, 89, 99, 101

Babin, Rex, 225
Baek, Judy Yoo-Kyung, 283
Baker, C. Lee, 213
Balaban, Noga, 227–229, 232
Baltin, Mark, 151, 164
Banda, Tracy, 234
Barbarczy, Anna, 41, 64
Barbiers, Sjef, 84
Bardel, Camilla, 41, 65
Barnwell, Katherine, 17
Bar-Shalom, Eva, 41, 70
Bartsch, Renate, 23
Bayer, Josef, 121
BBC News Africa, 233
Becker, Angelika, 40–41, 60–63, 65
Becker-Donner, Etta, 17
Bellugi, Ursula, 41, 50–51, 53, 55, 68

Benazzo, Sandra, 69
Bennis, Hans (Barbiers *et al.*), 84
Benveniste, Émile, 122, 216
Berman, Ruth, 71
Bernini, Giuliano, 65
Bever, Thomas G., 114
Biber, Douglas, 150, 154–155
Biberauer, Theresa, 80
Biq, Yung-O., 138
Birjulin, Leonid A., 90, 92
Blancquaert, E., 83
Bley-Vroman, Robert, 64–65
Blige, Nellie, 233, 245
Bloom, Lois, 41, 44, 46–47, 51–53
Bloom, Paul A., 227
Blutner, Reinhard, 198
Boef, Eefje (Barbiers *et al.*), 84
Bolinger, Dwight, 31, 187, 214
Bolinger, Dwight L., 31, 187, 214
Bond, Oliver, 17
Bosanquet, Bernard, 119
Bowerman, Melissa, 41, 65
Brammer, Michael J., 231
Bréal, Michel, 74, 140
Breitbarth, Anne, 84
Bremen, Klaus von, 203, 204, 205
Broschart, Jürgen, 91
Bryant, Margaret, 128
Bultinck, Bert, 101
Buridant, Claude, 77
Büring, Daniel, 63, 65, 150, 154
Burnstein, Eugene, 227
Bybee, Joan L., 15, 19

Cameron-Faulkner, Thea, 41, 43, 47, 50, 65
Cancino, Herlindo, 41, 60, 65
Capell, A., 86
Carden, Guy, 151, 158

Index of persons

Cerulli, Enrico, 26
Chafe, Wallace, 175–176, 179
Chamberlain, Larry, 234
Childers, Donald G., 227
Choi, Soonja, 41, 43, 46,–50, 65, 141
Chomsky, Noam, 171–172
Cinque, Guglielmo, 279
Clahsen, Harald, 41, 56, 60, 65, 67
Clancy, Patricia M., 41, 46, 66
Clark, Eve V., 41, 52, 66, 126
Clark, Herbert H., 41, 52, 66, 126
Cloarec-Heiss, France, 17
Comrie, Bernard (Haspelmath et al.), 10
Condoravdi, Cleo, 96, 121
Conrad, Susan (Biber et al.), 150, 154–155
Contini-Morava, Ellen, 18
Cook, Vivian, 63–64, 66
Corblin, Francis, 120
Corne, Chris, 90
Cornyn, William, 87
Crain, Stephen, 58, 69
Croft, William, 10, 28, 32, 77–78, 93–94
Curme, George O., 113
Curry, Haskell B., 12

Dahl, Östen, 2–3, 9–26, 76, 140
Damourette, Jacques, 131
David, Anthony S., 67–69, 71, 141, 160, 231
Dayal, Veneeta, 193
De Clerck, Bernard, 89
De Cuypere, Ludovic, 80, 83
De Swart, Henriëtte, 121, 190
De Vogelaer, Gunther (Barbiers et al.), 84
Deckert, Sharon (Yaeger-Dror et al.), 150
Déprez, Viviane, 53, 66
Derbyshire, Desmond C., 22
Dietrich, Rainer, 41, 60, 66
Dimroth, Christine, 4, 39–68
Dowty, David, 129, 141
Drenhaus, Heiner, 41, 66
Drozd, Kenneth F., 41, 46, 52, 54, 55, 66

Dryer, Matthew S., 3, 10, 12, 14, 16, 19, 21, 23–26, 75, 133
Du Bois, John W., 176, 230, 245, 252
Dylan, Bob, 130

Earle, Samuel, 112
Early, Robert, 46, 65, 69, 71, 83–84, 89, 90, 121, 257
Eckardt, Regine, 77, 121, 140
Eilam, Aviad, 230
Einenkel, Eugen, 100
Endo, Yoshio, 279
Eriksen, Pål Kristian, 27, 93
Espinal, M. Teresa, 121, 132–134
Falaus, Anamaria, 121
Farrar, Frederic William, 139
Farrar, Michael J., 45
Fauconnier, Gilles, 188
Fein, Ofer, 225–252
Felix, Sascha, 41, 49, 51–52, 66, 67
Fembeti, Samuel, 91
Ferguson, Heather J., 68, 227, 228
Finegan, Edward (Biber et al.), 150, 154–155
Firbas, Jan, 176
Fischer, Olga, 149, 191
Fischler, Ira, 227
Fitzmaurice, Susan M., 149
Floricic, Franck, 80, 121
Forges, Germaine, 92
Foss, Mark A., 233
Fowler, H. W., 116–117, 124, 128
Fraenkel, Tamar, 229, 232
François, Alexandre, 78
Frazier, Ian, 137
Freeman, Jason, 135
Frege, Gottlob, 118

Ganzi, Jonathan, 229
Gardiner, Alan H., 76
Gasparrini, Désirée, 150
Gast, Volker, 73, 100
Geach, Peter T., 193

Index of persons 339

Gennari, Silvia P., 58, 67
Gernsbacher, Morton Ann, 228, 235
Giampietro, Vincent C., 231
Giannakidou, Anastasia, 6, 121, 123, 189, 193, 208
Gil, David (Haspelmath et al.), 10
Giora, Rachel, 4, 225–252
Giuliano, Patricia, 41, 60–61, 67, 69
Givón, Talmy, 32, 75, 175–176
Glucksberg, Sam, 227, 232, 234–235, 252
Gopnik, Alison, 45, 67
Goussev, Valentin (van der Auwera et al.), 89
Green, Margaret M., 17
Greenbaum, Sidney (Quirk et al.), 159, 173
Greenberg, Joseph H., 9, 12, 23
Greenberg, Robert D., 91
Greenough, J. B., 113
Gretsch, Petra, 66
Grice, H. Paul, 115, 213
Grommes, Patrick, 41, 60, 66
Grossman, Eitan, 73, 95
Gualmini, Andrea, 70
GUCK, 246
Guerzoni, Elena, 121
Guidetti, Michèle, 44, 67

Haberzettl, Stefanie, 59, 67
Haegeman, Liliane, 131, 260
Hagaman, Joel A., 230
Hahn, Kyung-Ja P., 41, 67
Halliday, Michael A. K., 176
Hall-Lew, Lauren (Yaeger-Dror.et.al.), 150
Hansen, Björn, 79, 91
Hansen, Maj-Britt Mosegaard, 79, 91
Harris, Tony, 41, 50, 67, 130
Haspelmath, Martin, 10, 29–30, 96–100, 123
Hasson, Uri, 227, 232, 234
Heine, Bernd, 33, 75, 90, 191
Helder, Christina, 70
Hemforth, Barbara, 230
Herburger, Elena, 120

Heringer, James T., 151, 165
Heruti, Vered, 230
Hinch, H. E., 86
Hodgson, W. B., 127–128
Hoeksema, Jack, 5–6, 30–31, 77, 80, 121, 123, 139, 141, 187–218
Honda, Isao, 83, 259
Honda, Yoshihiko, 83, 259
Horn, Laurence R., 1–5, 7, 18, 20, 27, 31, 44, 46, 54–55, 67, 73, 77, 80, 97, 100, 111–141, 150–151, 154–155, 158, 164, 170–171, 176, 187, 193, 197–198, 209, 211–214, 226, 228–229, 232, 235, 238, 252, 262–264, 272, 279, 282
Hummer, Peter, 41, 44–46, 55, 67
Hyltenstam, Kenneth, 41, 60, 67

Igwe, Georgewill E., 17
Ingham, Richard, 149
Inkova, Olga, 121
Irene, 68, 245
Israel, Michael, 5, 187, 191, 196, 209, 228, 234, 252
Ito, Katsutoshi, 41, 67
Iyeiri, Yoko, 177

Jackson, Eric, 192–195, 197
Jäger, Agnes, 98
Jagger, Mick, 120
Janda, Richard D., 191
Jennings, R. E., 193
Jespersen, Otto, 1–3, 5, 9, 20, 23, 26, 31–33, 73–76, 80–81, 83, 85, 88–90, 94, 96–99, 101, 111, 114–115, 117, 120–123, 128–131, 140–141, 151–152, 154, 170–171, 177, 203, 228–229, 272
Joly, André, 121, 131–132
Jones, Lawrence E., 193, 207–208, 211–213, 233
Jordens, Peter, 41, 51, 56–58, 61–63, 65–68
Jung, Ingrid, 13
Just, Marcel A., 6, 127, 203, 226–227, 246, 268

Kadmon, Nirit, 187–189, 192, 197
Kahrel, Peter, 10, 14, 29–30
Kahrel, Pieter Johannus, 95–96, 100
Kallel, Amel, 149
Kato, Yasuhiko, 6, 141, 257, 264, 266, 272–274, 277, 279, 282
Kaup, Barbara, 227
Kawaguchi, Yuji, 79
Kayne, Richard S., 216
Keil, Frank C., 234
Keysar, Boaz, 235
Khanmagomedov, B. G.-K.
Kilpatrick, Carroll, 226
Kiparsky, Paul, 96, 121, 190
Kitagawa, Y., 271
Kittredge, G. L., 113
Kjellmer, Göran, 150
Klein, Henny, 213
Klein, Wolfgang, 51, 57, 61
Klima, Edward S., 9, 41, 50–51, 53, 55, 68, 98, 114, 277
König, Ekkehard, 190
Koopman, H., 271
Krämer, Irene, 59, 68
Krifka, Manfred, 189
Kroskrity, Paul V., 83
Kuipers, Aert H., 20
Kuroda, S.-Y., 259–260, 271
Kuteva, Tania, 33, 75, 90

Labov, William, 120, 151, 165
Ladd, D. Robert, 154
Ladusaw, William A., 5, 188, 192
Laing, Margaret, 149, 177
Landman, Fred, 187–189, 192, 197
Lange, Sven, 41, 68
Langendoen, D. T., 114
Larrivée, Pierre, 73, 121, 124, 126, 128, 141
Larsson, Kenneth, 41, 68
Lawler, John, 129, 135, 141
Lederer, Richard, 125
Lee, Thomas H.-T., 41, 68

Leech, Geoffrey, (Biber et al.) 154, (Quirk et al.) 159, 173
Lehmann, Winfred P., 23–24
Lejeune, Ludo (van der Auwera et al.), 27, 89, 96
Lepore, Ernest, 151
Leuthold, Hartmut, 227–228
Levi, Natalie Alkeslassy, 229
Levine, William H., 230
Levinson, Stephen C., 198
Levy, Gideon, 231
Li, Charles, 138
Li, Renzhi, 91
Liberman, Mark, 127, 134, 141, 154
Lichte, Timm, 191
Lidz, Jeffrey, 58, 68
Lieven, Elena, 65
Linebarger, Marcia, 6, 192, 194–195
Lowth, (Bishop) Robert, 111, 113
Lucas, Christopher, 74, 83
Lüdtke, Jana, 227

MacDonald, Maryellen C., 58, 67, 226–227
Mann, John, 251
Marchand, Hans, 114, 117
Marchese, Lynell, 32
Marklund, Thorsten, 18
Martin, Benjamin, 112
Martin, Samuel E., 285
Martineau, France, 79
Martins, Ana Maria, 120
Master, Alfred, 18
Mazzon, Gabriella, 2, 98, 149, 177
McCawley, James D., 111
McGill, Stuart, 91
McGregor, William B., 73, 94
McKenzie, Parker, 23
McNeill, David, 41, 46–47, 53, 68
McNeill, Nobuko B., 41, 46–47, 53, 68
Medford, Nicholas, 231
Meillet, Antoine, 75–76, 80, 83
Meisel, Jürgen, 40–41, 57, 60–65, 68, 71

Index of persons

Meltzoff, Andrew N., 67
Metuki, Nili, 225–252
Miestamo, Matti, 3, 10, 12–17, 19–22, 25, 73, 83, 86–89, 101
Mignon, Françoise, 121
Miller, R. Andrew, 140, 258
Milon, John P., 41, 68
Misgav, Uri, 231
Moeschler, Jacques, 101
Molinu, Lucia, 80
Montgomery, Michael, 136, 141
Motohashi, Tatsushi, 264, 282
Mougeon, Raymond, 79
Mukash Kalel, Timothée, 84
Muller, Claude, 78
Murray, Lindley, 113, 116
Musolino, Julien, 58, 68–69

Napoli, Donna Jo, 132–133, 190
Nash, Walter, 116–117
Nedjalkov, Igor, 21
Neeleman, Ad, 208
Nelson, Gerald, 150
Nespor, Marina, 132–133
Neuckermans, Annemie, 84, 99
Neukom-Hermann, Anja, 149, 159, 180
Nishio, Toraya, 285
Norde, Muriel (Fischer et al.), 191
Norman, Jerry, 91
Noveck, Ira A., 227

Ogawa, Sadayoshi, 266, 282
Ortony, Andrew, 233
Orwell, George, 113–114
Ota, Akira, 264
Otanes, Fe T., 100
Overdiep, G. S., 84

Pakendorf, Brigitte, 92
Palacios Martinez, Ignacio, 150
Pappas, Dino Angelo, 135
Paradis, Carita, 229, 232
Parodi, Teresa, 41, 61, 69

Parry, M. Mair, 84
Partee, Barbara H., 197
Patz, Elisabeth, 89
Pauwels, J. L., 80, 82–83
Payne, John R., 9–10, 12, 14, 16, 20–21, 29
Pea, Roy D., 39, 41, 43–47, 69, 216
Pearce, David, 228
Penner, Zvi, 39, 69
Perdue, Clive, 41, 60–62, 66, 68–69
Pereltsvaig, Asya, 190
Perridon, Harry (Fischer et al.), 191
Perry Jr., Nathan W., 227
Pichon, Édouard, 131
Pienemann, Manfred, 65
Pierce, Amy, 49, 53, 66
Pinker, Steven, 130
Pitkin, Harvey, 78
Pollock, J. Y., 271
Pons Bordería, Salvador, 138
Portner, Paul, 123, 132
Poser, William, 208
Postal, Paul M., 80, 121, 139, 141
Postma, Gertjan, 139
Pott, A. F., 77
Potts, Christopher, 134, 141
Prado, Jérôme, 227
Prévost, Philippe, 64, 69
Priest, Anne, 28
Priest, Perry, 28
Prince, Ellen F., 123, 133, 176
Pust, Lieselotte, 150

Quine, W. V. O., 197
Quirk, Randolph, 159, 173

Ramadoss, Deepti, 41, 69
Ramat, Paolo, 10, 74, 80, 82
Rausch, P. J.
Rautenberg, Wolfgang, 228
Reich, Shuli, 131
Reilly, Judy S., 41, 64
Reinhart, Tanya, 277

342 *Index of persons*

Richards, Keith, 120
Richter, Frank, 188
Rissanen, Matti, 149
Rizzi, Luigi, 279
Roberge, Paul T., 80, 82
Robertson, Rachel W., 235
Robusto, 251
Rooij, Robert van, 193
Rosansky, Ellen, 65
Ross, John Robert, 231, 273–274
Roucos, Salim E., 227
Rubio Fernández, Paula, 235
Rūķe-Draviņa, Velta, 41, 47, 69
Rullmann, Hotze (Hoeksema *et al.*), 31
Russell, Willis, 1, 151, 177

Sabah, Hadas, 229
Sahlin, Elisabeth, 189
Sailer, Manfred, 188, 191
Saito, Mamoru, 277
Sanchez-Valencia, Victor (Hoeksema *et al.*), 31
Sanford, Anthony J., 227–228
Sano, Tetsuya, 41, 69
Sapir, Edward, 115
Savage, Howard (Earle *et al.*), 112
Schachter, Paul, 100
Schalley, Ewa, 92
Schaner-Wolles, Chris, 41, 69
Schiffman, Harold F., 17
Schimke, Sarah, 40–41, 61, 64, 70–71
Schul, Yaacov, 227, 229, 232
Schumann, John, 65
Schwegler, Armin, 80–81
Schwenter, Scott A., 79, 121, 132–133, 138–140
Seanzky, 251
Seavey, Frank (Earle *et al.*), 112
Seright, Oren D., 115, 117
Seuren, Pieter A. M., 3, 213
Sharma, Dhirendra, 114
Shen, Yeshayahu, 235, 252
Shibuya, Katumi, 259

Shuval, Noa, 230
Silberstein, Dagmar, 41, 60, 62–63, 70
Slobin, Dan I., 41, 53–54, 56, 64, 66, 68, 70
Smith, Jennifer, 150
Smoczyńska, Magdalena, 41, 70
Smyth, H. W., 121
Snyder, William, 41, 70
Soehn, Jan-Philipp, 191
Sportiche, D., 271
Starren, Marianne, 66–67
Stauble, Anne-Marie, 60, 70
Stebbing, Lizzie Susan, 152
Stern, Clara, 39, 42, 70
Stern, Pnina, 225–252
Stern, William, 39, 42, 70
Stokes, Stephanie F., 41
Stokes, William, 151, 165
Stringaris, Argyris K., 231, 252
Stromswold, Karin, 41, 49, 52, 54, 70
Summerfield, Karen Anne, 234
Svartvik, Jan (Quirk *et al.*) 159, 173
Swart, Henriëtte de
Szabolcsi, Anna, 187, 192

Taeymans, Martine, 198
Tagliamonte, Sali, 150
Taglicht, Josef, 152–153, 157, 166
Takizawa, Naohiro, 150
Tam, Clara W.-Y., 41, 70
Tesnière, Lucien, 113, 131
Theakston, Anna, 65
thisfred, 246
Thornton, Rosalind, 58, 69
Tieken-Boon van Ostade, Ingrid, 112, 149
Tobler, Adolf, 5, 152
Tokieda, Motoki, 258
Tomasello, Michael, 45, 70
Tottie, Gunnel, 5–6, 149–150, 176
Toupin, Mike, 91
Tovena, Lucia, 120
Tracy, Rosemarie, 69
Traugott, Elisabeth Closs, 98, 191

Index of persons 343

Uesaka, Nobuo, 286
Unsworth, Sharon, 59, 70

Vaidyanathan, Raghunathan, 41, 46, 71
van den Berg, Helma, 93
van den Berg, René, 10
van der Auwera, Johan, 2–3, 10, 27, 32, 73, 76, 85, 89–90, 96, 99–100, 129, 140, 198, 272
van der Ham, Margreet (Barbiers *et al.*), 84
van der Wal, Sjoukje, 41, 71
van der Wouden, Ton, 115, 121, 129, 141, 187, 198, 209
van der Wurff, Wim (Tieken-Boon van Ostade *et al.*) 149
Van Gelderen, Elly, 75
Van Olmen, Daniel, 90
Van Valin, Robert, 53–54, 56, 62, 71
Vasishth, Shravan, 25
Vendler, Zeno, 193
Vendryès, Joseph, 122, 126, 134
Vennemann, Theo, 23
Verhagen, Josje, 40–41, 61, 64, 71
Véronique, Daniel, 41, 60, 71
Verrips, Maike, 41, 71
Veselinova, Ljuba, 28, 73, 93–95
Victoria, 251
Visconti, Jacqueline, 133
Volterra, Virginia, 41, 45, 71
Vondruska, Richard J., 233

Wakabayashi, Setsuko., 41, 71
Walker, James A., 150
Wallage, Phillip, 84
Ward, Gregory, 229
Warner, Anthony, 118, 150
Wason, Peter, 126, 131
Watanabe, Akira, 121, 260, 282

Watanabe, Minoru, 286
Watkins, Laurel J., 23
Weissenborn, Jürgen, 41, 49, 57, 69, 71
Werner, Necia K., 235
Westergren-Axelsson, Margareta, 150
Westerståhl, Dag, 194
Wexler, Kenneth, 41, 50–51, 67, 71
White, Lydia, 64, 69, 225, 88
Whitman, John, 257, 260, 264, 266, 272, 274, 279, 282
Whitman, Neal, 127
Willems, Klaas (De Cuypere *et al.*), 80, 83
Willners, Caroline, 229, 232
Wimmer, Heinz, 67
Wode, Henning, 41, 49–51, 53, 56, 58, 71
Wouden, Ton van der, 31, 188, 215
Wray, Alison, 162

Xrakovskij, Victor S., 90, 92

Yaeger-Dror, Malcah, 150
Yamada, Takao, 258
Yanagida, Yuko, 265–266, 278, 281, 282
Yoon, Suwon, 132, 133, 141
York, Byron, 65–69, 71, 119, 173, 197, 225
Yoshimura, Akiko, 286

Zanuttini, Raffaella, 120, 123, 132–133, 260
Zeijlstra, Hedde, 121, 132
Zeshan, Ulrike, 10
Zguris, Stephanie, 234
Ziegeler, Debra, 138
Zimmer, Karl E., 11, 111, 115, 141
Zimmermann, Kai, 41, 49, 54, 70
Zwaan, A. Rolf, 227
Zwarts, Frans, 187, 188, 192, 195–196, 208, 218